From Genotype to Phenotype

The HUMAN MOLECULAR GENETICS series

Series Advisors

D.N. Cooper, *Charter Molecular Genetics Laboratory, Thrombosis Research Institute, University of London, UK*

S.E. Humphries, *Division of Cardiovascular Genetics, University College London Medical School, London, UK*

T. Strachan, *Department of Human Genetics, University of Newcastle-upon-Tyne, Newcastle-upon-Tyne, UK*

Human Gene Mutation
From Genotype to Phenotype

Forthcoming titles

The Human Genome: a Functional Analysis
Molecular Genetics of Cancer
Environmental Mutagenesis

From Genotype to Phenotype

Steve E. Humphries
Division of Cardiovascular Genetics, University College London Medical School, London, UK

Sue Malcolm
Division of Biochemistry and Genetics, Institute of Child Health, University of London, London, UK

BIOS
SCIENTIFIC
PUBLISHERS

First published 1994

A CIP catalogue record for this book is available from the British Library.

ISBN 1 872748 62 7

BIOS Scientific Publishers Ltd
St Thomas House, Becket Street, Oxford OX1 1SJ, UK.
Tel. +44 (0)865 726286. Fax +44 (0)865 246823

DISTRIBUTORS

Australia and New Zealand
DA Information Services
648 Whitehorse Road, Mitcham
Victoria 3132

India
Viva Books Private Limited
4346/4C Ansari Road
New Delhi 110002

Singapore and South East Asia
Toppan Company (S) PTE Ltd
38 Liu Fang Road, Jurong
Singapore 2262

USA and Canada
Books International Inc
PO Box 605, Herndon, VA 22070

Typeset by Ann Buchan, Shepperton, UK.
Printed by Information Press Ltd, Eynsham, UK.

Contents

Contributors

Baas, F. Academisch Medisch Centrum, Universiteit van Amsterdam, Meiberdreef 9, 1105 AZ Amsterdam, The Netherlands

Boerwinkle, E. Center for Demographic and Population Genetics, University of Texas Graduate School of Biomedical Sciences, Health Science Center, PO Box 20334, Houston, TX 77225, USA

Bolhuis, P.A. Academisch Medisch Centrum, Universiteit van Amsterdam, Meiberdreef 9, 1105 AZ Amsterdam, The Netherlands

Chan, L. Department of Cell Biology and Medicine, Baylor College of Medicine, Houston, TX 77030, USA

Cowell, J.K. Oncology Group, Imperial Cancer Research Fund, Institute of Child Health, 30 Guilford Street, London WC1N 1EH, UK

Dalgleish, R. Department of Genetics, Adrian Building, University of Leicester, University Road, Leicester LE1 7RH, UK

Fennessy, M. The London Hospital Medical College, University of London, The Royal London Hospital, Whitechapel, London E1 1BB, UK

Hammans, S.R. Department of Neurology, National Hospital for Neurology and Neurosurgery, Queen Square, London WC1N 3BG, UK

Hirst, M.C. Molecular Genetics Group, Institute of Molecular Medicine, John Radcliffe Hospital, Headington, Oxford OX3 9DU, UK

Hitman, G.A. The London Hospital Medical College, University of London, The Royal London Hospital, Whitechapel, London E1 1BB, UK

Horowitz, M. Gaucher Clinic, Shaare-Zedek Medical Center, PO Box 3235, Jerusalem 91031, Israel

Humphries, S.E. Centre for Genetics of Cardiovascular Disorders, University College London Medical School, The Rayne Institute, University Street, London WC1E 6JJ, UK

Johnson, K.J. Genetics Unit, Department of Anatomy, Charing Cross and Westminster Medical School, University of London, Fulham Palace Road, London W6 8RF, UK

Malcolm, S. Division of Biochemistry and Genetics, Institute of Child Health, University of London, 30 Guilford Street, London WC1N 1EH, UK

Meijerink, P.H.S. Academisch Medisch Centrum, Universiteit van Amsterdam, Meiberdreef 9, 1105 AZ Amsterdam, The Netherlands

Metcalfe, K. The London Hospital Medical College, University of London, The Royal London Hospital, Whitechapel, London E1 1BB, UK

Pignatti, P.F. Istituto di Scienze Biologiche, Università di Verona, Strada le Grazie, 37134 Verona, Italy

Soutar, A.K. MRC Lipoprotein Team, Hammersmith Hospital, Du Cane Road, London W12 0HS, UK

Valentijn, L.J. Academisch Medisch Centrum, Universiteit van Amsterdam, Meiberdreef 9, 1105 AZ Amsterdam, The Netherlands

Wilkie, A.O.M. Institute of Molecular Medicine, John Radcliffe Hospital, Headington, Oxford OX3 9DU, UK

Winchester, C.L. Genetics Unit, Department of Anatomy, Charing Cross and Westminster Medical School, University of London, Fulham Palace Road, London W6 8RF, UK

Zimran, A. Gaucher Clinic, Shaare-Zedek Medical Center, PO Box 3235, Jerusalem 91031, Israel

Abbreviations

ABPA	allergic bronchopulmonary aspergillosis
AP	acute phase
apo	apolipoprotein
AS	Angelman syndrome
AVN	avascular necrosis
BMD	Becker muscular dystrophy
bp	base pairs
BWS	Beckwith–Wiedemann syndrome
CAD	coronary artery disease
CBAVD	congenital bilateral absence of the vas deferens
CBP	chronic *Pseudomonas* bronchitis
CE	cholesterol ester
CETP	cholesterol ester transfer protein
CF	cystic fibrosis
CFGAC	Cystic Fibrosis Genetic Analysis Consortium
CFTR	cystic fibrosis transmembrane regulator
CHD	coronary heart disease
CM	chylomicrons
CMT	Charcot–Marie–Tooth disease
CNP	2′,3′-cyclic nucleotide 3′-phosphodiesterase
COPD	chronic obstructive pulmonary disease
CV	chorionic villus
DDS	Denys–Drash syndrome
(C)DM	(congenital) myotonic dystrophy
DMD	Duchenne muscular dystrophy
EDS	Ehlers–Danlos syndrome
EGF	epidermal growth factor
EGR	early growth response
EMG	electromyography
FCPD	fibrocalculous pancreatic diabetes
FH	familial hypercholesterolaemia
FISH	fluorescence *in situ* hybridization
FMR1	fragile X mental retardation gene
FRAX	fragile site on the X chromosome
FraXA	fragile X syndrome
FraXE	fragile X mental retardation syndrome
FS	Frasier syndrome
G6P	glucose-6-phosphate
GCK	glucokinase
GU	genitourinary
HDL	high density lipoprotein

HDL-C	high density lipoprotein–cholesterol
HLA	human leukocyte antigen
HMSN	hereditary motor and sensory neuropathy
HNF	hepatic nuclear factor
HTGL	hepatic triglyceride lipase
ICA	islet cell antibody
IDDM	insulin-dependent diabetes mellitus
IDS	iduronate sulphatase
IGF	insulin-like growth factor
IGT	impaired glucose tolerance
ILNR	intralobular nephrogenic rests
IRT	immunoreactive trypsin
ISF	impaired sperm function
KSS	Kearns–Sayre syndrome
LCAT	lecithin–cholesterol acyl transferase
LDL	low density lipoprotein
LHON	Leber's hereditary optic neuropathy
LINES	long interspersed nuclear elements
LOH	loss of heterozygosity
Lp(a)	lipoprotein (a)
LPL	lipoprotein lipase
MAG	myelin-associated glycoprotein
MBP	myelin basic protein
MELAS	mitochondrial myopathy, encephalopathy, lactic acidosis and stroke-like episodes
MERRF	myoclonus epilepsy and ragged red fibres
MI	myocardial infarction
MIP	methylation induced premeiotically
MODY	maturity-onset diabetes of the young
MRDM	malnutrition-related diabetes mellitus
NARP	neurogenic muscle weakness, ataxia and retinitis pigmentosa
NBF	nucleotide-binding fold
ncv	nerve conduction velocity
NIDDM	non-insulin-dependent diabetes mellitus
NPHS	Northwick Park Heart Study
NTM	normal transmitting male
OI	osteogenesis imperfecta
OMGP	oligodendrocyte–myelin glycoprotein
P_0	protein zero
PCR	polymerase chain reaction
PDGF	platelet-derived growth factor
PDDM	protein-deficient diabetes mellitus
PEO	progressive external opthalmoplegia
PFG	pulse field gel analysis
PLNR	perilobular nephrogenic rests
PLP	proteolipid protein
PMD	Pelizaeus–Merzbacher disease
PNS	peripheral nervous system
PWS	Prader–Willi syndrome
RFLP	restriction fragment length polymorphism

RIP	repeat induced point mutation
RRF	ragged red fibres
SAP	sphingosine activator protein
SCA-I	spinocerebellar ataxia type I
SCE	sister chromatid exchange
snRNP	small nuclear ribonucleoprotein
SSCP	single strand conformational polymorphism technique
SSI	severity score index
STR	simple tandem repeat
TG	triglyceride
TGF-β	transforming growth factor β
TM	transmembrane domain
TNF	tumour necrosis factor
UAVD	unilateral absence of the vas deferens
VLDL	very low density lipoprotein
VNTR	variable number of tandem repeats
WAGR	Wilms'–aniridia, abnormal gonadal development and mental retardation syndrome
WHO	World Health Organization
WT	Wilms' tumour
XLMR	X-linked mental retardation
YAC	yeast artificial chromosome

Preface

It is now becoming widely recognized that the clinical manifestations of even the most simple single gene disorders are affected both by other genetic factors and by the environment experienced by the patient. For the complex disorders, such as diabetes or coronary artery disease, these factors may completely mask or mimic the underlying genetic aetiology — the classical problems of penetrance, expressivity and phenocopy.

In this book, 14 authors, each expert in their field, explore the problems of how the effects of different mutations — the 'genotype' of the individual — are modulated to produce variability in the clinical symptoms — the 'phenotype' shown by the patient. The mechanisms of these interactions are varied and, for different disorders, may be the varying effects of different mutations in the same gene (for example, in familial hypercholesterolaemia or cystic fibrosis) or mutations in different genes causing the same phenotype, as in the connective tissue disorders. Also, the effects may be modulated by other single or multiple genes, or by the patient experiencing or avoiding specific environmental risk factors, such as smoking for coronary artery disease or becoming obese for non-insulin-dependent diabetes. Imprinted genes on chromosome 15 will lead to different diseases (Prader–Willi or Angelman syndromes) depending on the parent of origin. Finally, even the impact of chance stochastic events may play a significant role, such as in the amplification of unstable DNA sequences in myotonic dystrophy or fragile X, or in the inheritance of two copies of a chromosome from one parent (uniparental disomy).

The broad coverage of these different mechanisms presented in this book gives a unique overview of the important interactions which determine the clinical expression of these and many other disorders, both common and rare, experienced by individuals in the population at large. A better understanding of these problems will be useful both for improved diagnosis and for determining specific therapeutic measures in the future.

Steve E. Humphries (*London*)
Sue Malcolm (*London*)

Foreword

From Genotype to Phenotype will be a most welcome contribution to the literature in the rapidly developing field of human and clinical genetics. This field has seen dramatic new developments, even during the last few years. Outstanding among these is the discovery that several diseases or disease predispositions are caused by increased number of repeats of short DNA sequences. Areas with such sequences are labile and may expand in number (or reduce) from one generation to the next. One of the most amazing discoveries has been that a phenomenon previously believed to be an artefact not only existed but had a most surprising biological explanation. This was the phenomenon of anticipation in myotonic dystrophy: the debut of the disease at a younger age in offspring than in parents. Anticipation has now turned out to be accompanied by expansion of a repeat region in the gene for myotonic dystrophy, from parent to offspring.

The fragile X syndrome is another example of a disorder caused by expansion of a labile DNA area. The transition from a 'premutation' to a 'full mutation' may have considerable consequences. At the research level, the major challenge is to arrive at a better understanding of the processes of expansion or restriction of labile areas.

In the earlier days, it was generally believed that it would be the rule that the mutation causing a given disease would be the same between families within the same ethnic group. Two of the disorders dealt with in this volume, hypercholesterolaemia, caused by defects in the low density lipoprotein receptor gene, and cystic fibrosis, have turned out to be disorders where an extremely high number of different mutations may cause the same disease.

It has, in recent years, become increasingly clear that such mutations are also of importance for many common disorders which were previously believed to be caused exclusively by environmental, lifestyle or dietary factors. Thus, it is the interaction between (several) genes and environmental or lifestyle factors that underlies the development of the disease, but the environmental factors cause disease preferentially in people who have a genetic predisposition. This important area is thoroughly discussed in this volume. Particular attention is paid to the 'variability gene concept', a concept that should be fruitful for the attempts to dissect the effect of genes and environment on, respectively, risk factor level and risk factor variability.

This volume also gives a comprehensive update on important biological phenomena, such as somatic mosaicism, chimerism and X chromosome inactivation, and excellent reviews on the genetics of diabetes and on the role of mitochondrial genes in human disease. Finally, the developing area of gene therapy is discussed, with respect to an extremely important group of diseases, dyslipidaemias, which cause accelerated atherosclerosis.

The excitement that comes from making new, often unexpected, observations in the area of clinical genetics is conveyed superbly by the editors and individual authors throughout this book. It is a pleasure to read a book that describes these recent advances so well and which is also so full of novelty.

Kåre Berg (*Oslo*)

1

Mutations and human disease

Susan Malcolm

1.1 Introduction

The chapters of this book describe, for a range of diseases, how different mutations of a gene can have quite different phenotypic effects. This introductory chapter explains some of the ways in which genes are mutated and how changes in DNA sequence alter gene expression.

It has been known for some time that only a minority of the human genome, perhaps 5%, consists of gene coding sequences. Gradually the structure and properties of the remaining DNA sequences have been elucidated and it has become increasingly obvious that, although the extra sequences may not, for the most part, contribute to the expressed coding regions of genes, they are highly significant in contributing to mutations in human DNA. Because the mutations frequently do not result in direct alterations to amino acid sequence in the coding regions, but affect the expression or processing of mRNA, they contribute greatly to the variable phenotypes observed in human disease.

The project to produce a high resolution map of the entire human genome has been made possible because of the high level of variability between individuals. Originally, restriction fragment length polymorphisms (RFLPs) caused by the presence or absence of a restriction site were used both for predictive genetic tests and as the basis of a genetic map. The restriction enzymes TaqI and MspI, with the recognition sites T**CG**A and C**CG**G, were found to be particularly useful in detecting polymorphisms. The dinucleotide doublet 5′CpG3′ is underrepresented in the human genome, except in the HTF islands near the beginning of many genes (Bird, 1986). The major site of methylation in human DNA is $5'^{me}$CpG3′ and the mechanism by which 5′CpG3′ dinucleotides have become depleted is believed to be deamination of ^{me}CG to TG. There is no mechanism to correct this and, following replication, it will be incorporated into the DNA sequence with a complementary 5′CpA3′ in the other strand. This leads to the high frequency of extremely useful polymorphisms detected by TaqI and MspI, but also is a major contributor to independent recurrent mutations in coding sequences.

It is an unfortunate fact that the same inherent instability of the genome which has led to the variation which has made the Human Genome Project possible also leads to many disease-causing mutations.

Polymorphisms based on two-allele RFLPs had limited usefulness for genetic analysis and were superceded by length polymorphisms based on stretches of DNA with variable numbers of tandem repeats (VNTRs) (Jeffreys *et al.*, 1985; Nakamura *et al.*, 1987). Over recent years, a complete genetic map of humans has been constructed based on the polymorphic nature of runs of dinucleotide repeats, particularly $(CA)_n$ (Weissenbach *et al.*, 1992; Todd, 1992). The length variability probably is caused principally by replication slippage, and the same instability in replication of repeated sequences appears to be another powerful drive towards mutation (see Section 1.3.1).

The majority of eukaryotic genes consist of coding regions interrupted by non-coding intervening sequences or introns. The introns are present in the primary transcript but are processed, or spliced, out of the final mRNA transcript. Most of the DNA sequence in the introns appears to have no function but some areas near the intron/exon boundaries are conserved and necessary for correct splicing. Mutations of these sequences can not only hamper correct splicing but mutations of other sequences, either intronic or exonic, can produce DNA matching the consensus sequences which will compete with the normal splice site. Variations of sequence involving the splicing mechanism are likely to be particularly important in complicating the correlation of genotype and phenotype, as the exact sequence environment will influence the relative splicing efficiencies and may be tissue specific or affected by other factors. A further complexity arises now that it is realized that exons containing premature stop mutations can be spliced out to leave a partially active protein (Dietz *et al.*, 1993).

Other biological processes which contribute to the production of mutations include active transposable elements which form a proportion of the interspersed repeats (e.g. Alu repeats and **l**ong **i**nterspersed **n**uclear **e**lements or LINES) found at high frequency throughout the genome, and failures of meiotic division which lead to a fetus receiving both copies of a particular chromosome from one parent, so-called uniparental disomy. These will be discussed in Section 1.4.3. A further cause of variation the expression of a disease phenotype is mosaicism; for example, as a result of variation in the pattern of X inactivation in females. This will be discussed in Chapter 10.

1.2 Mutations involving RNA splicing

1.2.1 Consensus sequences at splice junctions

Shapiro and Senepathy (1987) have surveyed sequences in the GenBank databank to ascertain the most common usage of sequences round intron/exon splice sites. The findings for primates are presented in Table 1.1. At the donor site, the strongest consensus is for the first two bases into the intron which are invariant GT. The only other strong bias is for G, found as the last base of the exon in 78%

of boundaries. The 3′ or acceptor end of the intron is generally more variable but contains a run of pyrimidines (Ts and Cs) followed by NC(74%)AG immediately before the exon. The AG is again invariant.

1.2.2 Mutations at splice sites and exon skipping

The simplest outcome of a mutation in the invariant GT or AG sites is that the exon is skipped during RNA processing. Sakuraba *et al.* (1992) describe a case of Fabry disease in which a G to T transversion of the invariant GT consensus in the 5′ splice site of intron 6 of the α-galactosidase gene consistently results in exon 6 elimination from the mRNA. Three different putative splicing mutations in the CFTR gene have been studied by analysing mRNA extracted from nasal epithelial cells from patients with cystic fibrosis (Hull *et al.*, 1993). A mutation which disrupts the 5′ splice donor site of exon 12, 1898+G→A, and one which disrupts the 3′ splice acceptor site of intron 10, 1717-1G→A, resulted in loss of exon 12 and 11 respectively. However, in one patient, a 621+1G→T mutation

Table 1.1. Nucleotide percentages at splice junctions in primates. Data derived from Shapiro and Senepathy (1987).

Position		A	C	G	T	Consensus
5′ splice site	−3	32	37	19	12	
	−2	58	13	15	15	A Exon
	−1	10	4	78	8	G
	+1	0	0	100	0	G
	+2	0	0	0	100	T Intron
	+3	57	2	39	2	AG
	+4	71	8	12	9	A
	+5	5	6	84	5	G
	+6	16	15	22	47	T
3′ splice site	−14	9	31	15	45	T/C
	−13	9	33	13	45	T/C
	−11	7	31	11	51	T/C
	−10	10	35	7	51	T/C
	−9	10	35	11	44	T/C
	−8	7	43	7	42	T/C
	−7	9	41	8	42	T/C
	−6	6	39	6	48	T/C
	−5	6	40	8	46	T/C
	−4	23	29	23	24	N
	−3	3	74	1	22	C
	−2	100	0	0	0	A Intron
	−1	0	0	100	0	G
	+1	28	13	49	10	G Exon

disrupting the 5′ splice donor site of exon 4 resulted in a proportion of the product lacking the last 93 bp of exon 4, as a result of activation of an alternative splice site within exon 4, as well as the normal product from the other chromosome. Another patient with the same mutation additionally showed product with complete skipping of exon 4. The reason for the same mutation resulting in different effects on mRNA splicing in the two patients is not known but may reflect further sequence variation within intron 4. Both patients had classical symptoms of cystic fibrosis but, at the more subtle level of the activity of chloride transport, there may be phenotypic differences.

1.2.3 Activation of splice site in coding sequence

Mutations of consensus splice sites may lead to sequences within an exon providing the next best match, in which case activation of the cryptic site will compete with exon skipping. In addition to the example above, Hoshide *et al.* (1993) described a child with carbamyl phosphate synthetase I deficiency who had a 9 bp in-frame deletion resulting from a G → C transversion of the last nucleotide of the exon in the splice donor site. Use of a cryptic donor site in the exon resulted in the deletion.

Mutations within the coding region itself can also create sites which compete with the naturally occurring splice site. Flomen *et al.* (1992) describe a mutation within the iduronate sulphate sulphatase (IDS) gene in a patient with Hunter syndrome in which a C→T transition in the coding sequence, resulting in a silent change in the codon for threonine 146, creates a strong donor splice site (Figure 1.1). The new splice site not only contains the consensus AG/GT found in the natural site but, at the new downstream positions, the sequence GAGA shows better homology with the consensus sequence than with the normal splice site.

1.2.4 Mutations within the intron

As may be anticipated, mutations within the intron may activate new splice sites. Moskowitz *et al.* (1993), studying the α-L-iduronidase gene, found a G→A mutation in intron 5 at a position –7 from exon 6, as one of the mutations in two

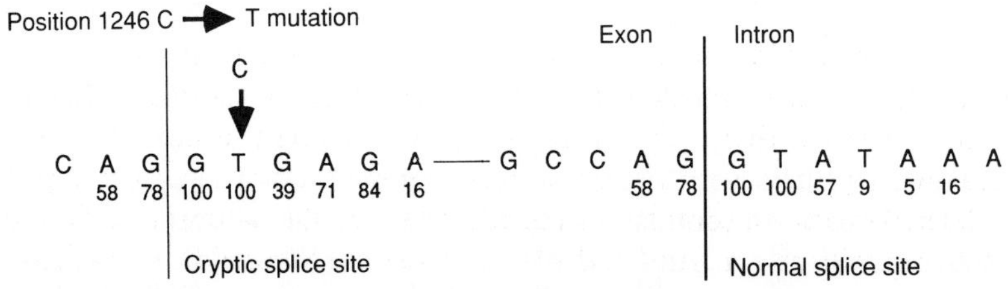

Figure 1.1. Activation of splice site within coding region of IDS gene. The numbers under the bases are taken from Shapiro and Senepathy (1987) and give the nucleotide percentage use.

compound heterozygote patients with Scheie syndrome. Scheie syndrome constitutes a much milder form of α-L-iduronidase deficiency than the allelic variant, Hurler syndrome. mRNA amplification from the patients showed a mixture of normally spliced species plus alternatively spliced cDNAs containing an extra five intronic nucleotides inserted into mRNA as a result of creation of a new alternative splice site. The additional nucleotides cause a frameshift which results in an almost immediate termination codon. The partial alternative splicing observed in these two cases probably explains the mild disease manifestation found in Scheie syndrome.

Gene mapping studies have shown that a locus for X-linked hydrocephalus, which is characterized by mental retardation and enlarged brain ventricles, maps to the same subchromosomal region as the gene for neural cell adhesion molecule L1, identifying this as a strong candidate. In cells from an affected individual, novel L1 mRNA species containing both deletions and insertions were found (Rosenthal *et al.*, 1992). Polymerase chain reaction (PCR) analysis of mRNA showed that there was a 69 bp insertion corresponding to the 3′ end of an intron. In addition, a single base change of adenine to cytosine was observed 19 bp upstream of the normal splice junction. It is most likely that the mutated A normally appears at the branchpoint found 10–50 bases upstream of the 3′ splice junction of an intron which forms a lariat structure during hnRNA splicing.

1.2.5 Partial correction of a severe phenotype

Duchenne and Becker muscular dystrophy. Mutations of the dystrophin gene, mapping to Xp21, give rise to both Duchenne muscular dystrophy (DMD) and Becker muscular dystrophy (BMD). The gene is of unique complexity and size, being over 2 Mb long, with 79 exons (Roberts *et al.*, 1993). The final product is a 427 kDa protein encoded by a 14 kb mRNA. Many cases of both DMD and BMD result from deletions of the dystrophin gene. The disease phenotype varies drastically from the most severe Duchenne form, resulting in death in the late teens, to very mild Becker forms. The explanation for most of this phenotypic difference was put forward by Monaco *et al.* (1988). Exon/intron borders can begin and end in any of the three positions of the triplet code for amino acids, and adjacent exons will have triplet codon breakpoints which maintain the correct translational open reading-frame during splicing. An intragenic deletion that, after joining, leaves adjacent exons in-frame will translate into a protein with an interstitial deletion of amino acids corresponding to the deleted exons. In contrast, intragenic deletions that alter the reading-frame after splicing will give a truncated protein when a stop codon is reached. The clinical phenotype appears to correlate with the amount of residual muscle dystrophin, being absent or drastically reduced in DMD patients but present in low but significant amount of shorter product in BMD patients. In general, Monaco and colleagues showed that deletions in DMD patients shifted the translational open reading-frame, whereas deletions in BMD patients maintained the translational open reading-frame.

Some exceptions were found to this general rule, for example patients with deletions of exons 3–7. Although the exon sequences predict that this would result in an out-of-frame mRNA, the patients often show a BMD-like phenotype. One possible explanation arose from studies of dystrophin mRNA in these patients (Chelly *et al.*, 1990) which, in addition to the major expected transcript joining exon 2 with exon 8, showed minor transcripts with splicing between both exon 1 and exon 8 and between exon 2 and exon 10. The juxtapositions of exons 1 and 8 and exons 2 and 10 restore the reading frame. This leads to low levels of a partially functional dystrophin protein and a milder phenotype. Other possible explanations exist, for example the presence of an alternative promoter.

Editing of premature termination codons. In general, the discovery of a point mutation within the coding region of a gene giving rise to a premature stop codon (TGA) has been considered sufficient evidence that the disease-causing mutation has been found. The assumption has been that no active protein will be produced and further study is unnecessary. The preceding discussion will have already shown the danger of this assumption, and recent reports have indicated that there may be systems for specifically activating alternative splicing to remove stop codons. Dietz *et al.* (1993) identified a patient with Marfan syndrome in which one allele of the fibrillin mRNA contained a 66-nucleotide deletion resulting from an in-frame skipping of an entire exon. The only identified sequence variation in the patient was a T→G change within the skipped exon, resulting in a premature TAG termination codon. Further screening revealed similar results for two nonsense mutations in the gene encoding δ-aminotransferase from patients with gyrate atrophy. Further examples of this phenomenon were observed in a survey of haemophilia A patients (Naylor *et al.*, 1993). Analysis of the factor VIII gene identified one patient who had mRNA that was missing exon 19, even though the splice junctions at both ends were normal. Within exon 19 there was a G→T mutation, resulting in premature termination. A second patient showed a mixture of normally processed mRNA and mRNA skipping exon 22, with a termination codon within exon 22 in genomic DNA. This may be of considerable clinical importance, as these patients are expected to produce significant amounts of defective protein. This will probably make them less likely to develop antibodies to factor VIII following treatment, as this complication occurs more often in patients with substantial loss of protein domains. None of the four previous patients with the nonsense mutation in exon 19 have developed inhibitors. Further evidence that nonsense mutations may alter splice site selection has been reported in the Fanconi anaemia C gene (Gibson *et al.*, 1993) and the AMP-deaminase gene (Morisaki *et al.*, 1993). Approximately 2% of Caucasians and African-Americans are homozygous for the nonsense mutation in exon 2 of the AMP-deaminase gene but, as many are asymptomatic, this result could have considerable importance for population screening and diagnosis.

Although searching genomic DNA for point mutations in candidate genes has been useful for diagnostic purposes, it is of little significance in understanding the course of disease, without RNA and allied protein or enzyme studies. This is because the complexity of possible alternative mRNA splicing and our poor understanding up to now of the signals involved means that the phenotype cannot be predicted accurately. Additionally, the effect of particular mutations on the amount of mRNA produced is unknown and this is also likely to be important in determining phenotype.

1.3 Mutations resulting from repeats in DNA

1.3.1 Non-homologous recombination and replication slippage

Gene duplication followed by gradual divergence in sequence of the genes has been a powerful force in evolution. However, the presence of two closely related sequences provides the opportunities for mutations arising by unequal interchromosomal crossing-over. This is a result of recombination occurring while the sequences are misaligned. The resultant product is either a duplication or a reciprocal deletion, depending on which chromosomal product survives. There are two copies of the α-globin gene 3.6 kb apart. If the two chromosomes misalign, so that the α_1-gene and its surrounding sequences on one chromosome misalign with the closely related α_2-gene on the other chromosome, then either a deletion or duplication can occur. The deletion is a common cause of α-thalassaemia and the duplication, resulting in three α-globin genes overall, has been found in many different populations. Recently, the presence of approximately 20 kb-long duplicated sequences, 1.5 Mb apart on chromosome 17 has been implicated in the duplication giving rise to hereditary motor and sensory neuropathy type I (Pentao *et al.*, 1992) and the reciprocal deletion found in hereditary neuropathy with liability to pressure palsies (Chance *et al.*, 1993; see Chapter 6).

A survey of all the reported rearrangements of the α-galactosidase gene found in patients with Anderson–Fabry disease illustrates well the high frequency with which deletions are associated with direct repeats at the breakpoints. Table 1.2 shows some typical examples. Of the 12 α-galactosidase rearrangements reported, six are deletions surrounded by direct repeats (Bernstein *et al.*, 1989; Ishii *et al.*, 1991; Kornreich *et al.*, 1990; de Jong, personal communication). One is a 8112 bp duplication flanked by a 5 bp repeat, which is likely to be the result of meiotic misalignment. One is a complex deletion/inversion associated with Alu repeats (Kornreich *et al.*, 1990). However, it is quite difficult to determine whether the mechanism is non-homologous recombination or replication slippage in any case. In case 2 in Table 1.2 the repeat is only 3 bp long, which reduces the chance of misaligning during meiosis. On the other hand, the deleted piece is 4651 bp long, which seems far too long for replication slippage (Bernstein *et al.*, 1989; Kornreich *et al.*, 1990). The remainder are all very small deletions/insertions. Two of them involve the insertion of an extra T (Ishii *et al.*, 1991;

Table 1.2. Rearrangements in the α-galactosidase A gene

Exons/ introns rearranged	Direct repeat gene	Length repeat (bp)	Outcome	Reference
Intron 2→intron 4	6548–6586 9746–9784	38	3197 bp del	Bernstein *et al.* (1990)
Exon 1→intron 2	1195–1197 5846–5848	3	4651 bp del	Bernstein *et al.* (1990)
Intron 1→exon 6	2589–2594 10 701–10 706	5	8112 bp dup	Bernstein *et al.* (1990)
Exons→3′ gene	Alu	~300	1710 bp del 151 bp inv	Bernstein *et al.* (1990)
Exon 5	1207	—	1 bp (T) ins	Davies *et al.* (1993)
Exon 7	11 050–11 051	2	2 bp (GA) del	de Jong, personal communication

del, deletion; ins, insertion; dup, duplication; inv, inversion.

Davies *et al.*, 1993) which, although in each case is adjacent to an existing T, this might be by chance. One involves the loss of GA from the sequence AG**GA**-**GA**AT and is most likely to be a replication slippage. The final 19 bp deletion is the only one in which there is no evidence for a repeat being involved in its derivation (S. Malcolm, unpublished results).

Considerable evidence exists in other genes for frameshifts occurring in sequences prone to slippage during replication. In factor VIII (Naylor *et al.*, 1993) three out of four were in such sequences, one was an insertion of an A in a run of six A residues and another a deletion of a single A in a run of eight. The third frameshift resulted from the loss of AAGA from the repeated sequence AAGAAAGA. In a case of retinitis pigmentosa, a deletion of an isoleucine was found in the rhodopsin gene by deletion of 3 bp from the sequence TCAT-CATCA (Inglehearn *et al.*, 1991) and the molecular defect underlying three unrelated cases of autosomal recessive chronic granulomatous disease is a dinucleotide deletion at GTGT tandem repeat (Casimir *et al.*, 1991).

The most drastic and clinically significant examples of deletions/insertions are the expansions of triplet repeats associated with fragile sites on the X chromosome and other diseases including myotonic dystrophy (Chapter 8) and Huntington's chorea. The mechanism by which these expansions occur is so far unknown, but they serve to reinforce the point that repetitive elements within human DNA are a profound source of instability.

1.3.2 Rearrangements caused by Alu repeats

Alu repeats are the best characterized repeats in human DNA. They are sequences of around 300 bp, repeated about 400 000 times, scattered throughout the human genome and comprising in total about 5% of human DNA. They are believed to have arisen originally by dimerization and have two halves, a right and a left half. The right half is homologous to the left half but has additionally a 31 bp insertion (Deininger *et al.*, 1981). As they exist on both DNA strands, adjacent pairs may have head-to-head, head-to-tail or tail-to-tail orientation. Examples have been found in several genes of deletions flanked by Alu repeats. These include the LDL receptor, globin (Lehrman *et al.*, 1987), apolipoprotein B (Huang *et al.*, 1989), hexosaminidase B (Neote *et al.*, 1990), adenosine deaminase (Markert *et al.*, 1988) and glycoprotein IIIa of the platelet fibrinogen receptor (Li and Bray, 1993). Many of these appear to result from recombination following misalignment between Alu sequences during meiosis, with a recombinant Alu sequence remaining after deletion (Markert *et al.*, 1988; Neote *et al.*, 1990). However, other deletions are flanked by Alu sequences which are not in a head-to-tail orientation (Lehrman *et al.*, 1987) or the rearrangement is more complicated, involving several breakpoints (Kornreich *et al.*, 1990; Li and Bray, 1993). Other possible mechanisms for Alu repeat involvement are reviewed in Lehrman *et al.* (1987).

1.3.3 Alu as a transposable element

Alu sequences have several additional structural features to those described above, which provide evidence that they have spread through the genome by insertion via the mechanism of retroposition. These features are:

(i) the first half of the Alu sequence contains a polymerase III promoter;
(ii) Alu insertions are flanked by direct repeats of the host sequence, suggesting that the Alu sequence has been inserted into staggered nicks of the host DNA which subsequently has been repaired; and
(iii) the terminal 3′ ends have a poly A tract, strongly suggesting that there is a history of reverse transcription of a polyadenylated mRNA.

As well as Alu retroposons, L1 elements are present in the genome of humans; they are present in roughly 100 000 copies and account for about 5% of genomic DNA. Full length L1 elements are 5–7 kb in length, have 3′ poly A tails and one or two open reading-frames, one of which encodes a reverse transcriptase. These structural features suggested that some L1 elements were also retrotransposons. Most copies are truncated at the 5′ end.

Both Alu repeats and L1 elements have been implicated in causing mutations by inserting into functional genes. Kazazian *et al.* (1988) found two patients with haemophilia A in whom *de novo* insertion of a truncated L1 element into the factor VIII gene caused the disease. The precursor to one of these was subsequently shown to be a full length L1 element on chromosome 22 with an intact open reading-frame coding for a reverse transcriptase activity in a yeast assay system (Dombroski *et al.*, 1991).

Insertions of Alu elements causing disease have been described for neurofibromatosis type 1 (Wallace *et al.*, 1991), haemophilia B (Vidaud *et al.*, 1993), and in a further family the cholinesterase gene was inactivated (Muritani *et al.*, 1991).

Sequence analysis of an Alu element found in two families with Huntington's disease, and therefore only recently transposed, allows a sub-group of Alu sequences still active in transposition to be identified (Hutchinson *et al.*, 1993).

1.4 Parental allele-specific gene expression

1.4.1 Inheritance modified by imprinting

Cytogenetically visible chromosomal aberrations, particularly deletions, have proved extremely useful in mapping both single gene disorders and contiguous gene syndromes (Tommerup, 1993). The level of resolution of the light microscope is fairly crude, particularly compared to molecular studies. However, two quite distinct syndromes, Angelman syndrome (AS) and Prader–Willi syndrome (PWS), both involving developmental delay in children, were mapped to the same chromosomal region 15q11–13 by the observation of *de novo* deletions in sporadic cases. The clinical features of the two disorders are quite distinct. PWS includes developmental delay and hypotonia in infancy but is particularly characterized by hyperphagia leading to gross obesity later. AS is characterized by ataxia, seizures and severe mental retardation with a particularly pronounced lack of speech. Molecular studies using polymorphic markers from the region confirmed the similarity of the two deletions but revealed that the deletions in PWS arose on the chromosome inherited from the father, whereas the deletions in AS arose on the chromosome inherited from the mother. This showed that the expression of at least some genes in the region are dependent on the parental origin, i.e. they are imprinted.

The genes causing PWS and AS have subsequently provided the best characterized example of imprinting in the human genome. Confirmation of the imprinting came from two sources. Nicholls *et al.* (1989) found a case of PWS arising through uniparental maternal disomy, i.e. the child had inherited both copies of the mother's chromosome 15s and, as in the deletion cases, lacked any paternal contribution. Correspondingly, AS cases were found resulting from uniparental paternal disomy (Malcolm *et al.*, 1991). Secondly, a family in which a chromosomal translocation t(15;22)(q13;q11) was segregating, produced a number of offspring in which the child inherited an unbalanced chromosome complement involving a deletion of the region 15q11–13 from a parent with a balanced chromosome translocation. The unbalanced translocation led to AS when inherited through the mother and PWS when inherited through the father (Hulten *et al.*, 1991). This provides clear-cut evidence for genomic imprinting, with unequivocal evidence that the same deletion produces different effects depending upon its parental origin.

A small number of families in which there is recurrence of AS syndrome have been reported, with two, or occasionally three, siblings. However, two extended pedigrees have been reported which, on first inspection, do not follow an obviously Mendelian pattern of inheritance (Meijers-Heijboer *et al.*, 1992; Wagstaff *et al.*, 1992, 1993) (Figure 1.2). The pattern of disease in family B can be understood by considering dominant inheritance modified by imprinting. Molecular studies showed that all the affected children in the third generation (III-1, III-3, III-5, III-7) had inherited 15q11–13 markers from their normal grandfather (I-1). He passed on the mutated gene to his daughters without phenotypic effect, as the gene is normally inactivated by imprinting on the paternally derived chromosome. However, when the same chromosome passed through his daughters, the imprinting was reversed so that the maternal chromosome becomes active. As this is the chromosome carrying the mutated gene, the children inheriting that copy of chromosome 15 are affected with AS. Family R also shows that an affected individual is only born when the gene has been inherited through the maternal line. In families with more than one affected

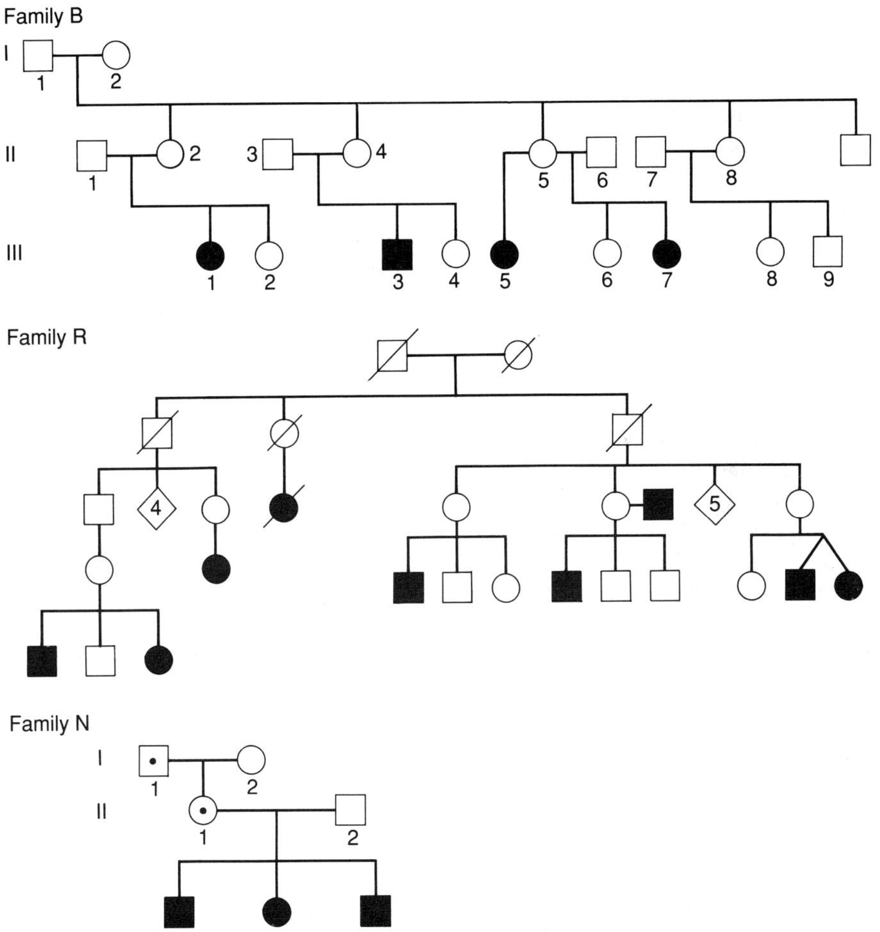

Figure 1.2. Pedigrees of extended families with multiple members affected by AS (Wagstaff *et al.*, 1992 (B); Meijer-Heijboer *et al.*, 1992 (R); Hamabe *et al.*, 1991 (N)).

child there is co-inheritance of the region 15q11–13 from the mother, but random inheritance from the father (Clayton-Smith *et al.*, 1992).

These unusual families indicate that the PWS and AS loci, although close and both within the normal cytogenetic deletion, are in fact separate. Further evidence comes from a unique family reported by Hamabe *et al.* (1991) (family N, Figure 1.2), in which a micro-deletion detectable by molecular genetic techniques was passed from a grandfather to his daughter with no phenotypic consequence. When the deleted chromosome was passed on to her three children they all suffered from AS. This family, and some rare translocation cases with PWS, have allowed preliminary mapping of the two disorders. A candidate gene, small nuclear ribonucleoprotein N (snRPN), which is expressed only from the paternal chromosome in mice, and is therefore imprinted, maps to the PWS region (Leff *et al.*, 1992; Ozcelik *et al.*, 1992). snRPN is expressed mainly in brain and heart and is involved in mRNA splicing. The relevance of this to the PWS phenotype is at present unclear.

1.4.2 Disorders on chromosome 11

Beckwith–Wiedemann syndrome (BWS) is characterized by overgrowth of numerous organs. It is often associated with the childhood nephroblastoma, Wilms' tumour. Most cases (85%) are sporadic, but both familial forms and some associated with abnormalities of chromosome 11p15 are found. It provides another candidate for a disorder transmitted by autosomal dominant inheritance modified by imprinting (Niikawa *et al.*, 1986; Junien, 1993). Linkage analysis mapped the disorder to 11p15.5 in familial cases. In families there is a reduced risk of being affected if chromosome 11 is inherited from the father. In the chromosomal duplication cases the extra material is of paternal origin. Sporadic cases of BWS are associated with uniparental disomy, in this case uniparental paternal isodisomy (i.e. two copies of the same chromosome) of a short region of chromosome 11, sometimes as a somatic mosaic. This is a situation analogous with AS, where a maternal contribution is required because the paternal allele has been inactivated by imprinting.

A further gene on chromosome 11, in the region 11q23–ter, subject to imprinting has been implicated in hereditary paragangliomas, slow-growing tumours of the head and neck (Heutink *et al.*, 1992). In a large pedigree with 16 affected individuals, all affected individuals inherited the disease gene through their father, expression is not seen in the children of affected females until the gene has been transmitted through a male carrier.

1.4.3 Uniparental disomy

Effects of uniparental disomy. The examples of uniparental disomy described above have all come to light because they involve regions of the human genome containing imprinted genes. There seem to be relatively few such regions in humans and it is quite possible for uniparental disomy to go unobserved. A rare consequence of uniparental disomy can be that an individual is homozygous for a recessive disorder when both chromosomes have been inherited from one parent. The first description of this by Spence *et al.* (1988) involved a female with cystic fibrosis who was being investigated for short stature. Molecular markers indicated that the father had contributed no DNA for the chromosome 7 region near the cystic fibrosis gene; markers from other chromosomes ruled out non-paternity. The presence of disease could be interpreted by the inheritance of two copies of chromosome 7 from the carrier mother. The patient was homozygous for a nonsense mutation G452X in exon 11 of the CFTR gene (Beaudet *et al.*, 1991). An example of uniparental disomy of chromosome 6 was found in a child with complete deficiency of complement 4, associated with systemic lupus erythematosus (Welch *et al.*, 1990). This deficiency is so rare that it is usually only found in children of related parents, but this patient had inherited two identical paternal chromosome 6 complements for all alleles tested. In this example there were no other phenotypic consequences, indicating that there are no imprinted genes on chromosome 6. The relevance of the short stature in the patient with cystic fibrosis is not clear but, in general, a child with a recessive genetic condition showing additional features not normally associated with the disease makes a good candidate for testing for uniparental disomy.

Origins of uniparental disomy. Engel (1980) proposed uniparental disomy as a new genetic mechanism several years before the first example, detailed above, was described. Based on the cytogenetic study of spontaneous abortion products, he calculated that aneuploidy is sufficiently frequent that there should be a finite number of conceptions arising from fertilization of a nullisomic gamete by a disomic gamete. This could give rise to either iso- or heterodisomy, depending on the stage of meiosis at which the non-disjunction occurred in the disomic gamete.

Results to date suggest many different mechanisms whereby uniparental disomy may arise (Figure 1.3). Studies in a series of PWS patients show non-disjunction at maternal meiosis I (i.e. maternal heterodisomy), associated with a raised maternal age, to be the most common mechanism (Mascari *et al.*, 1992). On the other hand, in AS, where the less common paternal non-disjunction must be postulated, uniparental disomy is both much less common and frequently appears to result in two identical copies of the paternal chromosome 15 (unpublished observations). The most likely explanation of this is duplication of a chromosome in order to rescue a monosomic conception arising from fertilization of a nullisomic egg by normal sperm.

Two lines of evidence show that post-fertilization errors can also occur. The first is the presence of somatic mosaics and the uniparental disomy of only part of chromosome 11 described for BWS above. The second comes from prenatal chorionic villus samplings carried out for maternal age. In one subject (Cassidy *et al.*, 1992), the CVS showed trisomy 15, but a later amniocentesis carried out to confirm this result showed a normal chromosome complement in each cell. Subsequently the baby was born and diagnosed as having Prader–Willi syn-

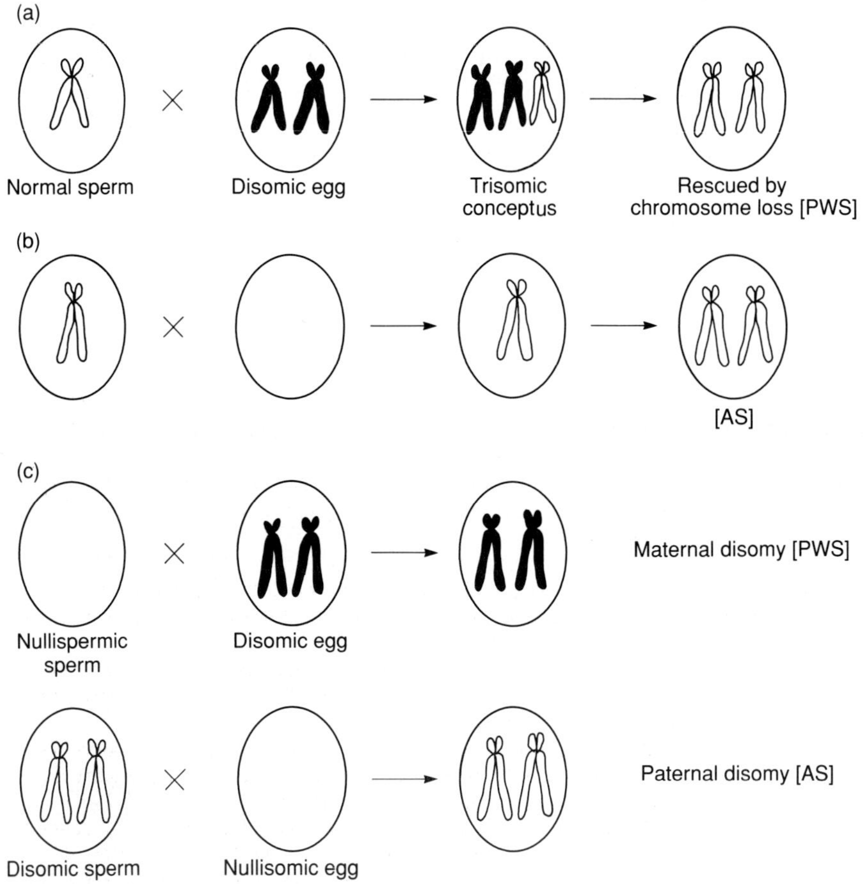

Figure 1.3. Mechanisms by which uniparental disomy (in this case, of chromosome 15) may arise.

drome, with uniparental maternal disomy. This case provides evidence that an original trisomic conception, as reflected in the CVS, has been rescued, unfortunately by loss of the single paternal chromosome. Further intriguing possibilities arise from the work of Robinson *et al.* (1993) who studied both an AS patient and a PWS patient who carried an additional small inverted chromosome 15s (+inv dup915)(pter→q11:q11→pter) in some of their cells. In both cases, the additional chromosome was found to be secondary to uniparental disomy causing the disease. However, the coincidence of the presence of chromosome 15 fragments suggests that their presence is left over from the event causing the uniparental disomy, possibly a by-product of 'rescuing' a trisomic fertilization.

1.5 Concluding remarks

This chapter has described many of the mechanisms leading to mutations in the human genome as a first step to understanding their effect. Splicing mutations have been considered in some detail as they show such extreme variability at the phenotypic level, even when the mutation is apparently the same. In contrast, disorders involving imprinted genes have been described where a largely constant phenotype is caused by very different events at the gene level.

References

Beaudet AL, Perciaccante RG, Cutting GR. (1991) Homozygous nonsense mutation causing cystic fibrosis with uniparental disomy *Am. J. Hum. Genet.* **48:** 1213.

Bernstein H, Bishop DF, Astrin KH, Kornreich R, Eng CM, Sakuraba H, Desnick RJ. (1989) Fabry disease: six gene rearrangements and an exonic point mutation in the alpha-galactosidase gene. *J. Clin. Invest.* **83:** 1390–1399.

Bird AP. (1986) CpG rich islands and the function of DNA methylation. *Nature* **321:** 209–213.

Casimir CM, Bu-Ghanim HN, Rodaway ARF, Bentley DL, Rowe P, Segal AW. (1991) Autosomal recessive chronic granulomatous disease caused by deletion at a dinucleotide repeat. *Proc. Natl Acad. Sci. USA* **88:** 2753–2757.

Cassidy SB, Lai L-W, Erickson RP, Magnuson L, Thomas E, Gendron R, Herrman J. (1992) Trisomy 15 with loss of the paternal 15 as a cause of Prader–Willi syndrome due to maternal disomy. *Am. J. Hum. Genet.* **51:** 701–708.

Chance PF, Alderson MK, Leppig KA, Lensch MW, Matsumari N, Smith B, Swanson PD, Odelberg SJ, Disteche CM, Bird TD. (1993) DNA deletion associated with hereditary neuropathy with liability to pressure palsies. *Cell* **72:** 143–151.

Chelly J, Gilgenkrantz H, Lambert M, Hamard G, Chafey P, Recan D, Katz P, de la Chapelle A, Koenig M, Ginjaar IB, Fardeau M, Tome F, Kahn A, Kaplan J-C. (1990) Effect of dystrophin gene deletions on mRNA levels and processing in Duchenne and Becker muscular dystrophies. *Cell* **63:** 1239–1248.

Clayton-Smith JA, Webb T, Robb SA, Dijkstra I, Willems P, Lam S, Cheng X-J, Pembrey ME, Malcolm S. (1992) Further evidence for dominant inheritance at the chromosome 15q11–13 locus in familial Angelman syndrome. *Am. J. Med. Genet.* **44:** 256–260.

Davies JP, Winchester BG, Malcolm S. (1993) Mutation analysis in patients with the typical form of Anderson–Fabry disease. *Hum. Mol. Genet.* **2:** 1051–1053.

Deininger PS, Jolly DJ, Rubin CM, Friedmann T, Schmid CW. (1981) Base sequence studies of 300 nucleotide renatured repeated human DNA clones. *J. Mol. Biol.* **151:** 17–33.

Dietz HC, Valle D, Francomano CA, Kendzior RJ, Pyeritz RE, Cutting GR. (1993) The skipping of constitutive exons *in vivo* induced by nonsense mutations. *Science* **259:** 680–683.

Dombroski BA, Mathias SL, Nanthakumar E, Scott AF, Kazazian HH. (1991) Isolation of an active human transposable element. *Science* **254:** 1805–1808.

Engel E. (1980) A new genetic concept: uniparental disomy and its potential effect, isodisomy. *Am. J. Med. Genet.* **6:** 137–143.

Flomen RH, Green PM, Bentley DR, Gianelli F, Green EP. (1992) Detection of point mutations and a gross deletion in six Hunter syndrome patients. *Genomics* **13:** 543–550.

Gibson RA, Hajianpour A, Murer-Orlando M, Buchwald M, Mathew CG. (1993) A nonsense mutation and exon skipping in the Fanconi anaemia group C gene. *Hum. Mol. Genet.* **2:** 797–799.

Hamabe J, Kuroki Y, Imaizumi K, Sugimoto T, Fukushima Y, Yamaguchi A, Izumikawa Y,

Niikawa N. (1991) DNA deletion and its parental origin in Angelman syndrome patients. *Am. J. Med. Genet.* **41:** 64–68.

Heutink P, van der Mey AGL, Sandkuijl LA, van Gils AP-G, Bardoel A, Breedveld GJ, van Vliet M, van Ommen GJB, Cornelisse CJ, Oostra BA, Weber JL, Devilee P. (1992) A gene subject to genomic imprinting and responsible for hereditary paragangliomas maps to chromosome 11q23–ter. *Hum. Mol. Genet.* **1:** 7–10.

Hoshide R, Matsuura T, Haraguchi Y, Endo F, Yoshinaga M, Matsuda I. (1993) Carbamyl phosphate synthetase I deficiency. One base substitution in an exon of the CPS I gene causes a 9-basepair deletion due to aberrant splicing. *J. Clin. Invest.* **91:** 1884–1887.

Huang L-S, Ripps ME, Korman SH, Deckelbaum RJ, Breslow JL. (1989) Hypobetalipoproteinemia due to an apolipoprotein B gene exon 21 deletion derived by Alu–Alu recombination. *J. Biol. Chem.* **264:** 11394–11400.

Hull J, Shackleton S, Harris A. (1993) Abnormal mRNA splicing resulting from three different mutations in the CFTR gene. *Hum. Mol. Genet.* **2:** 689–692.

Hulten M, Armstrong S, Challinor P, Gould C, Hardy G, Leedham P, Lee T, McKeown C. (1991) Genomic imprinting in Angelman and Prader–Willi translocation family. *Lancet* **338:** 638–639.

Hutchinson GB, Andrew SE, McDonald H, Goldberg YP, Graham R, Rommens JM, Hayden MR. (1993) An Alu element retroposition in two families with Huntington disease defines a new active Alu subfamily. *Nucleic Acids Res.* **21:** 3379–3383.

Inglehearn CF, Bashir R, Lester DH, Jay M, Bird AC, Bhattacharya SS. (1991) A 3-bp deletion in the rhodopsin gene in a family with autosomal dominant retinitis pigmentosa. *Am. J. Hum. Genet.* **48:** 26–30.

Ishii S, Sakuraba H, Shimmoto M, Minamikawa-Tachino MS, Suzuki T, Suzuki Y. (1991) Fabry disease : detection of 13 bp deletion in alpha-galactosidase A gene and its application to gene diagnosis of heterozygotes. *Ann. Neurol.* **29:** 560–564.

Jeffreys AJ, Wilson V, Thein SL. (1985) Hypervariable 'minisatellite' regions in human DNA. *Nature* **314:** 67–73.

Junien C. (1993) Beckwith–Wiedemann syndrome, tumorigenesis and imprinting. *Curr. Opin. Genet. Devel.* **2:** 431–438.

Kazazian HH, Wong C, Youssoufian H, Scott AF, Phillips DG, Antonarakis SE. (1988) Haemophilia A resulting from *de novo* insertion of L1 sequences represents a novel mechanism for mutation in man. *Nature* **332:** 164–166.

Kornreich R, Bishop DF, Desnick RJ. (1990) Alpha-galactosidase A gene rearrangements causing Fabry disease. Identification of short direct repeats at breakpoints in the Alu-rich gene. *J. Biol. Chem.* **265:** 9319–9326.

Leff SE, Brannan CI, Reed ML, Ozcelik T, Francke U, Copeland NG, Jenkins NA. (1992) Maternal imprinting of the mouse Snrpn gene and conserved linkage homology with the Prader–Willi syndrome region. *Nature Genetics* **2:** 259–264.

Lehrman MA, Russell DW, Goldstein JL, Brown MS. (1987) Alu–Alu recombination deletes splice acceptor sites and produces secreted low density lipoprotein receptor in a subject with familial hypercholesterolemia. *J. Biol. Chem.* **262:** 3354–3361.

Li L, Bray PF. (1993) Homologous recombination among three intra-gene Alu sequences causes and inversion-deletion resulting in the hereditary bleeding disorder Glanzmann thrombasthenis. *Am. J. Hum. Genet.* **53:** 140–149.

Malcolm S, Clayton-Smith J, Nichols M, Robb S, Webb T, Armour JAL, Jeffreys AJ, Pembrey ME. (1991) Uniparental disomy in the Angelman syndrome. *Lancet* **337:** 694–697.

Markert WL, Hutton JJ, Wiginton DA, States JC, Kaufman RE. (1988) Adenosine deaminase (ADA) deficiency due to deletion of the ADA gene promoter and first exon by homologous recombination between two Alu elements. *J. Clin. Invest.* **81:** 1323–1327.

Mascari MJ, Gottlieb W, Rogan PK, Butler MG, Waller DA, Armour JA, Jeffreys AJ, Ladda RL, Nicholls RD. (1992) The frequency of uniparental disomy in Prader–Willi syndrome. *N. Engl. J. Med.* **326:** 1599–1607.

Meijers-Heijboer EJ, Sandkuyl LA, Brunner HG, Smeets HJM, Hoogeboom AJM, Deelan WH,

van Hemel JO, Nelen MR, Smeets DFCM, Niermeijer MF, Halley, DJJ. (1992) Linkage analysis with chromosome 15q11–13 markers shows genomic imprinting in familial Angelman syndrome. *J. Med. Genet.* **29:** 853–857.
Monaco AP, Bertelson CJ, Liechti-Gallati S, Moser H, Kunkel LM. (1988) An explanation for the phenotypic differences between patients bearing partial deletions of the DMD locus. *Genomics* **2:** 90–95.
Morisaki H, Morisaki T, Newby LK, Holmes EW. (1993) Alternative splicing: a mechanism for phenotypic rescue of a common inherited defect. *J. Clin. Invest.* **9:** 2275–2280.
Moskowitz SM, Tieu PT, Neufeld EF. (1993) Mutation in Scheie syndrome (MPS IS): a G→A transition creates new splice site in intron 5 of one IDUA allele. *Hum. Mutation* **2:** 141–144.
Muritani KT, Hada Y, Yamamoto T, Kaneko Y, Shigeto T, Ohue J, Furuyama J, Higashino K. (1991) Inactivation of the cholinesterase gene by Alu insertion: possible mechanism for human gene transposition. *Proc. Natl Acad. Sci. USA* **88:** 11315–11319.
Nakamura Y, Leppert M, O'Connell P, Wolff R, Holm T, Culver M, Martin C, Fujimoto E, Hoff M, Kumlin E, White R. (1987) Variable number of tandem repeat (VNTR) repeat markers for human gene mapping. *Science* **235:** 1616–1622.
Naylor JA, Green PM, Rizza CR, Gianelli F.(1993) Analysis of factor VII mRNA reveals defects in everyone of 28 haemophilia A patients. *Hum. Mol. Genet.* **2:** 11–17.
Neote K, McInnes B, Mahuran DJ, Gravel RA. (1990) Structure and distribution of an Alu-type deletion in Sandhoff disease. *J. Clin. Invest.* **86:** 1524–1531.
Nicholls RD, Knoll JHM, Butler MG, Karam S, Lalande M. (1989) Genetic imprinting suggested by maternal heterodisomy in non-deletion Prader–Willi syndrome. *Nature* **342:** 281–285.
Niikawa N, Ishikiriyama S, Takahashi S, Inagawa H, Ohta Y, Hasa N, Kamei T, Kajii T. (1986) The Wiedemann–Beckwith syndrome: pedigree studies on five families with evidence for autosomal dominant inheritance with variable expression. *Am. J. Med. Genet.* **24:** 41–55.
Ozcelik T, Leff S, Robinson WP, Donlon T, Lalande M, Sanjines E, Schinzel A, Francke U. (1992) Small nuclear ribonucleoprotein polypeptide N (SNRPN), an expressed gene in the Prader–Willi syndrome critical region. *Nature Genetics* **2:** 265–269.
Pentao L, Wise CA, Chinault AC, Patel PI, Lupski JR. (1992) Charcot–Marie–Tooth type 1a duplication appears to arise from recombination at repeat sequences flanking the 1.5 Mb monomer unit. *Nature Genetics* **2:** 292–300.
Roberts R, Coffey AJ, Bobrow M, Bentley DR. (1993) Exon structure of the human dystrophin gene. *Genomics* **16:** 536–538.
Robinson WP, Wagstaff J, Bernasconi F, Baccichetti C, Artifoni L, Franzoni E, Suslak L, Shih L-Y, Aviv H, Schinzel A. (1993) Uniparental disomy explains the occurrence of the Angelman or Prader–Willi syndrome with an additional small inv dup(15) chromosome. *J. Med. Genet.* **30:** 756–760.
Rosenthal A, Jouet M, Kenwrick S. (1992) Aberrant splicing of neural cell adhesion molecule L1 mRNA in a family with X-linked hydrocephalus. *Nature Genetics* **2:** 107–112.
Sakuraba H, Eng CM, Desnick RJ, Bishop DF. (1992) Invariant exon skipping in the human α-galactosidase A pre-mRNA: a G^{+1} to T substitution in a 5′ splice site causing Fabry disease. *Genomics* **12:** 643–650.
Shapiro MB, Senepathy P. (1987) RNA splice junctions of different classes of eukaryotes: sequence statistics and functional implication in gene expresson. *Nucleic Acids Res.* **15:** 7155–7174.
Spence JE, Perciccante RG, Greig GM, Willard HF, Ledbetter DH, Hejtmanci JF, Pollack MS, O'Brien WE, Beaudet AL. (1988) Uniparental disomy as a mechanism for human genetic disease. *Am. J. Hum. Genet.* **42:** 217–226.
Todd JA. (1992) La carte des microsatellites est arrivée. *Hum. Mol. Genet.* **1:** 663–666.
Tommerup N. (1993) Mendelian cytogenetics. Chromosome rearrangements associated with Mendelian disorders. *J. Med. Genet.* **30:** 713–727.
Vidaud DM, Vidaud BR, Bahnak V, Siguret SG, Sanchez Y, Laurian D, Meyer M, Goossens M, Lavergne JM. (1993) Haemophilia B due to a *de novo* insertion of a human-specific Alu sub-family member within the coding region of the factor IX gene. *Eur. J. Hum. Genet.* **1:** 30–36.

Wagstaff J, Knoll JHM, Glatt KA, Shugart YY, Sommer A, Lalande M. (1992) Maternal but not paternal transmission of 15q11–13 linked nondeletion Angelman syndrome leads to phenotypic expression. *Nature Genetics* **1:** 291–294.

Wagstaff J, Shugart YY, Lalande M. (1993) Linkage analysis in familial Angelman syndrome. *Am. J. Hum. Genet.* **53:** 105–112.

Wallace MR, Andersen LB, Saulino AM, Gregory TW, Collins FS. (1991) A *de novo* Alu insertion results in neurofibromatosis type I. *Nature* **353:** 864–866.

Weissenbach J, Gyapay G, Dib C, Vignal A, Morrissette J, Millasseau P, Vaysseix G, Lathrop M. (1992) A second generation linkage map of the human genome. *Nature* **359:** 794–802.

Welch TR, Beischel LS, Choi K, Balakrishnan K, Bishof NA. (1990) Uniparental isodisomy 6 associated with deficiency of the fourth component of complement. *J. Clin. Invest.* **86:** 675–678.

2

Cystic fibrosis

Pier Franco Pignatti

2.1 Introduction

Cystic fibrosis (CF) is the most common severe recessive disease in Caucasoid populations. The incidence of CF varies among different groups, and it is usually reported to be 1 in 2500 births. CF is characterized by progressive sinopulmonary airflow limitation and infections, and by a nutritional deficiency due to exocrine pancreatic impairment.

The CF gene was located to chromosome 7 in 1985, and it was finally identified at the end of 1989. It was named CFTR (cystic fibrosis transmembrane regulator) because of its possible function and location (Rommens *et al.*, 1989). The basic alteration consists of defective epithelial transport of chloride through the CFTR membrane channel, which is regulated by cAMP-dependent protein kinase phosphorylation and requires binding of ATP for opening (Tsui and Buchwald, 1991; Collins, 1992; Welsh and Smith, 1993). Many mutations in the CFTR gene have been identified by screening CF patients and also, more recently, patients affected by certain other diseases. Therefore, the relationship between genotype (i.e. the CFTR gene mutations in an individual) and phenotype (i.e. the clinical manifestations in the patient) and also the organ, tissue and cell characteristics of the genotype after its interaction with individual genetic background and the environment, are multiple and complex.

2.2 The CFTR gene, its mutations and population differences

The CF gene consists of 27 exons, distributed over 230 kb of genomic DNA on human chromosome 7. A scheme of the gene and of the location of the mutations found in CF patients is shown in Figure 2.1. The predicted protein is 1480 amino acids long, and it comprises two transmembrane domains containing 12 membrane-spanning regions (TM), two ATP-binding domains or nucleotide-binding folds (NBF), similar to other ATP-binding transporter proteins (Hyde *et al.*, 1990), and a unique highly polar domain (R) with many possible phosphorylation sites, believed to have a regulatory role (Riordan *et al.*, 1989).

Figure 2.1 shows a CFGAC (Cystic Fibrosis Genetic Analysis Consortium)

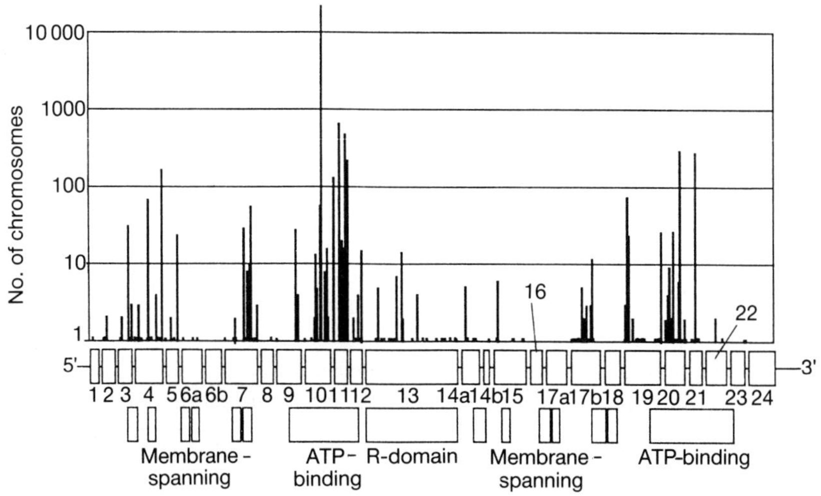

Figure 2.1. Frequency and location of CFTR gene mutations found in CF. Top: no. of chromosomes with mutations reported by the Cystic Fibrosis Genetic Analysis Consortium. Middle: scheme of the CFTR gene and its 27 exons and introns, plus untranslated 5′ and 3′ portions. Bottom: putative CFTR protein domains, showing the 12 transmembrane sequences forming the two transmembrane domains, the two ATP-binding domains and the single regulatory domain. Reproduced from Tsui (1992b) with permission from Elsevier Trends Journals.

mutation summary as of May 1992 (Tsui, 1992a): no important frequency distribution differences have since emerged, even if the most recent CFGAC report at the time of writing (November 1993), kindly provided by Dr Lap-Chee Tsui to members of the Consortium, lists more than 375 presumed mutations and 50 other DNA sequence variations in the gene. Figure 2.1 shows clusters of mutation frequencies found in CF patients, particularly in the ATP-binding domains and in the initial membrane-spanning regions. The tallest bar in exon 10 refers to the most common disease-producing mutation, ΔF508, a three-nucleotide deletion leading to the loss of phenylalanine in position 508, which was the first CF mutation identified (Kerem *et al.*, 1989b). The apparent 'hot spots' of mutation may, in part, be due to a biased search in selected regions of presumed functional importance, and they are also related to the different position of mutations in different clinical manifestations, as discussed later in this chapter.

In different populations, different frequencies of the various mutations have been reported. Table 2.1 gives the example of five different European populations arbitrarily selected for the purpose of showing the amount of diversity. A more comprehensive worldwide regional mapping of CFTR mutations found in CF patients is being prepared by the CFGAC. The mutations are designated following the nomenclature rules proposed by Beaudet and Tsui (1993). Briefly, for missense mutations, the amino acid position is indicated, preceded by the letter of the wild-type residue, and followed by the mutant one. For nonsense mutations, the number is followed by an X. Splice, and insertion/deletion mutations refer to the nucleotide number of the published exon sequence, with + or – symbols indicating the relative intron position substituted. It can be noted that ΔF508 is the most common mutation in all populations except in the Ashkenazi, and that some mutations are characteristically overrepresented in some populations, such as W1282X in Ashkenazi Jews, R1162X and 2183 AA→G in a north-east Italian cohort, 621+1G→T and 1898+1G→A in the Welsh, 1078 delT and G551D in Bretons, and R553X in eastern Germans. The frequency of the various CFTR mutations and the total fraction of mutations identified in CF patients has obvious consequences on the number of homo/heterozygote patients expected, and on the possibility of carrier screening in the population. Differences in the type and frequency of mutations raise interesting questions not only about their origin, diffusion and probable evolutionary advantage, but also about the possibility of a variable pattern of disease in diverse populations.

2.3 CFTR mutations and disease

Notwithstanding this wealth of information on the gene and its alterations, the pathophysiological explanation of the cell, tissue and organ effects of CFTR gene mutations is still lacking (Koch and Hoiby, 1993). The number of mutations indicates that many cases of CF and other CFTR gene-related diseases must be represented by individuals that carry two different mutations in their chromosomes and, in some populations in which different mutations are frequent, this

Table 2.1. Distribution of the most frequent CF mutations in some European populations

Population	Mutation (%)									
	ΔF508	W1282X	R1162X	2183AA→G	621+1G→T	1898+G→A	1078delT	G551D	R553X	Total
Ashkenazi*	23	**60**	—	—	—	—	—	—	—	92
Italian †	47.5	0.9	**9.8**	**9.3**	0.9	0.4	—	0.4	1.3	86.7
Welsh‡	71.6	—	—	—	**6.6**	**5.5**	2.2	2.2	1.1	99.5
Breton§	81.2	0.3	—	—	0.3	—	**4.9**	**4.1**	—	98
German¶	60.8	—	—	—	—	—	—	1.2	**5.7**	75.6

— Not found or not searched.
Data from: * Shoshani *et al.* (1992), † Gasparini *et al.* (1993) and our unpublished results, ‡ Cheadle *et al.* (1993), § Ferec *et al.* (1992), ¶ Coutelle *et al.* (1992).
Note: only selected mutations with exceptional incidence are shown.

may be the case for most patients. Regulatory effects between the two alleles during phenotypic expression are likely, and have to be considered in the analysis of compound heterozygotes and, therefore, when possible, true homozygotes are studied, even if they are rarer. Exceptionally, three different putative mutations have been found in CF patients, which further expands the genotype–phenotype correlation spectrum of possibilities. The task of relating gene mutations to disease outcome is complicated by the large clinical variability of CF manifestations, severity and age of onset. Moreover, information has emerged recently on several other conditions where the CFTR gene has already been shown to be definitely or probably involved, such as male infertility due to the absence of the vasa deferentia, or chronic obstructive pulmonary disease, respectively. If one considers that the incidence of congenital bilateral absence of the vas deferens may be about the same as that generally reported for cystic fibrosis, then one realizes how important it is to recognize the phenotypic manifestations of alterations in the CFTR gene for screening and prevention, and possibly for therapeutic purposes as well.

From the above considerations, it is therefore interesting and timely to review our present knowledge of the genotype–phenotype correlations in CF, even if the rapid progress of the field carries with it the certainty that this summary will soon be outdated (Pignatti, 1991). This chapter will treat the argument in a historical sequence as far as possible, CF first, and other diseases second, in order to give the feeling of the excitement of the recent developments, and an idea of some future research directions. The information has been drawn from the current literature and congress abstracts known to the author up to November 1993, and from our own work in progress. Another review of genotype–phenotype relationships in CF, which includes 1992 references, has appeared recently (Hamosh and Cutting, 1993).

2.4 Clinical manifestations and diagnosis of CF

CF has three main diagnostic criteria: chronic sinopulmonary disease, pancreatic insufficiency and elevated sweat electrolyte levels (Davis *et al.*, 1980). Other less common alterations may be present. After a brief summary in this section of the most important characteristics of the disease, each of them will then be treated singly in relation to CFTR genotype in the subsequent sections.

The clinical manifestations of CF (Boat *et al.*, 1989) are characterized by the respiratory involvement. Pulmonary obstruction is due to the presence of a thick viscous mucus and persistent and recurrent infections, especially with *Pseudomonas* and *Staphylococcus* strains, leading to respiratory failure and often to death. The term 'mucoviscidosis', now less commonly used, refers to this characteristic and it was introduced by Farber in 1945 (reviewed in Boat *et al.*, 1989). Most patients also have an intestinal involvement. In the majority of cases, exocrine pancreatic insufficiency is present, which begins *in utero* and causes steatorrhea and growth delay. The term 'cystic fibrosis of the pancreas' derives from the commonly observed small cysts formed by dilated ducts; it was first used by Anderson in 1938 (reviewed in Boat *et al.*, 1989) and it is now in general use, even if CF is not a disease of the pancreas, but a generalized disorder (Di Sant'Agnese and Powell, 1962). In about 10–15% of the cases, pancreatic function is sufficient, and the patients do not need dietary enzyme supplements for food digestion. Neonatal meconium ileus, a form of intestinal obstruction, is present in 5–10% of cases, and is typical of CF. Liver cirrhosis may be present in 2–5% of cases.

In CF patients, the 'sweat test' reveals an abnormally high level of sweat chloride (over 60 mM), and it is the most important diagnostic laboratory finding (Gibson and Cooke, 1959) for confirmation of the clinical diagnosis. In 1953 it was discovered that chloride, sodium and, to a lesser extent, potassium were increased in sweat and saliva of CF patients (reviewed in Di Sant'Agnese and Powell, 1962). As the overlap between normal and patients' values is minimal for chloride, increased sweat chloride concentration represents the commonly used CF diagnostic test, and the hallmark of the disease.

Adults with CF show infertility, which is present in 97–98% of males and is usually due to bilateral aplasia of the vas deferens. Obstructive azoospermia has in fact been proposed as a major criterion for diagnosis of the 'cystic fibrosis syndrome', together with the rise in sweat chloride and chronic obstructive pulmonary disease (Stern *et al.*, 1982). Infertility is also frequent in females.

Newborn screening has been done by using dried blood or meconium samples for quantitation of immunoreactive trypsin (IRT) and other enzymes (Boat *et al.*, 1989). It is now feasible also to screen directly for carriers of CFTR gene mutations in the general population by nucleic acid analysis (Williamson, 1993): the PCR-based procedure used is so simple that it is advertised currently for first-year medical student biochemistry laboratory training (Grody *et al.*, 1993).

2.4.1 CF exocrine pancreatic function and CFTR mutations

Pancreatic function is usually concordant among affected young (Corey *et al.*, 1989) or adult (Santis *et al.*, 1990), siblings, and among twins (Santis *et al.*, 1992), indicating a direct influence of genotype. Pancreatic sufficiency or insufficiency was associated tentatively with different CF mutations from the beginning, when the CF locus was first localized to chromosome 7q, and molecular marker RFLPs became available. The hypothesis of allelic heterogeneity underlying pancreatic function in CF had been advanced on the basis of observed differences between pancreatic-insufficient (Pl) and pancreatic-sufficient (PS) patients by CF locus-linked DNA marker haplotypes (Devoto *et al.*, 1989; Kerem *et al.*, 1989a; Ferrari *et al.*, 1990) and polymorphisms (Gasparini *et al.*, 1990a). When the gene was identified, Kerem *et al.* (1989b) observed that ΔF508 homozygotes were always PI, while ΔF508 heterozygotes, as well as patients with other, still unknown, mutations, could be PS. They hypothesized that mild mutations could provide some residual exocrine function (PS alleles), and could therefore exhibit a dominant phenotype over severe mutations with little or no function (PI alleles). Their hypothesis was soon confirmed by others with studies performed in different populations, and also in a population where the frequency of the ΔF508 mutation was much lower, and where therefore more, and possibly different, genotypes would produce the same pancreatic manifestations (Borgo *et al.*, 1990a). This analysis has continued successfully, so that several mutations have now been characterized as belonging to either the PI or the PS group, whereby the presence of at least one of the PS mutations is enough to confer sufficient pancreatic function.

Table 2.2 gives a list of mutations belonging to the two categories. It can be noted (Kristidis *et al.*, 1992) that severe mutations comprise all molecular types, and in particular those with presumably blocked or impaired synthesis of the CFTR protein, such as nonsense, frameshift and splice site mutations. By contrast, mild mutations are almost exclusively of the missense type. Also, the position of the mutation along the gene differs; most severe mutations reported in Table 2.2 are located in NBF domains, or just before NBF2 (R1162X and 3659delC), with the exception of mutations 612+1G→T and 1148T, which are in TM1. By contrast, most mild mutations are in TM domains, with the notable exception of a cluster located at the beginning (position 455), or at the end (positions 549, 551 and 574), of NBF1. The splice site mutation 1898+3A→G reported in the PS group occurs at a donor site position (+3) where sequence conservation is less stringent. This mutation was found in a PS compound heterozygote together with ΔF508 (Gasparini *et al.*, 1993), and initially it was

Table 2.2. Classification of CFTR gene mutations relative to pancreatic function in CF patients

Severe (two mutations necessary for PI)	
Deletions:	ΔF508*, Δ1507*, 2184delA
Nonsense:	Q493X*, G542X*, Q552X†, R553X*, R1162X‡, W1282X*
Frameshift:	457TAT→G‡, 556delA*, 2184delA§, 3659delC*, 3905insT¶
Missense:	1148T*, G480C*, V520F*, S549R¶, G551D*, R560T*, G1244E¶, G1255P¶, N1303K*‖, G1349D¶
Splice:	621+1G→T*, 1717–1G→T*, 1717–1G→A**, 1898+1G→A††
Mild (one mutation sufficient for PS)	
Missense:	R117H*, P205S‡‡, R334W*, T338I§§, R347P*, A455E*, S549N¶¶, G551S‖‖, P574H*
Splice:	1898+3A→G†, 3849+10 kb C→T***

Data from: * Kristidis *et al.* (1992); † Gasparini *et al.* (1993); ‡ Gasparini *et al.* (1992); § Lissens *et al.* (1993); ¶ Welsh and Smith (1993); ‖ Osborne *et al.* (1992); ** the Cystic Fibrosis Genotype–Phenotype Consortium (1993); †† Strong *et al.* (1992); ‡‡ Chillon *et al.* (1993); §§ Saba *et al.* (1993); ¶¶ Santis *et al.* (1990); ‖‖ Strong *et al.* (1991); *** Augarten *et al.* (1993) – note: 5/15 patients (33%) were PI.

Note: this table is not complete. Caution: pancreatic genotype–phenotype correlations are not absolute: some exceptions have been described. The use of this table for prediction of clinical outcome is not recommended.

reported as a rare variant in a PS patient carrying an unknown mutation in the other chromosome (Cremonesi *et al.*, 1992). The mutation may allow the synthesis of enough correctly spliced transcript to preserve the amount of function necessary for determining the PS phenotype, producing the partial loss of exon 12 from the spliced product. This exon corresponds to the last part of NBF1, where the above-discussed 574 PS mutation is also located, confirming its possible PS role. Interestingly, exon 12 complete skipping occurred with 1898+1G→A, a PI mutation, in two patients who were compound heterozygotes for ΔF508 (Strong *et al.*, 1992); the defect is predicted to result in the in-frame translation of a CFTR protein lacking a highly conserved region, and is therefore predicted to be defective in ATP binding and hydrolysis.

Some exceptions to the rule of PS/PI mutations explained above have been noted, most disturbingly some ΔF508 homozygotes being PS (Kopelman and Rozen, 1990; Santis *et al.*, 1990; Stuhrmann *et al.*, 1990; Lanng *et al.*, 1991; Hamosh *et al.*, 1993; Cystic Fibrosis Genotype–Phenotype Consortium, 1993). Also, a G85E 11-year-old homozygote boy was reported to be PS (Chalkey and Harris, 1991), while two G85E/ΔF508 patients were PI (Gasparini *et al.*, 1993). R117H/ΔF508 patients were reported as PS in 20 out of 23 cases (Cystic Fibrosis Genotype–Phenotype Consortium, 1993). Several explanations have been offered for these exceptions; certainly, there may be problems related to the accuracy of the diagnosis of pancreatic function which may have to be based just on the decision to supplement the diet with pancreatic enzyme, due to the difficulty of implementing invasive techniques in every patient. The age effect, whereby a progression from sufficiency to insufficiency is common (Borgo *et al.*, 1990b) may also help to explain some of these exceptions; for example, Kristidis *et al.* (1992) report that 15 of their currently PI patients had been PS earlier in life. Also, the possibility of other, modifying, mutations in the gene has been indicated. Two ΔF508 homozygote siblings with delayed onset of PI at puberty were found to carry a second CFTR gene mutation V1212I on one ΔF508 allele (Macek *et al.*, 1993); the authors suggest that the second mutation might play a role in mitigating the deleterious effect of ΔF508.

An accurate assessment of exocrine secretory function, based on a direct study by duodenal intubation and a pancreatic stimulation test, has allowed the analysis of a second, and different, functional parameter: the bicarbonate output level. This was reduced most severely in ΔF508 homozygotes, and therefore followed the same model presented above for the digestive function (Borgo *et al.*, 1990a). In conclusion, with the current state of knowledge, pancreatic secretory function is the CF symptom most easily related to the different gene mutations and to the patient's genotype.

2.4.2 Pancreatitis and CFTR mutations in CF patients

Symptoms of pancreatitis are occasionally but rarely present in adolescent and adult CF patients, especially in those who have retained some exocrine pancreatic function (Boat *et al.*, 1989). Recently, the genotypes of five CF patients with pancreatitis were reported (Hamosh *et al.*, 1993). These included R117H/ΔF508, R560T/R334W, and three carriers of mutations ΔF508 (two cases) and 444delA, respectively. Further studies are necessary to determine if particular mutations are more common in pancreatitis.

2.4.3 Diabetes in CF patients and CFTR mutations

Sometimes diabetes develops in CF patients, particularly in older ones, and abnormal glucose tolerance tests may be quite frequent (Boat *et al.*, 1989). The endocrine pancreatic function was unrelated to ΔF508 homozygosity or compound heterozygosity in a group of 211 patients, 15% of which were diabetic and 12% of whom had impaired glucose tolerance (Lanng *et al.*, 1991). By contrast, a meeting abstract has reported a greater frequency of diabetes among ΔF508 homozygote adults (Hamdi *et al.*, 1992).

2.4.4 CF meconium ileus and CFTR mutations

An early report on gene-linked molecular markers indicated that a particular J3.11 restriction polymorphism allele was more often associated with meconium ileus (MI), and suggested that multiallelism at the CF locus might explain the observation (Mornet *et al.*, 1988). Another group did not find such an effect (Curtis *et al.*, 1989), although the pooled data still showed a difference (Simon-Bouy *et al.*, 1989). Others confirmed this result only partially (Auvinet *et al.*, 1989). In a larger study, MI was present in 158 (13%) of a series of 1175 patients; in families in which the first child had MI, 29% of subsequent siblings also had MI, compared with 6% of siblings born to families in which the first child did not have MI, indicating the importance of genetic factors. This notwithstanding, allelic frequencies and haplotypes linked to the CF locus were similar in CF families with or without MI (E. Kerem *et al.*, 1989d).

MI occurred in 80 (21%) of the 375 PI patients reported by Kristidis *et al.* (1992), and in none of the 19 PS patients, indicating that the residual CFTR function conferring PS also prevents the development of MI. In the same study, MI appeared to be overrepresented (9/18) in patients who had the ΔF508/G542X genotype, while in ΔF homozygotes it was about 18% (49/277). MI was underrepresented in G551D/ΔF508 patients (5/75, or 7%), compared with ΔF508 homozygotes (15/62, or 24%) in another study (Hamosh *et al.*, 1992). MI occurred in every genotype group reported in the recent Cystic Fibrosis Genotype–Phenotype Consortium study (1993), except in 23 R117H/ΔF508 patients, who were mostly PS (20/23). The frequencies of MI varied from 6% in 17 W1282X/ΔF508 patients to 24% in 148 G542X/ΔF508 patients (this series includes the patients previously reported by Kristidis *et al.*, 1992) and, in this study, MI was observed in 57/394 (14.5%) ΔF508 homozygotes.

2.4.5 CF severity and CFTR mutations

Considering the complexity of the clinical manifestations of CF, which affect several organs, and the variable expression of the same genotype even among siblings, it is probably more instructive to consider the phenotypic manifestations separately. These considerations notwithstanding, and as a prognostic idea of disease severity would be of great practical interest for the patient and the physician, several attempts have been made at characterizing this most general form of phenotype correlation with genotype. Disease parameters given vary in different studies, indicating that the operational definition of 'severity' is not uniform. A correlation with the severity of CF had been attempted even before the discovery of the CF gene, by using linkage analysis of DNA markers. One such study indicated that the MP6d-9 genotype was associated significantly with very mild clinical manifestations, including pancreatic sufficiency, absence of meconium ileus and absence of *Pseudomonas* colonization (Gasparini *et al.*, 1990a).

After the identification of the CF gene, correlations with several different mutations were sought. In a study of 293 patients, ΔF508 homozygotes had received a diagnosis at an earlier age and had a greater frequency of PI (E. Kerem *et al.*, 1989). Later age at diagnosis and pancreatic sufficiency was found in CF siblings carrying ΔF508/R117H and ΔF508/R347P genotypes (Dean *et al.*, 1990). A study of 218 patients indicated that, compared to ΔF508 heterozygotes compounded with unknown mutations, ΔF508 homozygotes had earlier onset of symptoms and age at diagnosis, required greater pancreatic enzyme substitution, and showed poorer lung function, increased incidence of chronic *Pseudomonas aeruginosa* infection and higher mortality rates (Johansen *et al.*, 1991). Severe disease with PI, high incidence of MI, early age at diagnosis, poor nutritional status and variable pulmonary function, has been reported for mutation W1282X homozygotes, or W1282X compound heterozygotes with ΔF508 (Shoshani *et al.*, 1992). A milder clinical course in 12 ΔF508/R553X compound heterozygotes, and a more severe disease in 13 ΔF508/3905insT patients, compared to 45 ΔF508 homozygotes, all treated in the same centre, was found (Liechti-Gallati *et al.*, 1992).

A cohort study on 123 patients from 8.5 to 10 years has indicated a slightly higher mortality rate for ΔF508 homozygotes, and a better nutritional status for PS patients (Borgo *et al.*, 1993). A study by Augarten *et al.* (1993) compared the clinical course of 15 patients who were compound heterozygotes for mutation 3849+10 kb C→T with homozygotes or compound heterozygotes for ΔF508 and W1282X mutations, all of whom were treated similarly. In this study, patients with the 3849 mutation were older, diagnosed at a later age, and with a better nutritional status; none had MI, liver disease, or diabetes mellitus; sweat chloride values were normal in five of them, and pancreatic function was normal in 10. Age-adjusted pulmonary function did not differ between the two groups. Recently, 399 patients who were compound heterozygotes for ΔF508 and other mutations were matched with ΔF508 homozygotes of the same sex and closest age in 32 centres worldwide in a large analysis by the Cystic Fibrosis Genotype–

Phenotype Consortium (1993). In this series, decreased severity of the disease in the R117H/ΔF508 patients was reported, and these patients usually were PS, never had MI, were diagnosed at an older age and had lower sweat chloride concentrations. No other significant differences were found in this and any other tested genotype for a large series of variables, including pulmonary function, *Pseudomonas* colonization, nasal polyps, pancreatitis, diabetes, MI and liver cirrhosis.

In conclusion, and generally speaking, ΔF508 homozygotes seem to have a more severe disease than compound heterozygotes, as is discussed above for the exocrine pancreatic function; also other observed genotypes relate to disease in the same way as reported in Table 2.2 for PI- or PS-producing mutations. The same considerations will be repeated for CF pulmonary function, in Section 2.4.8. These observations suggest that the nutritional status of the patient as a consequence of the effect of the mutation is the most significant characteristic for its general clinical implications.

2.4.6 CF adult cases and CFTR mutations

Occasionally, milder CF cases are diagnosed in adulthood, in individuals whose general condition is relatively good. Upper lobe infections, chest atelectasis and chronic and recurrent infections are the usual manifestations (Boye *et al.*, 1980). The emerging picture is that less severe disease, as found typically in adult CF patients, may be associated with compound heterozygosity. This was indicated in a recent study on eight patients over 35 years of age selected from a group of 512 patients, where all the genotyped patients were compound heterozygotes. Five of them carried a missense mutation in the TM region and either ΔF508 or G542X; two of them had the splice mutation 2789+5G→A and ΔF508 (Ferec *et al.*, 1993). Therefore, less severe phenotypes do seem to be genetically determined.

2.4.7 CF sweat chloride concentration and CFTR mutations

A sweat chloride concentration of over 60 mM by the pilocarpine iontophoresis test (Gibson and Cooke, 1959), or sweat sodium over 70 mM, is taken as an indication of CF. Values increase with age, so that, in adults, the borderline chloride concentration is 80 mM, although intermediate values from 40 to 60 mM are sometimes observed. In general, patients with PS have a significantly lower sweat sodium and chloride concentration than do patients with PI. Occasionally, patients with low sweat chloride levels are found: 1–2% between 50 and 60 mM and, exceptionally, less than 50 mM (Davis *et al.*, 1980).

An early study found an association between CF locus-linked KM.19 marker alleles and sweat chloride concentration above or below the median value in 31 CF patients not carrying ΔF508, and indicated that an unknown CFTR gene mutation could alter electrolyte concentrations as detected in the sweat test (Gasparini *et al.*, 1991a). A study by Strong *et al.* (1991) described two G551S homozygote sisters, one of whom had borderline normal sweat electrolyte levels, the other one normal values, both being PS and adult. A meeting abstract reported one patient who was a compound heterozygote for the R117H/G542X mutations who had normal sweat chloride values, while his brother had increased chloride concentration (Highsmith *et al.*, 1991). These observations indicate that other factors interact with the CFTR genotype in determining chloride concentrations in the sweat. A ΔF508 homozygous patient with typical symptoms of gastrointestinal and pulmonary disease presented with almost normal chloride concentration in sweat tests, and an extra R553Q alteration was found associated with one ΔF508 gene, so that he had a total of three CFTR gene mutations! The authors suggest that the substitution of arg553 by glutamine modifies the effect of ΔF508 (Dork *et al.*, 1991).

2.4.8 CF pulmonary disease and CFTR mutations

Contrary to the correlation observed between pancreatic function and gene alteration, pulmonary disease, which is constant in CF, and represents the main cause of death, is not so clearly associated with CFTR mutations. A general problem in the evaluation of the degree of pulmonary disease is the difficulty of defining variables such as age progression and the influence of treatment, especially when few cases are analysed. The procedure most often adopted is to match the patients under study with others with a defined genotype and homogeneous for age, treatment and environment. Genetic factors are important in determining the severity of lung disease, as shown by a correlation between CF siblings in the forced expiratory volume in 1 sec given as a percentage of predicted normal values for age, height, sex and weight (% FEV 1 values; Santis *et al.*, 1990), which is probably the most useful parameter of lung function being used. Significant concordance in pulmonary disease severity between three monozygotic and four dizygotic sets of CF twins was reported, while the CFTR genotype was not the major determinant of lung disease. Thus genetic factors other than CFTR were suggested as influencing lung disease (Santis *et al.*, 1992). Digestive function seems to play a role in determining pulmonary status in CF patients. In a large study on 241 ΔF508 genotyped patients, PS individuals had better pulmonary function than PI ones, after adjustment for age (Kerem *et al.*, 1990). Accordingly, pulmonary function was better in 72 patients with normal fat absorption (Gaskin *et al.*, 1992). These observations indicate some influence of CFTR mutations on pulmonary disease progression, but that there is a weaker correlation between genotype and pulmonary than pancreatic function.

Given the clinical importance of the respiratory disease in CF, correlations with mutations have been attempted in several cases. A summary of some interesting results is given in Table 2.3. It should be noted that Table 2.3, contrary to Table 2.2, reports only genotypes and not single CFTR mutations as, in contrast to pancreatic function, no clear dominance picture emerges for pulmonary function. Poorer lung function of ΔF508 homozygotes was reported by some authors (Johansen *et al.*, 1991), while no difference was found by others (Al Jader *et al.*, 1992; Santamaria *et al.*, 1992). ΔF508 homozygote infants, identified through CF neonatal screening, have early

Table 2.3. A tentative classification of some CFTR gene mutations relative to pulmonary disease in CF patients with defined genotypes

More severe	
Deletions	ΔF508/ΔF508*
Splice:	621+1G→T/ΔF508[†]
Milder	
Nonsense:	G542X/G542X[‡], R553X/R553X[§], R1162X/R1162X[¶], S1255X/G542X[‖], W1316X/R553X[‖]
Frame shift:	2184delA/2184delA**
Missense:	G85E/G85E[††], R117H/ΔF508[†], R334W/G542X[‡‡], I336K/ΔF508[‡‡], G551S/G551S[§§], H1054D/ΔF508[‡‡]
Splice:	2789+5G→A/ΔF508[‡‡]

Data from: * Johansen *et al.* (1991); [†] Al Jader *et al.* (1992); [‡] Cuppens *et al.* (1990), Beaudet *et al.* (1991), Bonduelle *et al.* (1991); [§] Bal *et al.* (1991), Cheadle *et al.* (1992); [¶] Gasparini *et al.* (1992); [‖] Cutting *et al.* (1990); ** Lissens *et al.* (1993); [††] Chalkey and Harris (1991); [‡‡] Ferec *et al.* (1993); [§§] Strong *et al.* (1991).
Note: the table is not complete.
Caution: pulmonary function is variable (see text): use of this table for predicting the course of disease is strongly discouraged.

evidence of airways obstruction during the asymptomatic period (Mohon *et al.*, 1993). Studies with patients carrying the same mutation on both chromosomes are deemed to be of particular importance, as gene expression in such patients would not be modified by allelic heterogeneity. Nonsense mutation homozygotes or compound heterozygotes, while being PI, repeatedly have been reported to have a milder respiratory disease than ΔF508 homozygotes in several studies, as detailed in Table 2.3. Several nonsense mutations, all classified as severe for pancreatic function in Table 2.2, appear as mild for pulmonary function in Table 2.3. By contrast, W1282X homozygotes show variability in pulmonary function (Shoshani *et al.*, 1992). The reasons why termination mutations would determine milder respiratory than digestive symptoms is not known. The CFTR gene is expressed at a low level in the lung (Crawford *et al.*, 1991; Bremer *et al.*, 1992), and a small amount of residual wild-type transcript may be sufficient for CFTR function (Chu *et al.*, 1992; Slomski *et al.*, 1992). This in turn could be achieved by nonsense mutation-determined alternative splicing and exon skipping (Dietz *et al.*, 1993). Accordingly, in G553X homozygote respiratory cells, no CFTR mRNA has been detected (Will *et al.*, 1993). Another possibility may be that other tissue-specific cellular mechanisms substitute for absence of CFTR in the respiratory tract. A milder pulmonary, as well as pancreatic, disease, was reported for two R117H/ΔF508 siblings, and a trend to a more severe pulmonary disease in five 15–20-year-old patients with the 621+1G→T/ΔF508 genotype (Al Jader *et al.*,

1992). Other CFTR genotypes possibly associated with a milder pulmonary disease are also reported in Table 2.3.

In conclusion, variation in respiratory disease among patients with the same genotype is commonly observed (Gasparini *et al.*, 1992; Shoshani *et al.*, 1992; Borgo *et al.*, 1993; Macek *et al.*, 1993), indicating the importance of other genetic or environmental factors still to be determined. The observations on the correlation between some mutations and severity of pulmonary disease have not been supported by the recent Cystic Fibrosis Genotype–Phenotype Consortium study (1993), which pointed out that for none of the genotypes studied, which were the most common ones, could predictions be made about the severity or the course of pulmonary disease. Probably, the accessibility of the bronchiolar tissues to the environment makes them more prone than the pancreas to occasional external influences.

2.4.9 CF airways colonization and CFTR mutations

Lung infection is a major feature of CF, and several studies have considered it in relation to the patient's genotype. In a study on 235 patients, homozygous ΔF508 individuals had greater early incidence of chronic *P. aeruginosa* infection than did ΔF508 compound heterozygotes (Johansen *et al.*, 1991). In a more recent report, 267 CF children and adolescents were assessed for the age of onset of chronic airways colonization with *P. aeruginosa*, and it was observed that the age-specific colonization rates differentiated by pancreatic function (the same as seen above for pulmonary function). These rates were significantly lower in PS than in PI patients seen at the same centre, with the percentage of colonized patients increasing most rapidly in compound heterozygotes carrying ΔF508 and a nonsense mutation (R553X, G542X, W1282X or R1162X) in the nucleotide binding fold-encoding exons (Kubesh *et al.*, 1993). In the Cystic Fibrosis Genotype–Phenotype Consortium study (1993), *Pseudomonas* colonization varied from a minimum of 30% in 23 R117H/ΔF508 patients to a maximum of 82% in 17 W1282X/ΔF508 patients; ΔF508 homozygotes had 56% colonization.

2.4.10 CF liver disease and CFTR mutations

A study found that a quarter of adults with CF (57/233) had abnormal liver function, and suggested that intrahepatic impairment of biliary drainage may be important in the pathogenesis of liver disease (Nagel *et al.*, 1989). In 47 CF patients with hepatobiliary involvement, ΔF508 frequency was increased, but the difference was not significant compared to the general CF population (Ferrari *et al.*, 1991). By contrast, the risk of liver disease among 20 adult CF patients who were homozygotes for ΔF508 was found to be double that of compound heterozygotes (De Arce *et al.*, 1992). In seven CF patients who were transplanted for end stage biliary cirrhosis, ΔF508 was present in 11/14 chromosomes (79%), slightly more than expected on the basis of its general frequency in CF patients from the same population (Traystman *et al.*, 1993). Mild pulmonary, but severe pancreatic and hepatic, symptoms were described in a CF patient homozygous for a frameshift mutation (2184delA) in the regulatory domain of the CFTR (Lissens *et al.*, 1993).

2.5 A functional classification of CFTR mutations at the cell level, and relation to phenotype

Recent studies have elucidated several aspects of the CFTR protein function in apical chloride conductance in secretory epithelia. Several classes of mutations were identified, in which the protein either does not position properly in the cell membrane, or is not regulated properly by cAMP stimulation, or has altered ionic pore properties (Tsui *et al.*, 1992b). A more recent classification (Welsh and Smith, 1993) makes further distinctions within the first group, in which transport to the membrane is altered, between protein synthesis defects due to nonsense, frameshift and splice mutations (class I), and protein-processing defects (class II). Mutations G542X and 621+1G→T are examples of the first class, and ΔF508 and N1303K are examples of the second class. All of these mutations do not allow the proper insertion of CFTR into the cell membrane, and all belong to the PI group (see Table 2.2). The third class comprises mutations with diminished or severely impaired ATP response; examples are G551S and G551D, respectively, which are located in NBF1. PS or PI may develop, which is related to the degree of impaired regulation in 551S or 551D, respectively. The fourth class comprises mutations which affect the channel pore and its anion conductance. Examples are arginine substitutions in the first transmembrane domain, such as R117H, R334W and R347P; in these cases, the amount of cAMP-mediated chloride current is reduced (Sheppard *et al.*, 1993). Mutation R347P, in particular, determines a loss of multiple anion occupancy of the CFTR membrane pore (Tabcharani *et al.*, 1993). These 'mild' mutations confer PS in CF, and we can now add that they also represent the class most commonly found in several other diseases in which the gene seems to play an important role. The sections following describe the involvement of CFTR in different conditions, which presently include isolated male reproductive system and pulmonary alterations. Isolated intestinal diseases remarkably are still unlinked to CFTR. The classification of CFTR gene mutations into classes of CFTR protein function therefore helps in understanding the relationship between the functional alteration and disease phenotype, both in CF and in other diseases. These relationships are not absolute however, as other genetic or environmental factors may play a role during the long series of pathogenetic events that occur from the inheritance of the gene defect to its full expression in the patient.

2.6 Congenital bilateral absence of the vas deferens and CFTR mutations

Congenital bilateral absence of the vas deferens (CBAVD) is an obstructive azoospermia due to the absence of both scrotal vasa. Infertility is caused by CBAVD in approximately 97% of CF male patients, so that it represents an important diagnostic criterion for CF. Isolated CBAVD represents about 1–2% of male infertility cases (editorial, Lancet, 1992). As spermatogenesis is normal, CBAVD is now treated by microsurgical epididymal (Silber *et al.*, 1990) or testicular (Craft *et al.*, 1993) sperm aspiration, followed by microassisted insemination (Schoysman *et al.*, 1993) or *in vitro* fertilization (Hirsh *et al.*, 1993a). Autosomal recessive inheritance of CBAVD is suspected but not proven (no. 277180, McKusick, 1992). Recently, CFTR gene mutations have been looked for and found in CBAVD patients with no clinical manifestations of CF. Initially, an increased frequency of ΔF508 heterozygotes was reported in CBAVD, compared to the carrier rate expected from the mutation incidence observed in CF patients. In the first such study, seven heterozygotes were reported out of 17 patients: sweat chloride concentration was abnormal in six of them, therefore classifying them as CF patients, chronic sinusitis was present in three, chronic bronchial hypersecretion in one and diabetes in one (Dumur *et al.*, 1990a). In an update of these results, eight patients out of 19 were heterozygous for ΔF508, and seven of them had chronic sinusitis (Rigot *et al.*, 1991). Other CFTR mutations were then looked for and found by using the single strand conformation polymorphism (SSCP) technique (Orita *et al.*, 1989) for localizing sequence alterations in *in vitro*-amplified segments of the CFTR gene; 16 out of 25 cases (64%) had at least one detectable CFTR mutation (Anguiano *et al.*, 1992). Also, and most notably, three of these patients were compound heterozygotes carrying both ΔF508 and G576A (two cases) or D1270N, mutations, respectively. One out of three of the compound heterozygotes and two out of six of the carriers tested had elevated sweat electrolytes. These observations demonstrate that at least some healthy men with CBAVD have a primarily genital form of CF or an altogether different disease, without the pulmonary, pancreatic and electrolytic manifestations. The mutation D1270N has not been described previously in patients with CF, and the mutation G576A was described as a benign polymorphism not causing CF in combination with a known mutation, raising the possibility that different mutations in the same gene might be typical of this condition. Several other CFTR mutations have since been found in isolated CBAVD, as detailed below and summarized in Table 2.4.

A high frequency of the mutation R117H, already implicated in PS-CF and in 'mild' CF cases, was observed by SSCP screening in 4/23 cases of CBAVD, compounded with ΔF508 (three cases) or with 2322delG (Gervais *et al.*, 1993a). Four R117H heterozygotes compounded with ΔF508 (3) or R553X (1) have been found in a series of 26 patients (Patrizio *et al.*, 1993) with at least one identified mutation out of a total of 44 examined (59%). No sweat tests were performed in these patients. Carriers of one copy of a CFTR gene mutation are not infertile, as shown in CF families, in carrier fathers of CBAVD patients, and in their children, obtained by *in vitro* fertilization, whose vasa deferentia were normal (Patrizio *et al.*, 1993). Three more cases of ΔF508/R117H compound heterozygotes, one R117H/G551D and one ΔF508/R75Q, have been described in an update of their original series, taking the frequency of R117H in CBAVD to almost 10% (Oates and Amos, 1993). Among 35 men participating in a

Table 2.4. CFTR gene mutations found in CBAVD patients

(a) Most frequent mutations* (no. of chromosomes)

	Reference					
Mutation	1	2	3	4	5	6
ΔF508	16	25	11	8	1	20
R117H	4	4	4	—	—	5
W1282X	6	—	—	—	—	—
R75Q	—	3	—	—	—	—
G551D	—	2	—	—	—	—
R553X	2	—	—	1	—	—
G576A	—	2	—	—	—	—
Total identified	26/88 (30%)	40/82 (49%)	16/46 (35%)	12/52 (23%)	2/2 (100%)	25/70 (36%)

(b) Genotypes characterized (no. of patients)

	Reference					
Genotype	1	2	3	4	5	6
ΔF508/R117H	3	3	3	—	—	5
ΔF508/G576A	—	2	—	—	—	—
ΔF508/D1270N	—	1	—	—	—	—
ΔF508/R75Q	—	1	—	—	—	—
ΔF508/R347Q	—	—	—	1	—	—
ΔF508/F508C	—	—	—	—	1	—
R117H/G551D	—	1	—	—	—	—
R117H/R553X	1	—	—	—	—	—
R117H/2322delG	—	—	1	—	—	—
S549N/R1070Q	—	—	—	1	—	—
Total identified	4/44 (9%)	8/41 (20%)	4/23 (17%)	2/26 (8%)	1/1 (100%)	5/35 (14%)

— Mutation not found or not searched.
* Note: mutations reported once are not indicated.
References: 1. Patrizio *et al.* (1993); 2. Oates and Amos (1993); 3. Gervais *et al.* (1993a); 4. Osborne *et al.* (1993); 5. Meschede *et al.* (1993); 6. Williams *et al.* (1993).

microscopic epididymal sperm aspiration programme, 57% were carriers of ΔF508 and five of them were compound heterozygotes with R117H (Williams *et al.*, 1993). More information about R117H, which is the most characteristic CBAVD mutation, is now available. A normal woman with genotype R117H/ΔF508 was identified by screening 4000 women during pregnancy for CFTR mutations. She had no pulmonary or pancreatic disease, and normal sweat chloride concentration (Lee *et al.*, 1992). Also, in a pilot carrier-screening programme of 1710 individuals, R117H and G551D carriers were much more frequent (11% and 74%, respectively) than expected on the basis of CF patient determinations (0.1% and 2.3%, respectively) (Handelin *et al.*, 1992). These findings indicate that the R117H mutation is more common in mild CF cases and in other diseases, and raise questions about the spectrum of mutations to be screened during CF carrier tests in the population. The R117H mutation has therefore been found in compound heterozygotes expressing three different phenotypes: CBAVD, PS-CF and normal. Recent evidence, moreover, indicates that the mutation R117H, which is located in exon 4, is followed in the gene by a polypyrimidine tract acceptor-site length variant, 5T or 7T, which alters the splicing efficiency of exon 9, and results either in a patient with PS-CF, or with CBAVD and normal individuals, respectively (Kiesewetter *et al.*, 1993). The authors suggest that the general genetic context in which a mutation such as R117H occurs, can play a role in determining the type of illness.

In conclusion, and as shown in Table 2.4, ΔF508 and R117H are the most common mutations, and ΔF508/R117H is the most common genotype, in isolated CBAVD; almost all compound heterozygotes include one or the other, indicating a prominent role for these two mutations in determining the disease. Out of a total of 170 isolated CBAVD males reported in Table 2.4, 106 (62%) have at least one CFTR identified mutation, and 24 (14%) have two. Several groups are finding CFTR mutations in CBAVD patients, and the impression is that, with the increasing efficiency allowed by the recent mutation screening methods, more and more genotypes will be found to carry different CFTR mutations. It is interesting to note that no CBAVD patient who is a true homozygote has yet been identified, as if homozygosity could produce only a CF genotype, although it is possible that homozygotes for a rare or novel CF mutation may eventually be identified. It is still questionable whether all CBAVD cases will carry CFTR mutations, as seems most likely at present. In this regard, one recent report indicated that seven CBVAD subjects heterozygous for a known CF mutation did not show a second CFTR gene mutation, which was excluded after determining the DNA sequence of all the 27 exons of the gene and intron boundaries, and after reverse transcription and amplification of their CFTR mRNAs (Osborne *et al.*, 1993). The authors conclude that most CBAVD cases are not compound heterozygotes for mutations within the CFTR gene, and suggest that still undiscovered defects in the CFTR promoter or other regulatory regions may decrease the synthesis of the CFTR protein and cause disease only in the most susceptible organ, the vas deferens.

Specific CFTR mutations therefore appear to predominate in CBAVD, compared to CF but, as the same genotype may be present in CF and in CBAVD cases, other modifying factors need to be considered. They could be genetic factors determined by the same locus, as suggested by Meschede *et al.* (1993) for F508C, and by Kiesewetter *et al.* (1993) for R117H. CBAVD would therefore constitute a primarily (Anguiano *et al.*, 1992), or exclusively, genital form of CF. The effect of CFTR mutations on the development of the epididymis from the mesonephric duct is not known. It must occur after the 7th week of gestation, when the duct has split into its reproductive and ureteral parts, and it may affect the intraluminal electrolyte composition of the developing vasal structures necessary to achieve patency during embryogenesis (Oates and Amos, 1993). A careful search for pulmonary and pancreatic symptoms should be undertaken in CBAVD patients, as disfunctions might become apparent with time. CFTR gene mutation analysis should be recommended for CBAVD patients and their partners in cases of assisted reproduction in order to evaluate the risk of having a CF child. CFTR mutation testing in appropriate family members should also be considered, as brothers may be affected by CBAVD, sisters may be carriers, and all siblings would be at an increased risk of having a CF child.

2.6.1 Unilateral absence of the vas deferens and CFTR gene mutations

Males with unilateral absence of the vas deferens (UAVD) frequently have ipsilateral renal agenesis or other urogenital anomalies, and may be fertile and incidentally ascertained. At a recent meeting, it was reported that 6/14 infertile UAVD cases (43%) were heterozygous for CFTR mutations ΔF508, R117H or R75Q (two each), and one of them had two brothers with CBAVD (Mickle *et al.*, 1993). Therefore UAVD also seems to be related to CFTR, as does CBAVD, and it may only be a question of time before compound heterozygotes will be found.

2.6.2 Male infertility due to impaired sperm function or azoospermia, and CFTR mutations

At a recent meeting, communications were presented on the increased presence of CFTR mutations in men with infertility due to causes other than CBAVD. In 12 of 62 impaired sperm function cases, due to astheno-, oligo-, terato-zoospermias, or combinations thereof, heterozygotes for common CFTR mutations were found, compared to none in 27 normospermic men, with one impaired sperm function subject being a ΔF508/G551D compound heterozygote. In 3/21 azoospermic men, mutations were also found, one of them being a R117H/G551D compound heterozygote (van der Ven *et al.*, 1993). The authors indicate that it is likely that a second CFTR mutation will be found in these individuals (as seen above for CBAVD), and they conclude that CFTR gene mutations may result in male infertility due to causes different from those in CBAVD. These observations suggest that the CFTR gene may be more generally involved in normal sperm maturation, and not only in epididymis differentiation.

2.6.3 Young's syndrome and CFTR mutations

Young's syndrome is characterized by chronic sinopulmonary infections starting in early childhood, and obstructive azoospermia in adult males; infertility is usually, but not always, present, and spermatogenesis is normal (Handelsman *et al.*, 1984). The midepididymis ductules are obstructed by inspissated secretions, and no atresia of the vasa deferentia is seen. Normal pancreatic and sweat gland function is present. Autosomal recessive inheritance is consistent with family data (Kueppers, 1992).

Two out of seven patients examined for four common CFTR mutations were carriers of ΔF508 (Hirsh *et al.*, 1993b). Twelve patients with Young's syndrome have been genotyped for 30 CFTR gene mutations, with one ΔF508 and one R75Q mutation being identified in 24 chromosomes. This was not significantly different from the carrier frequency expected in the population. The authors conservatively conclude it is unlikely that the majority of Young's syndrome patients have a form of CF, but that further data are needed (Friedman *et al.*, 1993).

2.6.4 Chronic obstructive pulmonary disease and CFTR mutations

There are several pulmonary disorders that share some features with CF, and must therefore be considered as possible candidates for CFTR gene mutations which may affect disease susceptibility or even determine its development. A link between 'cystic fibrosis of the pancreas as it occurs in the pediatric age group to some of the common chronic pulmonary, gastrointestinal, and metabolic disorders of adults' was already suggested by several groups in the late 1950s and early 1960s, on the basis of sweat test positivity (Di Sant'Agnese and Powell, 1962). Chronic obstructive pulmonary disease (COPD) is characterized by chronic airways limitation to airflow, due to chronic bronchitis or emphysema. In many cases, bronchiectasis is also present. Familial aggregation has been observed, and involvement of the α-1-antitrypsin gene has been demonstrated (Kueppers, 1992). An early linkage disequilibrium and ΔF508 mutation analysis in an Italian population sample carried out in 16 COPD patients from 25 to 80 years of age could not demonstrate a CF gene involvement (Gasparini *et al.*, 1990b). An increased frequency of ΔF508 was found in six out of 65 French patients with chronic bronchial hypersecretion aged from 26 to 67 years; in one, a CF diagnosis was made, two had sweat chloride concentrations close to the normal borderline value of 60 mM and three had normal chloride concentrations (Dumur *et al.*, 1990b). Out of 10 German patients from 18 to 45 years old with disseminated bronchiectatic lung disease, two ΔF508 homozygotes (one PI and one PS) were classified as CF patients, and four ΔF508 heterozygous patients were found (Poller *et al.*, 1991). These data, together with others that follow, are reported in Table 2.5, which indicates the possible influence of CFTR gene mutations in COPD.

ΔF508 was not found in 21 adult Japanese chronic bronchitis patients with severe obstruction (Akai *et al.*, 1992). The French study of ΔF508 mutations recently was extended to a larger number of patients with chronic bronchitis or bronchiectasis. In total, 20 ΔF508 carriers were found out of 273 patients five ΔF508 carriers were found in a subset of 16 bronchiectasis patients (31%) with sweat chloride concentrations of 60 mM or more, a significant increase over the percentage expected. A slight increase was also found in bronchiectasis patients with normal sweat chloride (4/47), and in chronic bronchitis patients with high

Table 2.5. CFTR gene mutations found in COPD patients

	Reference		
Mutation	1	2	3
ΔF508	4	9	1
R117H	—	—	1
R1066C	—	—	1
Total	4/16 (25%)	9/126 (7%)	7/56 (12%)

— Mutation not searched.
Numbers indicate chromosomes with or without mutation.
References: 1. Poller *et al.* (1991) (selected data given); 2. Gervais *et al.* (1993b) (data from bronchiectatic patients); 3. Pignatti *et al.* (1993), and our unpublished results.

sweat chloride (4/49); in the chronic bronchitis patients with low sweat chloride, the frequency (7/61) was the same as in the normal population (Gervais *et al.*, 1993b). Taking into account these results, and considering that ΔF508 constitutes only about 48% of CF chromosomes in north-eastern Italy (Gasparini *et al.*, 1993), and is not found in Japan (Akai *et al.*, 1992), and that many other phenotypically mild or uncommon CFTR mutations had been described in the meantime, we have started a more extended survey of CFTR mutations in a larger COPD series. This has been carried out by screening for 40 known mutations and by looking for rare and unknown mutations with denaturing gradient gel electrophoresis analysis of the gene (Fisher and Lerman, 1983). In 28 COPD patients aged from 21 to 76 years, with normal sweat electrolytes and pancreatic function, seven carriers of CFTR mutations have been found to date, with ΔF508, R75Q, R117H, R1066C, two other rare and one novel mutation respectively (Pignatti *et al.*, 1993; and unpublished data from our group). This is a significant increase over the expected population carrier frequency. Moreover, most of the identified mutations detailed, except ΔF508 and one other mutation, are located in the internal or external ends of the membrane-spanning sequences of the CFTR, and three of them are arginine substitutions. Some of these mutations belong to class IV, have been found also in patients with mild CF and in CBAVD cases as described above and reported in Table 2.4, and might produce a less severe disease because they determine a reduction of the amount of cAMP-regulated apical chloride current (Sheppard *et al.*, 1993). These findings indicate that a particular type of CFTR mutation is present typically in COPD patients. Table 2.5 summarizes some mutations found to date in COPD patients.

2.6.5 Allergic bronchopulmonary aspergillosis or chronic **Pseudomonas** *bronchitis and CFTR mutations*

CFTR mutations have also been looked for in isolated allergic bronchopulmonary aspergillosis (ABPA) and chronic *Pseudomonas* bronchitis (CPB) adult patients with sweat chloride concentrations under 50 mM. ABPA was defined by specific IgE and IgG precipitins to *Aspergillus fumigatus*, elevated total IgE, eosinophilia, asthma, pulmonary infiltrates or bronchiectasis. One compound heterozygote ΔF508/R347H, and one ΔF508 carrier were found in eight ABPA patients, indicating a possible influence of the CFTR gene on the disease, but no common CFTR mutation was found in CPB patients (Miller *et al.*, 1993). In an expanded series of patients, a third ΔF508 carrier was identified out of a total of 13 ABPA patients (G. Cutting, personal communication).

2.7 Neonatal transitory hypertrypsinaemia and CFTR mutations

A commonly used newborn CF screening test employs immunoreactive trypsin (IRT) level determination in dried blood spots. In 149/28 000 babies born in 1990 in the Caen region of France, elevated IRT over 600 $\mu g\ l^{-1}$ was detected on day 4, which decreased on retesting at day 21; nine of them were ΔF508 carriers (9/149=6%), higher than expected in the population (Laroche and Travert, 1991). The observation was confirmed by a similar study performed over the same period of time in the Paris region: 16 babies out of 51 assumed not to have CF carried ΔF508 (Lucotte *et al.*, 1991). No information on the frequency of other mutations is available.

2.8 Nasal polyposis and CFTR mutations

The CF phenotype often includes nasal polyposis. A study on nasal polyps removed from patients with no clinical history of CF, but no sweat test done, reported CFTR mutations in seven out of 56 patients examined: G551D was present in four, ΔF508 in two and R553X in one, respectively. As G551D was significantly overrepresented, the authors suggest a causal role for this particular mutation in determining isolated polyposis nasi (Burger *et al.*, 1991). In the recent Cystic Fibrosis Genotype–Phenotype Consortium study (1993), nasal polyps were present in all defined genotypes characterized, including both PI and PS groups, with rates ranging from 14% in 30 1717–1G→A/ΔF508 to 15% in 34 R553X/ΔF508 to 40% in 15 W1282X/ΔF508 compound heterozygotes; ΔF508 homozygotes developed nasal polyposis in 16% of the cases; G551D genotypes were not reported.

2.9 Conclusions

Exceptionally productive work carried out in the last 4 years since the discovery of the CFTR gene, has determined the function of the protein, and the nature of the defects responsible for CF and other diseases. Mutations characteristic of pancreatic function impairment have been determined successfully in digestive function sufficient and insufficient patients. Pulmonary disease is less well correlated with the genetic defect, possibly because of a tissue-specific variation in the expression of the gene, and other genetic and environmental factors contributing to airway function and bacterial colonization. According to a recent review by Tsui (1992b), CF phenotypes can be classified at three levels. The first comprises manifestations common to all CF patients, such as the increase in sweat chloride concentration. The second shows a direct relation to genotype, and is exemplified by pancreatic function. The third level of clinical manifestations is influenced strongly by other factors, and the symptoms are highly variable, even within genotypes or families; this is the case for pulmonary disease. More research is needed on the cellular, tissue and organ manifestations of different mutations to clarify further the pathophysiological mechanisms of disease. The involvement of the CF gene in other diseases represents a challenge for further study. Some male sterility cases without signs of CF, such as CBAVD, where several cases of compound heterozygosity have already been observed, and perhaps also other azoospermias, are now ascribable to CFTR mutations, indicating a recessive inheritance of the disease gene. Mutations in the CFTR gene are beginning to be found also in lung disease of the adult with no other manifestation of CF, as in COPD and ABPA. Only one patient with identified mutations in both genes has been described up to now, but perhaps in a short time others will follow, suggesting that in some of these cases recessive inheritance of CFTR gene mutations may also play a role.

Thus one has now to think in terms of a continuum of CFTR-derived phenotypes, which may vary among different diseases as well as in different manifestations of the same disease (Figure 2.2). In an extension of the natural selection hypothesis proposed for CF, whereby the diminished intestinal chloride and fluid secretion (Quinton, 1983) would increase carrier resistance to diarrhoeas mediated by bacterial toxins, once common in Europe (Hansson, 1988), we would now propose, in the light of all these findings, that other CFTR gene-related diseases, such as CBAVD and perhaps some cases of COPD, may be the present day result of heterozygote advantage. A wider picture is emerging, which encompasses not only CF, but single manifestations of the reproductive, airway, and possibly also the digestive, systems. The number of CFTR-related diseases keeps expanding, as does the number of CFTR mutations, and the

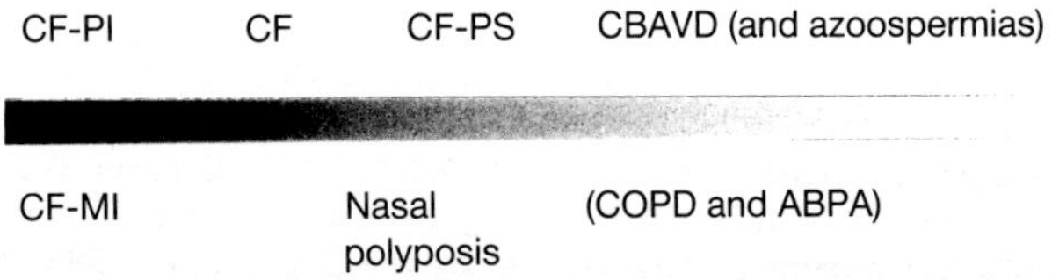

Figure 2.2. A model of CFTR-related diseases, tentatively ordered by decreasing severity. CF is shown in the middle, with its various manifestations. PI and PS are partly separated, because of the different mutations involved. Male infertility and pulmonary disease areas are shown while other, possibly related diseases are not, because more observations have been made of these.

consequences of these observations have far reaching implications for the pathogenesis, prevention and therapy of these illnesses.

Acknowledgements

The author thanks R. Williamson for the initial suggestion and encouragement, L.-C. Tsui, the Cystic Fibrosis Genetic Analysis Consortium and G. Cutting, for providing published and unpublished information and for useful discussions, and A. Varlien for patient typing.

Our work was supported by grants from the Italian C.N.R. Target projects 'Biotechnology and Bioinstrumentation' and 'Genetic Engineering', and from M.U.R.S.T. 40% and 60%.

References

Akai S, Okayama H, Shimura S, Tanno Y, Sasaki H, Takishima T. (1992) Delta F508 mutation of cystic fibrosis gene is not found in chronic bronchitis with severe obstruction in Japan. *Annu. Rev. Respir. Dis.* **146:** 781–783.

Al-Jader LN, Meredith AL, Ryley HC, Cheadle JP, Maguire S, Owen G, Goodchild MC, Harper PS. (1992) Severity of chest disease in cystic fibrosis patients in relation to their genotypes. *J. Med. Genet.* **29:** 883–887.

Anguiano A, Oates RD, Amos JA, Dean M, Gerrard B, Stewart C, Maher TA, White MB, Milunsky A. (1992) Congenital bilateral absence of the vas deferens – a primarily genital form of cystic fibrosis. *J. Am. Med. Assoc.* **267:** 1794–1797.

Augarten A, Kerem BS, Yahav Y, Noiman S, Rivlin Y, Tal A, Blau H, Ben-Tur L, Szeinberg A, Kerem E, Gazit E. (1993) Mild cystic fibrosis and normal or borderline sweat test in patients with the 3849+10 kb C→T mutation. *Lancet* **342:** 25–26.

Auvinet M, Morel Y, Chambon V, Andre J, Vidaud M, Goossens M, Bellon G, Gilly R. (1989) Cystic fibrosis with and without meconium ileus. *Lancet* **i:** 160–161.

Bal J, Stuhrmann M, Schlosser M, Schmidtke J, Reiss J. (1991). A cystic fibrosis patient homozygous for the nonsense mutation R553X. *J. Med. Genet.* **28:** 715–717.

Beaudet A, Tsui LC. (1993) A suggested nomenclature for designating mutations. *Hum. Mutat.* **2:** 245–248.

Beaudet A, Perciaccante R, Cutting G. (1991) Homozygous nonsense mutation causing cystic fibrosis with uniparental disomy. *Am. J. Hum. Genet.* **48:** 1213.

Boat TF, Welsh MJ, Beaudet AL. (1989) Cystic fibrosis. In: *The Metabolic Basis of Inherited Disease*, Vol. 2, 6th Edn (eds CL Scriver, AL Beaudet, WS Sly, D Valle). McGraw-Hill, New York, pp. 2649–2680.

Bonduelle M, Lissens W, Malfroot A, Liebaers I, Dab I. (1991) Mild cystic fibrosis in a child homozygous for G542 nonsense mutation in CF gene. *Lancet* **338:** 189.

Borgo G, Mastella G, Gasparini P, Zorzanello A, Doro R, Pignatti PF. (1990a) Pancreatic function and gene deletion F508 in cystic fibrosis. *J. Med. Genet.* **27:** 665–669.

Borgo G, Mastella G, Gasparini P, Pignatti PF. (1990b) Genotype analysis and pancreatic function classification in cystic fibrosis. *Lancet* **335:** 1601.

Borgo G, Gasparini P, Bonizzato A, Cabrini C, Mastella G, Pignatti PF. (1993) Cystic fibrosis: the delta F508 mutation does not lead to an exceptionally severe phenotype: a cohort study. *Eur. J. Pediatr.* **152:** 1006–1011.

Boye N, Skarpaas I, Fausa O. (1980) Cystic fibrosis in adult patients. *Eur. J. Respir. Dis.* **61:** 227–232.
Bremer S, Hoof T, Wilke M, Busche R, Sholte B, Riordan JR, Maass G, Tummler B. (1992) Quantitative expression of multidrug-resistance P-glycoprotein (MDR1) and differentially spliced CFTR mRNA transcripts in human epithelia. *Eur. J. Biochem.* **206:** 137–149.
Burger J, Macek M, Stuhrmann M. (1991) Genetic influences in the formation of nasal polyps. *Lancet* **337:** 974.
Chalkey G, Harris A. (1991) A cystic fibrosis patient who is homozygous for the G85E mutation has very mild disease. *J. Med. Genet.* **28:** 875–877.
Cheadle J, Al-Jader L, Goodchild M, Meredith A. (1992) Mild pulmonary disease in a cystic fibrosis child homozygous for R553X. *J. Med. Genet.* **29:** 597.
Cheadle J, Goodchild M, Meredith A. (1993) Direct sequencing of the complete CFTR gene: the molecular characterization of 99.5% of CF chromosomes in Wales. *Hum. Mol. Genet.* **2:** 1551–1556.
Chillon M, Casals T, Nunes V, Gimenez J, Perez Ruiz, E, Estivill X. (1993) Identification of a new missense mutation (P205S) in the transmembrane domain of the CFTR gene associated with a mild cystic fibrosis phenotype. *Hum. Mol. Genet.* **2:** 1741–1742.
Chu C-S, Trapnell BC, Curristin SM, Cutting GR, Crystal RG. (1992) Extensive posttranscriptional deletion of the coding sequences for part of nucleotide-binding fold 1 in respiratory epithelial mRNA transcripts of the CFTR gene is not associated with the clinical manifestations of CF. *J. Clin. Invest.* **90:** 785–790.
Collins F. (1992) Cystic fibrosis: molecular biology and therapeutic implications. *Science* **256:** 774–779.
Corey M, Durie P, Moore D, Forstner G, Levison H. (1989) Familial concordance of pancreatic function in cystic fibrosis. *J. Pediatr.* **115:** 274–277.
Coutelle C, Bruckner R, Grade K, Behrens F, Gedschold J, Hein J, Szibor R, Bauer I, Brock J, Graupner I, Urner U, Leucht B. (1992) Prevalence of cystic fibrosis mutations in the East German population. *Hum. Mutat.* **1:** 109–112.
Craft I, Bernett V, Nicholson N. (1993) Fertilizing ability of testicular spermatozoa. *Lancet* **342:** 864.
Crawford I, Maloney PC, Zeitlin PL, Guggino WB, Hyde SC, Turley H, Gatter KC, Harris A, Higgins CF. (1991) Immunochemical localization of the cystic fibrosis gene product CFTR. *Proc. Natl Acad. Sci. USA* **88:** 9262–9266.
Cremonesi L, Ferrari F, Belloni E, Magnani C, Seia M, Ronchetto P, Rady M, Russo MP, Romeo G, Devoto M. (1992) Four new mutations of the CFTR gene (541 del C, R347H, R352Q, E585X) detected by DGGE analysis in Italian CF patients, associated with different clinical phenotypes. *Hum. Mutat.* **1:** 314–319.
Cuppens H, Marynen P, De Boeck C, De Baets F, Eggermont E, Van den Berghe H, Cassiman JJ. (1990) A child, homozygous for a stop codon in exon 11, shows milder cystic fibrosis symptoms than her heterozygous nephew. *J. Med. Genet.* **27:** 717–719.
Curtis A, Jackson J, Keston M, Brock D. (1989) Genetic differences between cystic fibrosis with and without meconium ileus. *Lancet* **i:** 1078–1079.
Cutting GR, Kasch LM, Rosenstein BJ, Tsui L-C, Kazazian HH, Antonorakis SE. (1990) Two patients with cystic fibrosis, nonsense mutations in each cystic fibrosis gene, and mild pulmonary disease. *N. Engl. J. Med.* **323:** 1685–1689.
Cystic Fibrosis Genetic Analysis Consortium. (1990) Worldwide survey of the delta F508 mutation: report from the Cystic Fibrosis Genetic Analysis Consortium. *Am. J. Hum. Genet.* **47:** 354–359.
Cystic Fibrosis Genotype–Phenotype Consortium. (1993) Correlation between genotype and phenotype in patients with cystic fibrosis. *N. Engl. J. Med.* **329:** 1308–1313.
Davis P, Hubbard V, Di Sant'Agnese P. (1980) Low sweat electrolytes in a patient with cystic fibrosis. *Am. J. Med.* **69:** 643–646.
De Arce M, O'Brien S, Hegarty J. (1992) Deletion F508 and clinical expression of cystic fibrosis related liver disease. *Clin. Genet.* **42:** 271–272.

Dean M, White MB, Amos J, Gerrard B, Stewart C, Khaw KT, Leppert M. (1990). Multiple mutations in highly conserved residues are found in mildly affected cystic fibrosis patients. *Cell* **61:** 863–870.

Devoto M, De Benedetti L, Seia M, Piceni Sereni L, Ferrari M, Bonduelle ML, Malfroot A, Lissens W, Balassopoulou A, Adam G, Loukopoulos D, Cochaux P, Vassart G, Szibor R, Hein J, Grade K, Berger W, Wainwright B, Romeo G. (1989) Haplotypes in cystic fibrosis patients with or without pancreatic insufficiency from four European populations. *Genomics* **5:** 894–898.

Di Sant'Agnese PA, Powell GF. (1962) The eccrine sweat defect in cystic fibrosis of the pancreas (mucoviscidosis). *Ann. N.Y. Acad. Sci.* **93:** 555–559.

Dietz H, Kendzior R, Eldadah Z. (1993) Nonsense mutations act in *cis* to alter splice-site selection: verification in a heterologous expression system. *Am. J. Hum. Genet.* **53** (Suppl. 3): abstract 675.

Dork T, Wulbrand U, Richter T, Neumann T, Wolfes H, Wulf B, Maass G, Tummler B. (1991) Cystic fibrosis with three mutations in the cystic fibrosis transmembrane conductance regulator gene. *Hum. Genet.* **87:** 441–446.

Dumur V, Gervais R, Rigot JM, Lafitte J-J, Manouvrier S, Biserte J, Mazeman E, Roussel P. (1990a) Abnormal distribution of the CF delta F508 allele in azoospermic men with congenital aplasia of epididymis and vas deferens. *Lancet* **336:** 512.

Dumur V, Lafitte J-J, Gervais R, Debaecker D, Kesteloot M, Lalau G, Roussel P. (1990b) Abnormal distribution of cystic fibrosis delta F508 allele in adults with chronic bronchial hypersecretion. *Lancet* **335:** 1340.

Editorial. (1992) Congenital bilateral absence of the vas deferens and cystic fibrosis. *Lancet* **339:** 1328–1329.

Ferec C, Audrezet MP, Mercier B, Guillermit H, Moullier P, Quere I, Verlingue C. (1992) Detection of over 98% cystic fibrosis mutations in a Celtic population. *Nature Genetics* **1:** 188–191.

Ferec C, Verlingue C, Guillermit H, Querè I, Raguenes O, Felgelson J, Audrezet M-P, Moullier P, Mercier B. (1993) Genotype analysis of adult cystic fibrosis patients. *Hum. Mol. Genet.* **2:** 1557–1560.

Ferrari M, Antonelli M, Bellini F, Borgo G, Castiglione O, Curcio L, Dallapiccola B, Devoto M, Estivill X, Gasparini P, Giunta A, Marianelli L, Mastella G, Novelli G, Pignatti PF, Romano L, Romeo G, Seia M, Williamson R. (1990) Genetic differences in cystic fibrosis patients with and without pancreatic insufficiency: an Italian collaborative study. *Hum. Genet.* **84:** 435–438.

Ferrari M, Colombo C, Sebastio G. (1991) Cystic fibrosis patients with liver disease are not genetically distinct. *Am. J. Hum. Genet.* **48:** 815–816.

Fisher S, Lerman L. (1983) DNA fragments differing by single base-pair substitutions are separated in denaturing gradient gels: correspondence with melting theory. *Proc. Natl Acad. Sci. USA* **80:** 1579–1583.

Friedman K, Teichtal H, Robinson J. (1993) Screening Young's syndrome patients for CFTR mutations. *Pediatr. Pulmonol.* **15** (Suppl. 9): abstract 125.

Gaskin K, Gurwitz D, Durie P, Corey M, Levison H. (1992) Improved respiratory prognosis in patients with cystic fibrosis with normal fat absorption. *J. Pediatr.* **100:** 857–862.

Gasparini P, Novelli G, Estivill X, Olivieri D, Savoia A, Ruzzo A, Nunes V, Borgo G, Antonelli M, Williamson R, Pignatti PF, Dallapiccola B. (1990a) The genotype of a new linked DNA marker, MP6d-9, is related to the clinical course of cystic fibrosis. *J. Med. Genet.* **27:** 17–20.

Gasparini P, Savoia A, Luisetti M, Peona V, Pignatti PF. (1990b) The cystic fibrosis gene is not likely to be involved in chronic obstructive pulmonary disease. *Am. J. Respir. Cell. Mol. Biol.* **2:** 297–299.

Gasparini P, Pignatti PF, Borgo G, Mastella G. (1991a) Sweat chloride concentration in cystic fibrosis patients varies with KM.19 genotype but not with the presence or absence of the common F508 deletion. *Am. J. Med. Genet.* **39:** 230–231.

Gasparini P, Nunes V, Savoia A, Dognini M, Morral N, Gaona A, Bonizzato A, Chillon M, Sangiuolo F, Novelli G, Dallapiccola B, Pignatti PF, Estivill X. (1991b) Search for South

European cystic fibrosis mutations: identification of two new mutations, four variants and intronic sequences. *Genomics* **10:** 193–200.

Gasparini P, Borgo G, Mastella G, Bonizzato A, Dognini M, Pignatti PF. (1992) Nine cystic fibrosis patients homozygous for the CFTR nonsense mutation R1162X have mild or moderate lung disease. *J. Med. Genet.* **29:** 558–562.

Gasparini P, Marigo C, Bisceglia G, Nicolis E, Zelante L, Bombieri C, Borgo G, Pignatti PF, Cabrini G. (1993) Screening of 62 mutations in a cohort of cystic fibrosis patients from north eastern Italy: their incidence and clinical features of defined genotypes. *Hum. Mutat.* **2:** 389–394.

Gervais R, Dumur V, Rigot J-M, Lafitte J-J, Roussel P, Claustres M, Demaille J. (1993a) High frequency of the R117H cystic fibrosis mutation in patients with congenital absence of the vas deferens.

Gervais R, Lafitte J-J, Dumur V, Kesteloot M, Lalau G, Houdret N, Roussel P. (1993b) Sweat chloride and delta F508 mutation in chronic bronchitis or bronchiectasis. *Lancet* **342:** 997.

Gibson L, Cooke R. (1959) A test for concentration of electrolytes in sweat in cystic fibrosis of pancreas utilizing pilocarpine by iontophoresis. *Pediatrics* **23:** 545–549.

Grody WW, Kronquist KE, Lee EU, Edmond J, Rome LH. (1993) PCR-based cystic fibrosis (CF) carrier screening in a first-year medical student biochemistry laboratory. *Am. J. Hum. Genet.* **53:** 1352–1355.

Hamdi I, Payne S, Barton D, McMahon R, Green M, Shneerson J, Hales C. (1992) Genotype analysis for delta F508 allele in cystic fibrosis in relation to the occurrence of diabetes melitus. *J. Med. Genet.* **29:** 280.

Hamosh A, Cutting GR. (1993) Genotype/phenotype relationships in cystic fibrosis. In: *Current Topics in Cystic Fibrosis*, Vol. 1 (eds JD Dodge, DJH Brook, JW Widdicompe). John Wiley & Sons, New York, pp. 69–92.

Hamosh A, King TM, Rosenstein BJ, Corey M, Levison H, Durie P, Tsui L-C, McIntosh I, Keston M, Brock DJH, Macek M Jr, Zemkova D, Krasnicanova H, Vavrova V, Macek M Sr, Golder N, Schwarz MJ, Super M, Watson EK, Williams C, Bush A, O'Mahoney SM, Humphries P, De Arce MA, Reis A, Burger J, Stuhrmann M, Schmidtke J, Wulbrand U, Dork T, Tummler B, Cutting GR. (1992) Cystic fibrosis patients bearing both the common missense mutation Gly→Asp at codon 551 and the delta F508 mutation are clinically indistinguishable from delta F508 homozygotes, except for decreased risk of meconium ileus. *Am. J. Hum. Genet.* **51:** 245–250.

Hamosh A, Macek M, Nash E, Curristin SM, Rosenstein BJ, Cutting GR. (1993).Mutation analysis in cystic fibrosis patients with pancreatic sufficiency, pancreatitis, borderline sweat chloride concentrations or isolated congenital bilateral absence of the vas deferens. *Am. J. Hum. Genet.* **53** (Suppl. 3): abstract 1169.

Handelin B, Witt, D, Skoletsky J, Shuber A. (1992) Unexpected prevalence of R117H and G551D CF mutations in a randomly screened population. *Am. J. Hum. Genet.* **51** (Suppl. 4): abstract 858.

Handelsman D, Conway A, Boylan L, Turtle J. (1984) Young's syndrome, Obstructive azoospermia and chronic sinopulmonary infections. *N. Engl. J. Med.* **310:** 3–9.

Hannson G. (1988) Cystic fibrosis and chloride secreting diarrhoea. *Nature* **333:** 711.

Highsmith W, Barch L, Silverman L, Knowles M. (1991) Identification of an unusual genotype (R117H/G542X) in two adult siblings with mild cystic fibrosis but discordant sweat ductal function. *Pediatr. Pulmonol.* **13** (Suppl. 6): abstract 73.

Hirsh A, Montgomery J, Mohan P, Mills C, Bekir J, Tan S-L. (1993a) Fertilisation by testicular sperm with standard IVF techniques. *Lancet* **342:** 1237–1238.

Hirsh A, Williams C, Williamson B. (1993b) Young's syndrome and cystic fibrosis mutation delta F508. *Lancet* **342:** 118.

Hyde SC, Emsley P, Hartshorn MJ, Mimmack MM, Gileadi U, Pearce SR, Gallegher MP, Gill DR. (1990) Structural model of ATP-binding proteins associated with cystic fibrosis, multidrug resistance and bacterial transport. *Nature* **346:** 362–365.

Johansen HK, Nir M, Hoiby N, Koch C, Schwartz M. (1991) Severity of cystic fibrosis in patients homozygous and heterozygous for delta F508 mutation. *Lancet* **337:** 631–634.

Kerem BS, Buchanan J, Durie P, Corey M, Levison H, Rommens J, Buchwald M, Tsui L-C. (1989a) DNA marker haplotype association with pancreatic sufficiency in cystic fibrosis. *Am. J. Hum. Genet.* **44:** 827–834.

Kerem BS, Rommens J, Buchanan J, Markiewicz D, Cox T, Chakravarti A, Buchwald M, Tsui L-C. (1989b) Identification of the cystic fibrosis gene: genetic analysis. *Science* **245:** 1073–1080.

Kerem E, Corey M, Kerem BS, Durie P, Tsui L-C, Levison H. (1989) Clinical and genetic comparisons of patients with cystic fibrosis, with or without meconium ileus. *J. Pediatr.* **114:** 767–773.

Kerem E, Corey M, Kerem BS, Rommens J, Markiewicz D, Levison H, Tsui L-C. (1990) The relation between genotype and phenotype in cystic fibrosis – analysis of the most common mutation (delta F508). *N. Engl. J. Med.* **323:** 1517–1522.

Kiesewetter S, Macek M, Davis C, Curristin SM, Chu C-S, Graham C, Shrimpton AE, Cashman SM, Tsui L-C, Mickle J, Amos J, Highsmith WE, Shuber A, Witt DR, Crystal RG, Cutting GR. (1993) A mutation in CFTR produces different phenotypes depending on chromosomal background. *Nature Genetics* **5:** 274–278.

Kock C, Hoiby N. (1993) Pathogenesis of cystic fibrosis. *Lancet* **341:** 1065–1069.

Kopelman H, Rozen R. (1990) Genetic analysis and pancreatic function in cystic fibrosis. *Lancet* **335:** 1601.

Kristidis P, Bozon D, Corey M, Markiewicz D, Rommens J, Tsui L-C, Durie P. (1992) Genetic determination of exocrine pancreatic function in cystic fibrosis. *Am. J. Hum. Genet.* **50:** 1178–1184.

Kubesh P, Dork T, Wulbrand U, Kalin N, Neumann T, Wulf B, Geerlings H, Weissbrodt H, von der Hardt H, Tummler B. (1993) Genetic determinants of airways colonization with *Pseudomonas aeruginosa* in cystic fibrosis. *Lancet* **341:** 189–193.

Kueppers F. (1992) Chronic obstructive pulmonary disease. In: *The Genetic Basis of Common Diseases* (eds R King, J Rotter, A Motulsky). Oxford University Press, Oxford, pp. 222–239.

Lanng S, Schwartz M, Thorsteinsson B, Koch C. (1991) Endocrine and exocrine pancreatic function and the delta F508 mutation in cystic fibrosis. *Clin. Genet.* **40:** 345–348.

Laroche D, Travert G. (1991) Abnormal frequency of delta F508 mutation in neonatal transitory hypertrypsinaemia. *Lancet* **337:** 55.

Lee R, Nemzer L, Witt DR, Farmer G, Fitzgerald P, Fishbach A, Holtzman J, Kornfeld S, Palmer R, Shay G. (1992) Identification of previously undiagnosed cystic fibrosis patients: an unexpected outcome of prenatal carrier screening. *Am. J. Hum. Genet.* **51** (Suppl. 4): abstract 16.

Liechti-Gallati S, Bonsall I, Malik N, Schneider V, Kraemer LG, Ruedeberg A, Moser H, Kraemer R. (1992) Genotype/phenotype association in cystic fibrosis: analyses of the delta F508, R553X, and 3905 insT mutations. *Pediatr. Res.* **32:** 175–178.

Lissens W, Desmyttere S, Bonduelle M, Dab I, Liebaers I, Mercier B, Audrezet MP, Ferec C. (1993) Mild pulmonary, but severe hepatic disease in a cystic fibrosis patient homozygous for a frameshift mutation in the regulatory domain of the CFTR. *J. Med. Genet.* **30:** 446.

Lucotte G, Perignon JL, Lenoir G. (1991) Transient neonatal hypertrypsinaemia as test for delta F508 heterozygosity. *Lancet* **337:** 988.

Macek M Jr, Vavrova V, Davis C, Hamosh A, Macek M, Cutting GR. (1993) Two delta F508 homozygote siblings with delayed onset of pancreatic insufficiency carry a second cystic fibrosis gene mutation in one allele. *Am. J. Hum. Genet.* **53** (Suppl.3): abstract 83.

McKusic VA. (1992) *Mendelian Inheritance in Man*, 10th Edn. John Hopkins University Press, Baltimore, MD.

Meschede D, Eigel A, Horst J, Nieschlag E. (1993) Compound heterozygosity for the delta F508 and F508C CFTR mutations in a patient with congenital bilateral aplasia of the vas deferens. *Am. J. Hum. Genet.* **53:** 292–293.

Mickle J, Oates R, Colin A, Maher TA, Milunsky A, Amos JA. (1993) Increased frequency of cystic fibrosis mutations in males with unilateral absence of the vas deferens (UAVD). *Am. J. Hum. Genet.* **53** (Suppl. 3): abstract 1204.

Miller C. (1993) Sickly channels in mild disease. *Nature* **362:** 106.

Miller P, Macek M Jr, Hamosh A, Walden S, Loury MC, Cutting GR. (1993) Identification of CFTR mutations in adult patients with allergic bronchopulmonary aspergillosis and chronic *Pseudomonas* bronchitis. *Pediatr. Pulmonol.* **15** (Suppl. 9): abstract 128.

Mohon R, Wagener J, Abman S, Seltzer W, Accurso F. (1993) Relationship of genotype to early pulmonary function in infants with cystic fibrosis identified through neonatal screening. *J. Pediatr.* **122:** 550–555.

Mornet E, Serre J, Farral M, Boue J, Simon-Bouy B, Estivill X, Williamson R, Boue A. (1988) Genetic differences between cystic fibrosis with and without meconium ileus. *Lancet* **i:** 376–378.

Nagel RA, Javaid A, Meire HB, Wise A, Westaby D, Kavani J, Lombard MG, Williams R, Hodson ME. (1989) Liver disease and bile duct abnormalities in adults with cystic fibrosis. *Lancet* **ii:** 1422–1425.

Oates R, Amos J. (1993) Congenital bilateral absence of the vas deferens and cystic fibrosis. A genetic commonality. *World J. Urol.* **11:** 82–88.

Orita MY, Suzuki Y, Sekiya T, Hayashi K. (1989) Rapid and sensitive detection of point mutations and DNA polymorphisms using the polymerase chain reaction. *Genomics* **5:** 874–879.

Osborne L, Santis G, Schwarz M, Klinger K, Dork T, McIntosh I, Schwartz M, Nunes V, Maceck M Jr, Reiss J, Highsmith WE Jr, McMahon R, Novelli G, Malik N, Burger J, Anvret M, Wallace A, Williams C, Mathew C, Rozen R, Graham C, Gasparini P, Bal J, Cassiman JJ, Balassopoulou A, Davidow L, Raskin S, Kalaydjieva L, Kerem B, Richards S, Simon-Bouy B, Super M, Wulbrand U, Keston M, Estivill X, Vavrova V, Friedman KJ, Barton D, Dallapiccola B, Stuhrmann M, Beards F, Hill AJM, Pignatti PF, Cuppens H, Angelicheva D, Tummler B, Brock DJH, Casals T, Macek M, Schmidtke J, Magee AC, Bonizzato A, DeBoeck C, Kuffardjieva A, Hodson M, Knight RA. (1992) Incidence and expression of the N1303K mutation of the cystic fibrosis (CFTR) gene. *Hum. Genet.* **89:** 653–658.

Osborne LR, Lynch M, Middleton PG, Alton EWFW, Geddes DM, Pryor JP, Hodson ME, Santis GK. (1993) Nasal epithelial ion transport and genetic analysis of infertile men with congenital bilateral absence of the vas deferens. *Hum. Mol. Genet.* **2:** 1605–1609.

Patrizio P, Asch R, Handelin B, Silber S. (1993) Aetiology of congenital absence of vas deferens: genetic study of three generations. *Hum. Reprod.* **8:** 215–220.

Pignatti PF. (1991) Cystic fibrosis gene mutations, and correlation with clinical manifestations. *Path. Biol.* **39:** 582–584.

Pignatti PF, Bombieri C, Marigo C, Luisetti M. (1993) Cystic fibrosis gene mutations found in chronic obstructive pulmonary disease patients. *Am. J. Hum. Genet.* **53** (Suppl. 3): abstract 1213.

Poller W, Faber JP, Scholz S, Olek K, Muller KM. (1991) Sequence analysis of the cystic fibrosis gene in patients with disseminated bronchiectatic lung disease. *Klin. Wochenschr.* **69:** 657–663.

Quinton P. (1983) Chloride impermeability in cystic fibrosis. *Nature* **301:** 421–422.

Rigot JM, LaFfitte J-J, Dumur V, Gervais R, Manouvrier S, Biserte J, Mazeman E, Roussel P. (1991) Cystic fibrosis and congenital absence of the vas deferens. *N. Engl. J. Med.* **325:** 64–65.

Riordan J, Rommens J, Kerem BS, Rozmahel R, Grzelczack Z, Zielenski J, Lok S, Plavsic N, Chou JL, Drumm M, Iannuzzi M, Collins F, Tsui L-C. (1989) Identification of the cystic fibrosis gene: cloning and characterization of complementary DNA. *Science* **245:** 1066–1073.

Rommens J, Iannuzzi M, Kerem BS, Drumm M, Melmer G, Deab M, Rozmahel R, Cole Kennedy D, Hidaka N, Zsiga M, Buchwald M, Riordan J, Tsui L-C, Collins F. (1989) Identification of the cystic fibrosis gene: chromosome walking and jumping. *Science* **245:** 1059–1065.

Saba L, Leoni GB, Meloni A, Faà V, Cao A, Rosatelli MC. (1993) Two novel mutations in the transmembrane domains of the CFTR gene in subjects of Sardinian descent. *Hum. Mol. Genet.* **2:** 1739–1740.

Santamaria F, Salvatore D, Castiglione D, Raia V, deRitis G, Sebastio G. (1992) Lung involvement, the delta F508 mutation and DNA haplotype analysis in cystic fibrosis. *Hum. Genet.* **88:** 639–641.

Santis G, Osborne L, Knight R, Hodson M. (1990) Independent genetic determinants of

pancreatic and pulmonary status in cystic fibrosis. *Lancet* **336:** 1081–1084.
Santis G, Osborne L, Knight R, Smith M, Davison T, Hodson M. (1992) Genotype–phenotype relationship in cystic fibrosis: results from the study of monozygotic and dizygotic twins with cystic fibrosis. *Pediatr. Pulmonol.* **14** (Suppl. 8): abstract 239.
Schoysman R, Vanderzwalmen P, Nijs M, Segal L, Segal-Bertin G, Geerts L, van Roosendaal E, Schoysman D. (1993) Pregnancy after fertilisation with human testicular spermatozoa. *Lancet* **342:** 1237.
Sheppard DN, Rich DP, Ostedgaard LS, Gregory RJ, Smith AE, Welsh MJ. (1993) Mutations in CFTR associated with mild-disease form Cl-channels with altered pore properties. *Nature* **362:** 160–164.
Shoshani T, Augarten A, Gazit E, Bashan N, Yahav Y, Rivlin Y, Tal A, Yaar L, Kerem E, Kerem B-S. (1992) Association of a nonsense mutation (W1282X), the most common mutation in the Ashkenazi Jewish cystic fibrosis patients in Israel, with presentation of severe disease. *Am. J. Hum. Genet.* **50:** 222–228.
Silber SJ, Ord T, Balmaceda J, Patrizio P, Asch RH. (1990) Congenital absence of the vas deferens. The fertilizing capacity of human epididymal sperm. *N. Engl. J. Med.* **323:** 1788–1792.
Simon-Bouy B, Serre J-L, Mornet E, Tallandier A, Boué J, Boué A. (1989) Genetic differences between cystic fibrosis with and without meconium ileus. *Lancet* **i:** 102.
Slomski R, Schloesser M, Berg L-P, Wagner M, Kakkar VV, Cooper DN, Reiss J. (1992) Omission of exon 12 in CFTR gene transcripts. *Hum. Genet.* **89:** 615–619.
Stern R, Boat T, Doershuk C. (1982) Obstructive azoospermia as a diagnostic criterion for the cystic fibrosis syndrome. *Lancet* **i:** 1401–1404.
Strong TV, Smit LS, Turpin SV, Cole JL, TomHon C, Markiewicz D, Petty TL, Craig MW, Rosenow EC III, Tsui LC, Iannuzzi MC, Knowles MR, Collins FS. (1991) Cystic fibrosis gene mutation in two sisters with mild disease and normal sweat electrolyte levels. *N. Engl. J. Med.* **325:** 1630–1634.
Strong T, Smit L, Nasr S, Wood D, Cole J, Iannuzzi M, Stern R, Collins F. (1992) Characterization of an intron 12 splice donor mutation in the CFTR gene. *Hum. Mutat.* **1:** 380–387.
Stuhrmann M, Macek M Jr, Reis A, Schmidtke J, Tummler B, Dork T, Vavrova V, Macek M, Krawczak M. (1990) Genotype analysis of cystic fibrosis patients in relation to pancreatic sufficiency. *Lancet* **i:** 738–739.
Tabcharani JA, Rommens JM, Hou Y-X, Chang X-B, Tsui L-C, Riordan JR, Hanrahan JW. (1993) Multi-ion pore behaviour in the CFTR chloride channel. *Nature* **366:** 79–82.
Traystman M, Mack D, Cochran G, Zuvanich E, Calabro C, Antonson D, Markin R, Vanderhoof J, Langnas A, Shaw B, Sammut P, Colombo J. (1993) Mutational analysis in cystic fibrosis patients with biliary cirrhosis. *Am. J. Hum. Genet.* **53** (Suppl. 3): abstract 515.
Tsui L-C. (1992a) Mutations and sequence variations detected in the cystic fibrosis transmembrane conductance regulator (CFTR) gene: a report from the Cystic Fibrosis Genetic Analysis Consortium. *Hum. Mutat.* **1:** 197–203.
Tsui L-C. (1992b) The spectrum of cystic fibrosis mutations. *Trends Genet.* **8:** 392–398.
Tsui L-C, Buchwald M. (1991) Biochemical and molecular genetics of cystic fibrosis. *Adv. Hum. Genet.* **20:** 153–266.
van der Ven K, Rahman A, van der Ven H, Heilman S, Jeyendran RS, Ober C. (1993) CFTR mutations in men with impaired sperm function (ISF) and azoospermia without congenital absence of the vas deferens (CBAVD). *Am. J. Hum. Genet.* **51** (Suppl. 31): abstract 1246.
Welsh M, Smith A. (1993) Molecular mechanisms of CFTR chloride channel dysfunction in cystic fibrosis. *Cell* **73:** 1251–1254.
Will K, Reiss, J, Dean M, Schlosser M, Slomski R, Schmidtke J, Stuhrmann M. (1993) CFTR transcripts are undetectable in lymphocytes and respiratory epithelial cells of a CF patient homozygous for the nonsense mutation R553X. *J. Med. Genet.* **30:** 833–837.
Williams C, Mayall E, Williamson R, Hirsh A, Cookson H. (1993) A report on CF carrier

frequency among men with infertility owing to congenital absence of the vas deferens. *J. Med. Genet.* **30:** 973–974.

Williamson R. (1993) Universal community carrier screening for cystic fibrosis? *Nature Genetics* **3:** 195–201.

3

Mutations in type I and type III collagen genes

Raymond Dalgleish

3.1 Introduction

The collagens of connective tissues (Van der Rest and Garrone, 1991; Mayne and Brewton, 1993) are members of a complex superfamily (Hulmes, 1992). Some are ubiquitously expressed in the human body, while others have a more limited distribution. Each collagen type serves a specific function or set of functions and together the collagens are the most abundant proteins in the human body. As a consequence of the complexities of their biosynthesis and the precise functions which they must perform through interactions with other extracellular matrix components, the collagens are probably the most mutation-sensitive of biological systems. The clinical phenotypes resulting from these mutations are wide ranging in their manifestations and their severity (Kivirikko, 1993). This chapter will focus on mutations in just two of the 19 known collagens – types I and III – which are co-expressed in many tissues. It is not intended that this chapter be a comprehensive review of all known mutations. Rather, the intention is to convey a flavour of the complexity of the genotype–phenotype relationship.

3.2 Collagen structure and biosynthesis

The biosynthesis of collagen is complex (Kielty *et al.*, 1993). Its basic structure is unique among proteins and is characterized by a helical structure involving three α-chains consisting of repeating -Gly-X-Y- tripeptides, in which the X and Y positions are frequently occupied by the imino acids proline and hydroxyproline, respectively. In the case of collagen types I and III, the -Gly-X-Y- repeats span 1014 and 1029 amino acids, respectively. (By convention, the first glycine of the first repeat is designated as position 1.) Individually, the α-chains have little ordered structure. However, when perfectly aligned with one another, the adoption of a helical structure involving three chains is stabilized by hydrogen bonding and other charge interactions. The α-chains individually adopt a highly extended left-handed polyproline II helix and together they form a right-handed triple helix. The requirement for a glycine at every third amino acid is absolute, as this amino acid is the only one whose side-chain can be accommodated at the centre of the triple helix. The prolines and hydroxyprolines are essential for helix stability as, probably, is the presence of alanines in -Gly-Ala-Pro(Hyp)- and -Gly-Pro-Ala- motifs. Most of these tripeptide motifs are known to occur in regions of high thermal stability (Bächinger *et al.*, 1993), and a run of three of the former type is found in the helical region of the α1 chain of type I collagen at amino acid positions 868–876 immediately following a -Gly-Pro-Pro- motif.

Collagen α-chains are translated on ribosomes of the rough endoplasmic reticulum initially as proα-chains with long N- and C-terminal propeptides. The nascent chains then undergo post-translational modification. The hydroxyproline which is found in collagen is derived from proline through modification by the enzymes prolyl 4-hydroxylase and prolyl 3-hydroxylase in the rough endoplasmic reticulum. About 100 proline residues are required to be so modified for the α-chains to adopt a stable triple helix at physiological temperatures, and the completion of helix formation is delayed until sufficient residues are modified. Similarly, the enzyme lysyl hydroxylase modifies about 20 lysines to hydroxylysines, many of which are subsequently glycosylated with galactose and glucosylgalactose. The modifications of prolines and lysines only take place prior to helix formation, with the adoption of the triple helical structure preventing any further modification. As will be seen below, any delay in the formation of the triple helix leads to the overmodification of nascent α-chains which is characteristic of many cases of connective tissue disease.

The alignment of the α-chains is facilitated through the initial interaction of the C-terminal propeptides, which probably also play a role in ensuring the correct selection of chains where a heterotrimeric molecule is to be assembled. Such is the case in type I collagen, which consists of two α1 chains and one α2 chain. Some collagens, such as types II and III collagen, are homotrimers while others, such as types VI and IX, are heterotrimers consisting of three dissimilar chains. The formation of the helix is propagated from the C terminus towards the N terminus to give a product known as procollagen.

The procollagen is next translocated to the Golgi apparatus and subsequently is secreted via vacuoles into the extracellular space. The N-terminal and C-terminal propeptides are removed by the action of specific peptidases, leaving short telopeptides at each end of the triple helical α-chain region. The collagen triple helices then self-assemble laterally and linearly into characteristic quarter staggered arrays, which are stabilized by the formation of covalent cross-links involving critical lysine residues towards the ends of the triple helix. These arrays are the characteristic 300 nm banded fibrils seen by electron microscopy. Most collagen fibrils contain more than one collagen type (Van der Rest and Garrone, 1991) and those of skin contain both types I and III.

3.3 The genes encoding fibrillar collagens

Collagen α-chains are the products of distinct gene loci which are mostly dispersed throughout the human genome. The α1 and α2 chains of type I collagen and the α1 chain of type III collagen are encoded, respectively, by the loci COL1A1 (17q21–q22), COL1A2 (7q21–q22) and COL3A1 (2q24–q33). The genes for collagen types I and III are large and complex (Chu and Prockop, 1993), but an appreciation of one particular aspect of their structure sheds considerable light on the genotype–phenotype relationship. Whole numbers of -Gly-X-Y- repeats are encoded in the exons of the α-chain regions of the genes. These exons, which begin with the first base of a glycine codon and end with the last base of the amino acid in the Y position, are commonly 54 and 108 bp in length, encoding six and 12 repeats respectively. Mutations leading to the deletion of whole exons from mRNA, either because of exon skipping or the deletion of entire exons from the gene, will give rise to translation products in which the translation reading frame is maintained. A precise number of -Gly-X-Y- motifs will have been lost but the shortened α-chains will still be able to participate in helix assembly.

3.4 Collagen mutations result in connective tissue disorders

The biosynthesis of collagen only proceeds in the ordered manner described above if the procollagen α-chains are free from mutation. Substitution of a glycine with another amino acid usually has two effects. The first is that the bulkier amino acid replacing the glycine causes a temporary pause in the formation of the collagen triple helix. This allows increased access of the modifying hydroxylases to their substrate prolines and lysines. The over-hydroxylation of prolines may be of little consequence as, normally, most prolines are modified in this way. However, over-hydroxylation of lysines leads to a greater number of hydroxylysines, which are substrates for glycosylation reactions. A major effect of over-glycosylation is to reduce the thermal stability of the collagen, which typically melts at a temperature of around 41°C. Reductions in the denaturation temperature of 2–4°C are typical, though, in at least one case, a reduction of over 20°C has been noted (Westerhausen *et al.*, 1990). Even slight reductions in the melting temperature are likely to lead to collagen molecules that are unstable at physiological temperatures. Such unstable collagen is likely to be degraded largely by intracellular proteases, resulting in much reduced secretion into the extracellular space. Further, molecules that contain two abnormal α1(I) chains, rather than one, have a markedly lower thermal stability and secretion (Wallis *et al.*, 1990). There has even been the suggestion that, where a mutation only minimally disrupts helix stability and fibril assembly, over-glycosylation alone, resulting from a temporary pause in helix winding, may still result in a disease phenotype (Tenni *et al.*, 1993).

The second effect relates to the necessity to accommodate a bulkier than usual amino acid at the centre of the triple helix. Although the biosynthetic machinery is often able to do so, the result is to produce procollagen that contains imperfections, sometimes as severe as distinct kinks, in the normally rod-like structure which can interfere with the removal of the N-terminal propeptide due to incorrect registration of the chains at the cleavage site (Vogel *et al.*, 1988). It is probable that less severe kinks not affecting processing might still interfere with molecular packing and cross-linking. The effects of collagen mutations on molecular packing may be visualized by electron microscopy (Kobayashi *et al.*, 1990).

The fact that collagens are trimers has consequences for the 'amplification' of the primary defect. In type I collagen where there are two α1 chains and one α2 chain, a mutation in one allele of the gene encoding α1 chains will result in 75% of the collagen produced being defective (Figure 3.1). This effect, termed 'procollagen suicide' (Prockop, 1984) – though 'death by misadventure' would seem more apt since 'suicide' implies an active, rather than passive, process – comes about because, during chain selection to form a type I collagen fibril, there is the possibility of incorporating one, two or no defective α1 chains. There are two distinctly different ways in which to form a trimer incorporating a single defective α1 chain. If all molecules possessing one or two mutant α-chains are degraded, only 25% of the procollagen formed is defect free. In the case of α2 chain mutations, the outcome is more as expected – 50% of collagen molecules will be defective.

It is now well established that mutations in collagen genes, of the type described above, result in connective tissue disorders. Our present knowledge follows from the observations in the late 1970s that persons suffering from connective tissue diseases, such as osteogenesis imperfecta, produced defective collagen as judged by a number of biochemical criteria. It is now also evident that some forms of Ehlers–Danlos syndrome are due to mutations in collagen types I

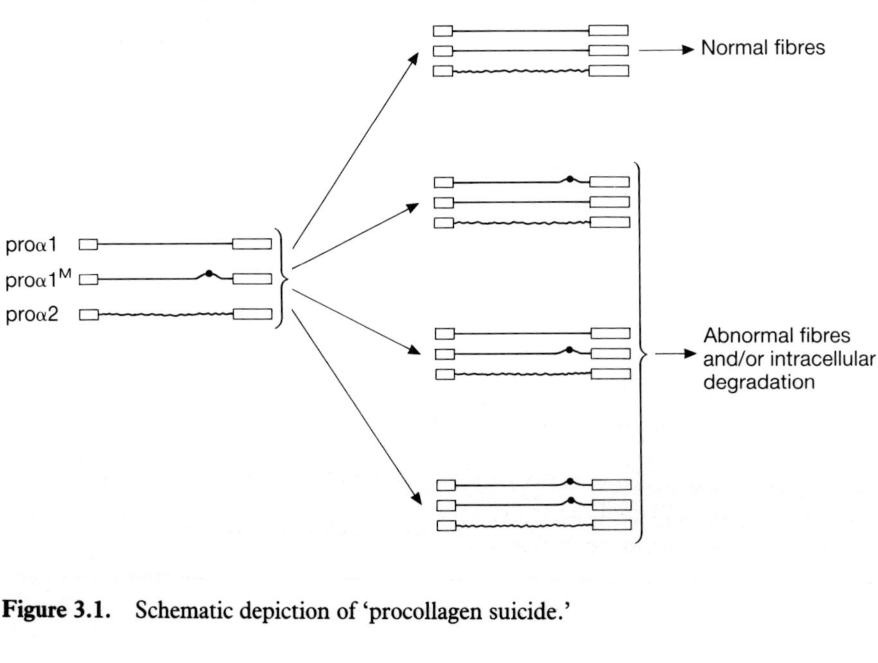

Figure 3.1. Schematic depiction of 'procollagen suicide.'

and III. In each case, the clinical phenotype results from a combination of reduced collagen secretion and the disruptive effects of defective molecules which do reach the extracellular matrix. Equally, the optimism of the early 1980s that collagen mutations would be shown to account for Marfan syndrome is now shown to have been misplaced.

3.5 Osteogenesis imperfecta

The main phenotypic features of osteogenesis imperfecta (OI) are bone fragility and reduced bone mass. It is a disorder with an overall prevalence of about 1 in 10 000 and is usually divided into four sub-types (Table 3.1) based on clinical criteria (Sillence *et al.*, 1979). OI is mostly either inherited from a parent as a dominant disorder or arises as a new dominant mutation, though in a few cases recessive inheritance is recognized. To help understand the genotype–phenotype relationship it is perhaps best to deal first with the new dominant mutations. These are found to be present in probands with the disease but are absent in either parent, indicating mutations that have arisen in the germline or early in fetal development. In the former case this can result in multiple affected offspring and, before the molecular pathology was better understood, gave rise to the notion that OI was inherited in a recessive manner. However, not all mutations

Table 3.1. Clinical features and patterns of inheritance for the different types of osteogenesis imperfecta (after Byers *et al.*, 1991)

OI type	Clinical features	Inheritance
I (mild)	Normal stature, little or no deformity; blue sclerae; hearing loss in 50%, dentinogenesis imperfecta is rare and may distinguish a subset	Autosomal dominant
II (lethal)	Lethal in the perinatal period, minimal calvarial mineralization, beaded ribs, compressed femurs, marked long bone deformity, platyspondyly	Autosomal dominant (new) Autosomal recessive (rare)
III (severe)	Progressively deforming bones, usually with moderate deformity at birth; sclerae variable in colour, often lighten with age; dentinogenesis imperfecta common, hearing loss common; stature very short	Autosomal dominant (new) Autosomal recessive (rare)
IV (moderate)	Normal sclerae, mild to moderate bone deformity and variable short stature; dentinogenesis imperfecta common and hearing loss occurs in some	Autosomal dominant

that occur in early development necessarily give rise to a disease phenotype. Somatic tissues may be unaffected (Kuivaniemi *et al.*, 1988) or mildly affected (Edwards *et al.*, 1992), with mutations in the germline giving rise to severely affected offspring.

Most cases of OI types II, III and IV result from new dominant mutations in COL1A1 and COL1A2 – a major exception, though, is a recessive form of type III OI in black southern African populations (Wallis *et al.*, 1993). They are dominant because of the 'procollagen suicide' effect, described in Section 3.4 above, where defective chains are incorporated into collagen triple helices. Deletions, insertions and amino acid substitutions in the α1 or α2 chains of type I collagen are known to account for each of these OI types (Kuivaniemi *et al.*, 1991; Byers, 1993).

Deletions can arise in two ways – either through partial gene deletions or because of mutations causing exon skipping during RNA processing. As discussed earlier, shortened α-chains produced by these mutations will still associate with full-length chains and interfere with helix formation and/or stability provided that the C-terminal propeptide is unaffected by the mutation. Many examples of such mutations exist, resulting in both lethal and non-lethal forms of the disease (Kuivaniemi *et al.*, 1991; Byers, 1993). Not all deletions are as large as entire exons – some are small but surprisingly devastating, and two are of particular interest. The first is the deletion of a -Gly-Ala-Pro- repeat from the C-terminal end of the α1(I) chain which causes type II OI (Hawkins *et al.*, 1991). As discussed in Section 3.2, this is one of three such tandem repeats between positions 868 and 876, and the deletion of one evidently has a dramatic effect. This is probably due, in part, to the loss of the stabilizing effect of such a motif. The second is the deletion of 4 bp in the C-terminal propeptide of the α2(I) chain, which creates a frameshift and causes type III OI (Pihlajaniemi *et al.*, 1984). This is the first characterized mutation that causes a recessively inherited OI.

Insertions are also possible, though they tend to be less common. A frameshift mutation in the α1(I) chain caused by a single base insertion results in an essentially non-functional chain and causes type II OI (Bateman *et al.*, 1989). This case is particularly interesting in that some mutant chains are incorporated into procollagen molecules but interfere with helix assembly, resulting in the degradation of such molecules. Only normal type I collagen was incorporated into the extracellular matrix in the proband but this was at a level of only 20% of that of normal, thus accounting for the phenotype.

Some insertions are very much larger. The recombination processes which brought about the present structure of the type I collagen genes have left behind the means for internal duplications as well as deletions to take place (Cohn *et al.*, 1993).

By far the more interesting type of mutation in OI is the single amino acid substitution. There are 338 glycines in each of the α1 and α2 chains of type I collagen. These are encoded by the codons GGA, GGC, GGG and GGT, and a single base substitution at either of the first two positions will result in the substitution of one of a limited number of other amino acids or the introduction of a stop codon (Table 3.2). Mutation to a stop codon is likely to result in type I OI for reasons that will be discussed later. As discussed above, substitution of a glycine results in delay of triple helix formation and all of its consequences.

Over the years, there have been attempts to correlate the severity of the OI produced by amino acid substitutions with their position and type. It was clear that there was a correlation between the position of a mutation and the start of overmodification of both the α1 and α2 chains (Bonadio and Byers, 1985). From this evolved a simple gradient model (Byers *et al.*, 1991) in which substitutions towards the N terminus of α-chains result in a more severe phenotype than those lying further C-terminal. This was based on the supposition that the phenotype was predominantly a consequence of the extent of overmodification and that the substitution of a glycine by any amino acid at any position would be equally poorly tolerated. Such a model appeared to account for the majority of α1(I) substitutions, especially for arginine and cysteine (Figure 3.2a). However, there were two problems. The first was that the model did not seem to hold so well for serine, and the second was that each amino acid had to have its own gradient in which the transition from one severity of phenotype to another took place at a different point along the α-chain. As more amino acid substitutions were

Table 3.2. Amino acid codons into which glycine codons may be mutated by a single base change

Wild-type glycine codons			
GGT	Gly		
GGC	Gly		
GGA	Gly		
GGG	Gly		
First position mutations		**Second position mutations**	
TGT	Cys	GTT	Val
TGC	Cys	GTC	Val
TGA	Stop	GTA	Val
TGG	Trp	GTG	Val
CGT	Arg	GCT	Ala
CGC	Arg	GCC	Ala
CGA	Arg	GCA	Ala
CGG	Arg	GCG	Ala
AGT	Ser	GAT	Asp
AGC	Ser	GAC	Asp
AGA	Arg	GAA	Glu
AGG	Arg	GAG	Glu

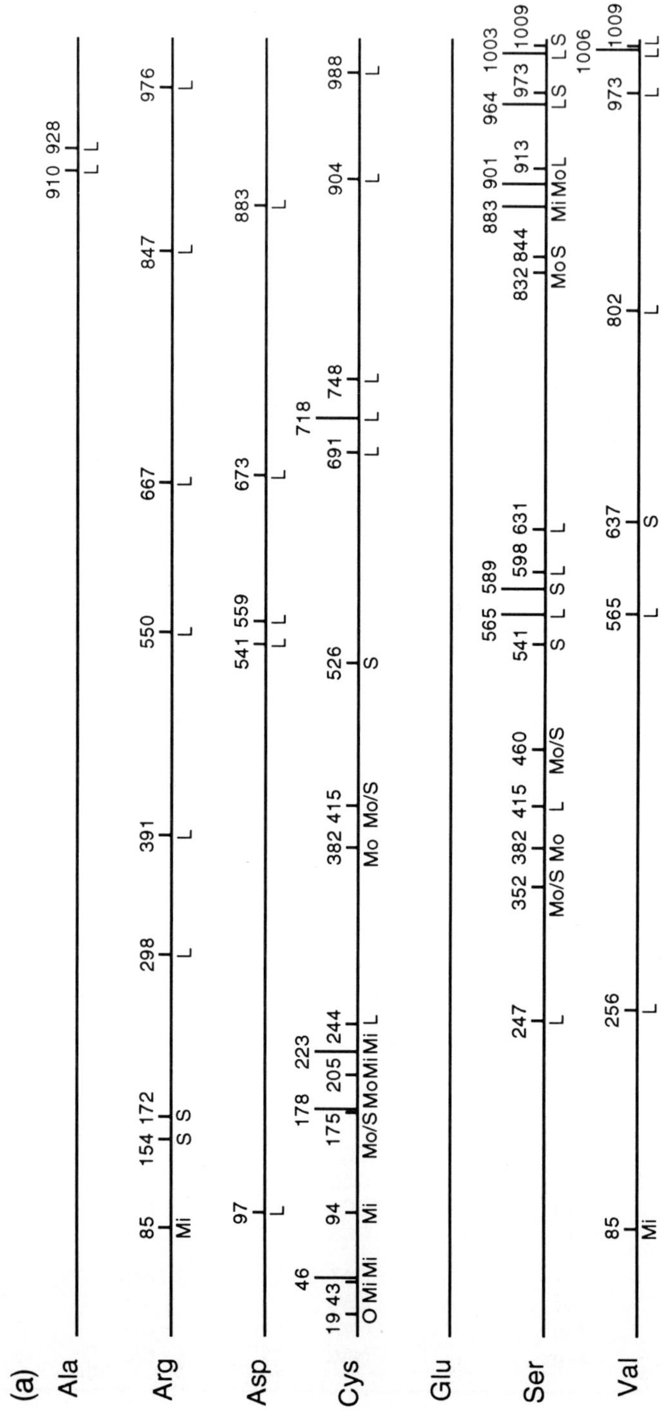
(a)
Ala 910 928 L L
Arg 85 Mi 154 172 S S 298 L 391 L 550 L 667 L 847 L 976 L
Asp 97 L 541 559 L L 673 L 883 L
Cys 19 43 46 O Mi Mi Mi 94 Mi 175 178 Mo/S Mo 205 223 Mo Mi Mi 244 L 382 415 Mo Mo/S 526 S 691 718 748 L L L 904 L 988 L
Glu
Ser 247 L 352 382 415 460 Mo/S Mo L Mo/S 541 565 589 598 631 S L S L L 832 844 Mo S 883 901 913 Mi Mo L 964 973 1003 1009 L S L S
Val 85 Mi 256 L 565 L 637 S 802 L 973 1006 1009 L L L

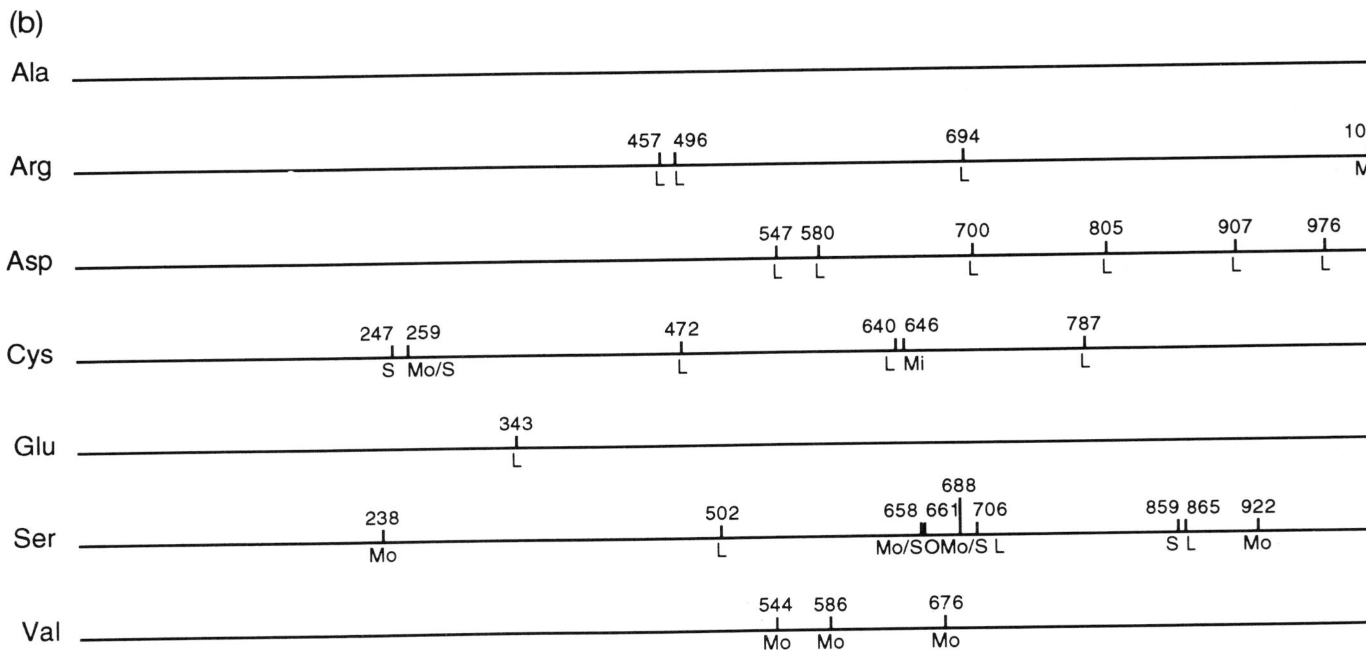

Figure 3.2. Substitutions of glycine in (a) the α1(I) and (b) the α2(I) chain. The position of each substitution is given, along with the severity of the osteogenesis imperfecta phenotype; mild (Mi), moderate (Mo), severe (S) and lethal (L). Two cases of osteoporosis (O) are indicated. Correct as of October 1993.

discovered, the model stood up less well to scrutiny and now a domain model is favoured (Deak *et al.*, 1991; Wenstrup *et al.*, 1991; Marini *et al.*, 1993) in which there are some regions where substitutions cause severe phenotypes and others where the effects are less pronounced. In particular, the α2 chain mutations can only be reconciled in this way (Figure 3.2b). It is interesting to note that, compared with the α1(I) chain, there are few amino acid substitutions in the N-terminal region of the chain. Perhaps substitutions do exist in this region but result in conditions with a sub-clinical phenotype. This seems rather unlikely, since most of the N-terminal substitutions which are known each result in conditions which are, at least, moderate in their severity. The data in Figure 3.2 indicate the status of the type I collagen glycine substitutions as of October 1993.

Why do glycine substitutions at different positions result in different phenotypes? Put simply, it must be that some are less disruptive of helix formation than others. It is presumed that neighbouring amino acids (probably most importantly those lying N-terminal to the substitution) determine which glycine substitutions are readily accommodated and which will be disruptive, and hence lead to greater overmodification. Hypotheses have been proposed to explain the effects of mutations on helix stability and folding (Bächinger *et al.*, 1993).

It is now becoming possible to address two more points with regard to the genotype–phenotype relationship. The first is whether or not the phenotype is determined solely by the substitution or whether it is modified by the effect of other genes. The identification of unrelated individuals harbouring the same mutation and having broadly the same disease phenotype supports the former. A glycine to arginine substitution at position 154 of the α1(I) chain causes type III OI in two unrelated individuals, and a glycine to serine substitution at position 1003 causes type II OI (Pruchno *et al.*, 1991). Other similar data also exist in support of this idea. The second point relates to whether different substitutions at the same point in a chain cause the same disease phenotype. Here the data are more scarce, but the glycine at position 541 of the α1(I) chain substituted with serine causes type III OI (Mackay *et al.*, 1993a), whereas an aspartic acid substitution at the same position causes type II OI (Zhuang *et al.*, 1991). However, substitution of the glycine at position 565 of the α1(I) chain with either valine or serine results in type II OI (Bateman *et al.*, 1993; Mackay *et al.*, 1993b).

Type I OI is the mildest form and is inherited in a truly autosomal dominant fashion, with the disease being inherited from an affected parent. It deserves to be considered separately from the other types in that the underlying cause of the disease in the majority of cases is fundamentally different from the other OI types. Linkage analysis has demonstrated, however, that the underlying cause is still mutations in the same genes, namely COL1A1 and COL1A2 (Sykes *et al.*, 1990). In a comprehensive study, Willing *et al.* (1992) have shown, in 21 unrelated families, that mRNA is detectable from only one allele of COL1A1. Affected persons produce half the normal amount of proα1(I) chains and their disease is explained simply in terms of reduced production of type I collagen. The excess proα2(I) chains do not associate into trimers and are degraded. This indicates that it is better to produce fewer collagen molecules than to produce defective ones. Presumably, null alleles of COL1A2 must also exist which would result in an excess of proα1(I) chains. Unlike their α2 counterparts, these can associate into homotrimers and it is likely that the resulting phenotype would be milder and perhaps sub-clinical. As yet, no data exist concerning the cause of the reduced levels of mRNA in OI, though they presumably include mutations in promoters, enhancers and glycine to termination codon substitutions, which interfere with mRNA maturation (McIntosh *et al.*, 1993).

Less commonly, the type I OI phenotype can result from situations in which defective chains are produced but are not incorporated into collagen molecules. A 5 bp deletion near the 3′ end of one COL1A1 allele causes a frameshift, resulting in elongated α1 chains which are unstable (Willing *et al.*, 1990).

Some cases of type I OI are attributable to structural mutations as in type II, III and IV. A number of substitutions by cysteine of α1 chain glycines have been characterized including that at the third position of the C-terminal propeptide (Cohn *et al.*, 1988) and at position 94 in the triple helical region (Starman *et al.*, 1989). It must be assumed that these substitutions only interfere with chain assembly and/or fibril formation to a very limited degree. In another type I OI family, deletion from the mRNA of the 99 bp exon 17 of COL1A1 is caused by a splicing defect (Willing *et al.*, 1993). This mRNA is present in only very small amounts and so results in the production of few defective α1(I) chains. Each of these instances in which an aberrant protein is produced in type I OI are exceptions rather than the rule (Wenstrup *et al.*, 1990).

3.6 Ehlers–Danlos syndrome type VII

The Ehlers–Danlos syndrome (EDS) is a heterogeneous group of heritable connective tissue disorders affecting skin, joints, ligaments and internal organs (Beighton, 1970; McKusick, 1972; Steinmann *et al.*, 1993). Type VII EDS is a congenital variant typified by loose jointedness, which may or may not be accompanied by skin changes. Patients typically produce type I collagen chains in which the N-terminal propeptide is retained but the C-terminal propeptide is normally cleaved. Initial studies of cultured skin fibroblasts from affected patients indicated that there was a marked reduction in activity of the peptidase responsible for the removal of the N-propeptides of the α1 and α2 chains of type I collagen. Later studies were to reveal a case in which α2(I) chains retaining the propeptide (pNα2(I) chains) were found to be present in a 1:1 ratio with normal α2(I) chains (Steinmann *et al.*, 1980). All of α1(I) chains were normal as were activity levels of the N-peptidase. On the basis of this and other supporting evidence, it was proposed that a structural mutation of type I procollagen was responsible for the disease. Subsequently, the presence of pNα1(I) chains was noted in the collagens produced by other EDS VII patients and this resulted in the subdivision of the disorder into types VIIA and VIIB, corresponding to the gene harbouring the mutation – COL1A1 and COL1A2, respectively – and both are inherited as autosomal dominants.

It is now recognized that the defects resulting in the presence of pNα1(I) and pNα2(I) in EDS VII patients have a common basis. All patients with EDS types VIIA or VIIB, that have been analysed at the molecular level, lack all or part of exon 6 in the mRNA of either the α1(I) or α2(I) chains The cleavage site for the peptidase responsible for the removal of the N-propeptides from these chains is encoded in exon 6 which, in addition, also encodes one of the lysines essential for the correct formation of inter-molecular cross links. The lack of exon 6 is predominantly due to single base mutations at the intron–exon boundaries causing exon skipping during the splicing of hnRNA (D'Alessio *et al.*, 1991; Nicholls *et al.*, 1991; Vasan *et al.*, 1991), though in one case a deletion of exon 6 has been noted (Steinmann *et al.*, 1993). A further case of EDS VII, described by Chiodo *et al.* (1992), is of particular note. Instead of the whole exon being skipped, a cryptic splice site in exon 6 is utilized, resulting in a truncated exon 6 which lacks the N-peptidase cleavage site but retains the lysine required for cross-linking. This finding points to the phenotype resulting from the loss of the peptidase cleavage site rather than that of the cross-linking lysine. However, there remains the possibility that the loss of sequences adjacent to the lysine may, in some way, interfere with its function in cross-linking and that this too contributes to the phenotype.

Recently, firm evidence has emerged for the existence of a variant of EDS VII truly attributable to absence of the N-peptidase and comparable to dermatosparaxis in cattle (Nusgens *et al.*, 1992; Smith *et al.*, 1992). The phenotype of so-called EDS VIIC differs somewhat from the A and B variants in that the skin is more severely affected. The genetic locus responsible for this variant has not been determined but the inheritance appears to be autosomal recessive.

3.7 Marfan syndrome

Many textbooks still cite type I collagen mutations as the cause of Marfan syndrome though it is now recognized that most cases are the result of fibrillin mutations (Dietz *et al.*, 1991). The confusion arises from the identification, in the early 1980s, of an atypical case in which a defect of type I collagen was noted (Byers *et al.*, 1981). Subsequently, the arginine at position 618 (a -Y- position) of half of the α2(I) chains of this patient was shown to be substituted by glutamine (Phillips *et al.*, 1990). The arginine at position 618 is highly conserved across species and in all the known human fibrillar collagen genes for which sequences are available. Its substitution probably interferes either with collagen fibril formation or, more likely, with the interaction of type I collagen with other components of the extracellular matrix. However, the precise reason why this substitution results in Marfan syndrome is obscure. This remains the only case attributable to a fully characterized mutation, though anomalous electrophoretic migration of α2(I) chains has also been noted in another Marfan patient who was shown not to have fibrillin abnormalities (Godfrey *et al.*, 1990). It is worth noting that the former patient presented with an atypical form of the disease in that her height (164.5 cm) would be considered to be in the normal range. Full clinical descriptions of this case may be found in Byers *et al.* (1981) and Phillips *et al.* (1990).

3.8 Ehlers–Danlos syndrome type IV

Type IV EDS is due to a deficiency of type III collagen and so affects tissues such as blood vessels, skin and hollow internal organs that are rich in type III collagen (Steinmann *et al.*, 1993). It is a life-threatening disorder that often results in the rupture of blood vessels, the gastro-intestinal tract and the gravid uterus. There is variation in the severity between families though no systematic sub-classification has been developed as in the case of OI. Although inherited as an autosomal dominant, about 50% of cases represent new mutations. The 'protein suicide' effect described for type I collagen applies more so for type III collagen as it is a homotrimer. In a patient with EDS IV, only 1/8 of the type III collagen will contain three normal chains. The other 7/8 is usually retained intracellularly and degraded resulting in the decreased type III collagen secretion that is characteristic of the disease.

The types of mutation known to cause EDS IV are the same as those for OI, though deletions and splicing mutations appear to be more common than amino acid substitutions. The relationship between genotype and phenotype in EDS IV can be confusing. Although there are certainly exceptions, mutations towards the C-terminal end of the α-chain produce a disease that is more severe than that caused by those lying more N-terminal. However in EDS IV, no amino acid substitution has been found any further N-terminal than position 400 (MacKay *et al.*, manuscript in preparation). As will be seen below, at least one substitution lying more N-terminal causes a distinctly different disorder, albeit involving blood vessels.

One type III collagen gene deletion is worth note. It is 9 kb in length and removes 15 exons encoding amino acids 586–999 (Vissing *et al.*, 1991). Remarkably, it results in a mild form of the disease, inviting speculation that collagen molecules containing shortened α-chains are partly functional.

3.9 Aortic aneurysms

Not all mutations of type III collagen result in a phenotype which is distinctly that of EDS IV. Kontusaari *et al.* (1990a) reported a splicing mutation in which, although there was some phenotypic overlap with EDS IV, the predominant feature was aortic aneurysms. In another family, a substitution of glycine-619 of the α1(III) chain with arginine also resulted in aortic aneurysms (Kontusaari *et al.*, 1990b). Further support for the notion that type III collagen mutations cause aortic aneurysms comes from the study of two patients whose cultured fibroblasts produced reduced amounts of type III collagen which had reduced thermal stability (Deak *et al.*, 1992). However, a more recent study of 50 unrelated patients suggests that, in general, mutations in the α-chain region of type III collagen are not responsible for aortic aneurysms (Tromp *et al.*, 1993). Nucleotide changes resulting in altered type III collagen were found in only two patients. One had a proline by threonine substitution at amino acid position 501 and the other a glycine by arginine substitution at position 136. The functional significance of the former is not clear, though the latter is clearly consistent with the overall picture of glycine substitutions.

3.10 Therapies for connective tissue disorders

In spite of the progress that has been made in understanding the genotype–phenotype relationship in diseases involving collagen types I and III, there has, as yet, been little impact on clinical treatment. There is currently no cure for any of the disorders discussed above and treatment is aimed at alleviating symptoms and lessening the likelihood of further complications. In the case of OI, the emphasis of treatment is appropriate orthopaedic management (Byers and Steiner, 1992) as no attempts to reduce the frequency of bone fracture or to increase bone density by drug or dietary means have been successful (Marini, 1988). Treatment for EDS IV and aortic aneurysms is also limited and restricted to symptomatic therapy, prophylaxis and counselling. Only in the case of type I OI (the mildest form) is any other form of intervention contemplated at present. Such treatment would have to be aimed at either increasing collagen production where null alleles were involved or in some way preventing the expression of abnormal chains whose incorporation into collagen results in the disease phenotype. Toward this latter goal, anti-sense oligonucleotides are being used to attempt to modulate collagen gene expression in cultured cells from OI patients (J. Marini, personal communication).

References

Bächinger HP, Morris NP, Davis JM. (1993) Thermal stability and folding of the collagen triple helix and the effects of mutations in osteogenesis imperfecta on the triple helix of type I collagen. *Am. J. Med. Genet.* **45:** 152–162.

Bateman JF, Lamande SR, Dahl H-HM, Chan D, Mascara T, Cole WG. (1989) A frameshift mutation results in a truncated nonfunctional carboxyl-terminal proα1(I) propeptide of type I collagen in osteogenesis imperfecta. *J. Biol. Chem.* **264:** 10960–10964.

Bateman JF, Lamande SR, Hannagan M, Moeller I, Dahl H-HM, Cole WG. (1993) Chemical cleavage method for the detection of RNA base changes: experience in the application to collagen mutations in osteogenesis imperfecta. *Am. J. Med. Genet.* **45:** 233–240.

Beighton P. (1970) *The Ehlers–Danlos Syndrome*. William Heinemann Medical Books, London.

Bonadio J, Byers PH. (1985) Subtle structural changes in the chains of type I procollagen produce osteogenesis imperfecta type II. *Nature* **316:** 363–366.

Byers PH. (1993) Osteogenesis imperfecta. In: *Connective Tissue and its Heritable Disorders* (eds PM Royce, B Steinmann). Wiley-Liss, New York, pp. 317–350.

Byers PH, Steiner RD. (1992) Osteogenesis imperfecta. *Annu. Rev. Med.* **43:** 269–282.

Byers PH, Siegel RC, Peterson KE, Rowe DW, Holbrook KA, Smith LT, Chang Y-H, Fu JCC. (1981) Marfan syndrome: abnormal α2 chain in type I collagen. *Proc. Natl Acad. Sci. USA* **78:** 7745–7749.

Byers PH, Wallis GA, Willing MC. (1991) Osteogenesis imperfecta: translation of mutation to phenotype. *J. Med. Genet.* **28:** 433–442.

Chiodo AA, Hockey A, Cole WG. (1992) A base substitution at the splice acceptor site of intron 5 of the COL1A2 gene activates a cryptic site within exon 6 and generates abnormal type I procollagen in a patient with Ehlers–Danlos syndrome type VII. *J. Biol. Chem.* **267:** 6361–6363.

Chu M-L, Prockop DJ. (1993) Collagen: gene structure. In: *Connective Tissue and its Heritable Disorders* (eds PM Royce, B Steinmann). Wiley-Liss, New York, pp. 149–165.

Cohn DH, Apone S, Eyre DR, Starman BJ, Andreassen P, Charbonneau H, Nicholls AC, Pope FM, Byers PH. (1988) Substitution of cysteine for glycine within the carboxyl-terminal telopeptide of the $\alpha 1(I)$ chain of type I collagen produces mild osteogenesis imperfecta. *J. Biol. Chem.* **263:** 14605–14607.

Cohn DH, Zhang X, Byers PH. (1993) Homology-mediated recombination between type I collagen gene exons results in an internal tandem duplication and lethal osteogenesis imperfecta. *Hum. Mutat.* **2:** 21–27.

D'Alessio M, Ramirez F, Blumberg BD, Wirtz MK, Rao VH, Godfrey MD, Hollister DW. (1991) Characterization of a COL1A1 splicing defect in a case of Ehlers–Danlos syndrome type VII: further evidence of molecular homogeneity. *Am. J. Hum. Genet.* **49:** 400–406.

Deak SB, Scholz PM, Amenta PS, Constantinou CD, Levi-Menzi SA, Gonzalez-Lavin L, Mackenzie JW. (1991) The substitution of arginine for glycine 85 of the $\alpha 1(I)$ procollagen chain results in mild osteogenesis imperfecta. The mutation provides direct evidence for three discrete domains of cooperative melting of intact type I collagen. *J. Biol. Chem.* **266:** 21827–21832.

Deak SB, Ricotta JJ, Mariani TJ, Deak ST, Zatina MA, Mackenzie JW, Boyd CD. (1992) Abnormalities in the biosynthesis of type III procollagen in cultured skin fibroblasts from two patients with multiple aneurysms. *Matrix* **12:** 92–100.

Dietz HC, Cutting GR, Pyeritz RE, Maslen CL, Sakai LY, Corson GM, Puffenberger EG, Hamosh A, Nanthakumar EJ, Curristin SM, Stetten G, Meyers DA, Francomano CA. (1991) Marfan syndrome caused by a recurrent *de novo* missense mutation in the fibrillin gene. *Nature* **352:** 337–339.

Edwards MJ, Wenstrup RJ, Byers PH, Cohn DH. (1992) Recurrence of lethal osteogenesis imperfecta due to parental mosaicism for a mutation in the COL1A2 gene of type I collagen. The mosaic parent exhibits phenotypic features of a mild form of the disease. *Hum. Mutat.* **1:** 47–54.

Godfrey M, Menashe V, Weleber RG, Koler RD, Bigley RH, Lovrien E, Zonana J, Hollister D. (1990) Cosegregation of elastin-associated microfibrillar abnormalities with the Marfan phenotype in families. *Am. J. Hum. Genet.* **46:** 652–660.

Hawkins JR, Superti-Furga A, Steinmann B, Dalgleish R. (1991) A 9 base pair deletion in COL1A1 in a lethal variant of osteogenesis imperfecta. *J. Biol. Chem.* **266:** 22370–22374.

Hulmes DJS. (1992) The collagen superfamily – diverse structures and assemblies. *Essays Biochem.* **27:** 49–67.

Kielty CM, Hopkinson I, Grant ME. (1993) Collagen: the collagen family: structure, assembly, and organization in the extracellular matrix. In: *Connective Tissue and its Heritable Disorders* (eds PM Royce, B Steinmann). Wiley-Liss, New York, pp. 103–147.

Kivirikko KI. (1993) Collagens and their abnormalities in a wide spectrum of diseases. *Ann. Med.* **25:** 113–126.

Kobayashi K, Hata R, Nagai S, Niwa J. (1990) Direct visualization of affected collagen molecules synthesized by cultured fibroblasts from an osteogenesis imperfecta patient. *Biochem. Biophys. Res. Commun.* **172:** 217–222.

Kontusaari S, Tromp G, Kuivaniemi H, Ladda RL, Prockop DJ. (1990a) Inheritance of an RNA splicing mutation ($G^{+1\ IVS20}$) in the type III procollagen gene in a family with aortic aneurysms and easy bruisability: phenotypic overlap between familial arterial aneurysms and the Ehlers–Danlos syndrome type IV. *Am. J. Hum. Genet.* **47:** 112–120.

Kontusaari S, Tromp G, Kuivaniemi H, Romanic AM, Prockop DJ. (1990b) A mutation in the gene for type III procollagen (COL3A1) in a family with aortic aneurysms. *J. Clin. Invest.* **86:** 1465–1473.

Kuivaniemi H, Sabol C, Tromp G, Sippola-Thiele M, Prockop DJ. (1988) A 19-base pair deletion in the pro$\alpha 2(I)$ gene of type I procollagen that causes in-frame RNA splicing from exon 10 to exon 12 in a proband with atypical osteogenesis imperfecta and in his asymptomatic mother. *J. Biol. Chem.* **263:** 11407–11413.

Kuivaniemi H, Tromp G, Prockop DJ. (1991) Mutations in collagen genes: causes of rare and some common diseases in humans. *FASEB J.* **5:** 2052–2060.

Mackay K, Byers PH, Dalgleish R. (1993a) An RT-PCR-SSCP screening strategy for detection of mutations in the gene encoding the α1 chain of type I collagen: application to four patients with osteogenesis imperfecta. *Hum. Mol. Genet.* **2:** 1155–1160.

Mackay K, Lund AM, Raghunath M, Steinmann B, Dalgleish R. (1993b) SSCP detection of a Gly565Val substitution in the proα1(I) collagen chain resulting in osteogenesis type II. *Hum. Genet.* **91:** 439–444.

Marini JC. (1988) Osteogenesis imperfecta: comprehensive management. *Adv. Pediatr.* **35:** 391–426.

Marini JC, Lewis MB, Wang Q, Chen KJ, Orrison BM. (1993) Serine for glycine substitutions in type I collagen in two cases of type IV osteogenesis imperfecta (OI). Additional evidence for a regional model of OI pathophysiology. *J. Biol. Chem.* **268:** 2667–2673.

Mayne R, Brewton RG. (1993) New members of the collagen superfamily. *Curr. Opin. Cell Biol.* **5:** 883–890.

McIntosh I, Hamosh A, Dietz HC. (1993) Nonsense mutations and diminished mRNA levels. *Nature Genetics.* **4:** 219.

McKusick VA. (1972) The Ehlers–Danlos Syndrome. In: *Heritable Disorders of Connective Tissue*, 4th edn. CV Mosby, St Louis, pp. 292–371.

Nicholls AC, Oliver J, Renouf DV, McPheat J, Palan A, Pope FM. (1991) Ehlers–Danlos syndrome type VII: a single base change that causes exon skipping in the type I collagen α2(I) chain. *Hum. Genet.* **87:** 193–198.

Nusgens BV, Verellen-Dumoulin C, Hermanns-Lê T, De Paepe A, Nuytinck L, Pièrard GE, Lapière CM. (1992) Evidence for a relationship between Ehlers–Danlos type VIIC in humans and bovine dermatosparaxis. *Nature Genetics.* **1:** 214–217.

Phillips CL, Shrago-Howe AW, Pinnell S, Wenstrup RJ. (1990) A substitution at a non-glycine position in the triple helical domain of proα2(I) collagen chains present in an individual with a variant of the Marfan syndrome. *J. Clin. Invest.* **86:** 1723–1728.

Pihlajaniemi T, Dickson LA, Pope FM, Korhonen VR, Nicholls A, Prockop DJ, Myers JC. (1984) Osteogenesis imperfecta: cloning of a pro-α2(I) collagen gene with a frameshift mutation. *J. Biol. Chem.* **259:** 12941–12944.

Prockop DJ. (1984) Osteogenesis imperfecta: phenotypic heterogeneity, protein suicide, short and long collagens. *Am. J. Hum. Genet.* **36:** 499–505.

Pruchno CJ, Cohn DH, Wallis GA, Willing MC, Starman BJ, Zhang X, Byers PH. (1991) Osteogenesis imperfecta due to recurrent point mutations at CpG dinucleotides in the COL1A1 gene of type I collagen. *Hum. Genet.* **87:** 33–40.

Sillence DO, Senn A, Danks DM. (1979) Genetic heterogeneity in osteogenesis imperfecta. *J. Med. Genet.* **16:** 101–116.

Smith LT, Wertelecki W, Milstone LM, Petty EM, Seashore MR, Braverman IM, Jenkins TJ, Byers PH. (1992) Human dermatosparaxis: a form of Ehlers–Danlos syndrome that results from the failure to remove the amino terminal propeptide of type I procollagen. *Am. J. Hum. Genet.* **51:** 235–244.

Starman BJ, Eyre D, Charbonneau H, Harrylock M, Weis MA, Weiss L, Graham JM, Byers PH. (1989) Osteogenesis imperfecta: the position of substitution for glycine by cysteine in the triple helical domain of the proα1(I) chains of type I collagen determines the clinical phenotype. *J. Clin. Invest.* **84:** 1206–1214.

Steinmann B, Tuderman L, Peltonen L, Martin GR, McKusick VA, Prockop DJ. (1980) Evidence for a structural mutation of procollagen type I in a patient with the Ehlers–Danlos syndrome type VII. *J. Biol. Chem.* **255:** 8887–8893.

Steinmann B, Royce PM, Superti-Furga A. (1993) The Ehlers–Danlos syndrome. In: *Connective Tissue and its Heritable Disorders* (eds PM Royce, B Steinmann). Wiley-Liss, New York, pp. 351–407.

Sykes B, Ogilvie D, Wordsworth P, Wallis G, Mathew C, Beighton P, Nicholls A, Pope FM, Thompson E, Tsipouras P, Schwartz R, Jensson O, Arnason A, Børresen A-L, Heiberg A, Frey D, Steinmann B. (1990) Consistent linkage of dominantly inherited osteogenesis imperfecta to the type I collagen loci: COL1A1 and COL1A2. *Am. J. Hum. Genet.* **46:** 293–307.

Tenni R, Valli M, Rossi A, Cetta G. (1993) Possible role of overglycosylation in the type I collagen triple helical domain in the molecular pathogenesis of osteogenesis imperfecta. *Am. J. Med. Genet.* **45:** 252–256.

Tromp G, Wu Y, Prockop DJ, Madhatheri SL, Kleinert C, Earley JJ, Zhuang J, Norrgård Ö, Darling RC, Abbott WM, Cole CW, Jaakkola P, Ryynänen M, Pearce WH, Yao JST, Majamaa K, Smullens SN, Gatalica Z, Ferrell RE, Jimenez SA, Jackson CE, Michels VV, Kaye M, Kuivaniemi H. (1993) Sequencing of cDNA from 50 unrelated patients reveals that mutations in the triple-helical domain of type III procollagen are an infrequent cause of aortic aneurysms. *J. Clin. Invest.* **91:** 2539–2545.

Van der Rest M, Garrone R. (1991) Collagen family of proteins. *FASEB J.* **5:** 2814–2823.

Vasan NS, Kuivaniemi H, Vogel BE, Minor RR, Wootton JAM, Tromp G, Weksberg R, Prockop DJ. (1991) A mutation in the proα2(I) gene (COL1A2) for type I procollagen in Ehlers–Danlos syndrome type VII: evidence suggesting that skipping of exon 6 in RNA splicing may be a common cause of the phenotype. *Am. J. Hum. Genet.* **48:** 305–307.

Vissing H, D'Alessio M, Lee B, Ramirez F, Byers PH, Steinmann B, Superti-Furga A. (1991) Multi-exon deletion in the procollagen III gene is associated with mild Ehlers–Danlos syndrome type IV. *J. Biol. Chem.* **266:** 5244–5248.

Vogel BE, Doelz R, Kadler KE, Hojima Y, Engel J, Prockop DJ. (1988) A substitution of cysteine for glycine 748 of the α1 chain produces a kink at this site in the procollagen I molecule and an altered *N*-proteinase cleavage site over 255 nm away. *J. Biol. Chem.* **263:** 19249–19255.

Wallis GA, Starman BJ, Schwartz MF, Byers PH. (1990) Substitution of arginine for glycine at postion 847 in the triple-helical domain of the α1(I) chain of type I collagen produces lethal osteogenesis imperfecta: molecules that contain one or two abnormal chains differ in stability and secretion. *J. Biol. Chem.* **265:** 18628–18633.

Wallis GA, Sykes B, Byers PH, Mathew CG, Viljoen D, Beighton P. (1993) Osteogenesis imperfecta type III: mutations in type I collagen structural genes, COL1A1 and COL1A2, are not necessarily responsible. *J. Med. Genet.* **30:** 492–496.

Wenstrup RJ, Willing MC, Starman BJ, Byers PH. (1990) Distinct biochemical phenotypes predict clinical severity in non-lethal variants of osteogenesis imperfecta. *Am. J. Hum. Genet.* **46:** 975–982.

Wenstrup RJ, Shrago-Howe AW, Lever LW, Phillips CL, Byers PH, Cohn DH. (1991) The effects of different cysteine for glycine substitutions within α2(I) chains: evidence of distinct structural domains within the type I collagen triple helix. *J. Biol. Chem.* **266:** 2590–2594.

Westerhausen A, Kishi J, Prockop DJ. (1990) Mutations that substitute serine for glycine α1–598 and glycine α1–631 in type I procollagen. The effects on unfolding of the triple helix are position-specific, and demonstrate that the protein unfolds through a series of cooperative blocks. *J. Biol. Chem.* **265:** 13995–14000.

Willing MC, Cohn DH, Byers PH. (1990) Frameshift mutation near the 3′ end of the COL1A1 gene of type I collagen predicts an elongated proα1(I) chain and results in osteogenesis imperfecta type I. *J. Clin. Invest.* **85:** 282–290.

Willing MC, Pruchno CJ, Atkinson M, Byers PH. (1992) Osteogenesis imperfecta type I is commonly due to a COL1A1 null allele of type I collagen. *Am. J. Hum. Genet.* **51:** 508–515.

Willing MC, Pruchno CJ, Byers PH. (1993) Molecular heterogeneity in osteogenesis imperfecta type I. *Am. J. Med. Genet.* **45:** 223–227.

Zhuang JP, Constantinou CD, Ganguly A, Prockop DJ. (1991) A single base mutation in type I procollagen (COL1A1) that converts glycine α1–541 to aspartate in a lethal variant of osteogenesis imperfecta – detection of the mutation with a carbodiimide reaction of DNA heteroduplexes and direct sequencing of products of the PCR. *Am. J. Hum. Genet.* **48:** 1186–1191.

4

Genotype–phenotype correlation in Gaucher disease

M. Horowitz and A. Zimran

4.1 Introduction

Gaucher disease is the most prevalent sphingolipid storage disease. It is inherited as an autosomal recessive deficiency of the lysosomal enzyme glucocerebrosidase (Barranger and Ginns, 1989). Glucocerebrosidase catalyses the hydrolysis of glucosyl ceramide into ceramide and glucose. It does so in the presence of a low molecular weight activator, designated saposin C or SAP C (sphingosine activator protein C), or sap 2. As a lysosomal enzyme, glucocerebrosidase is synthesised on polyribosomes as a 56 kDa peptide which translocates into the endoplasmic reticulum via its signal peptide. It undergoes N-glycosylation of four asparagines to produce several intermediate forms with higher molecular weights. It is then transported through the Golgi network into the lysosomes, where a protein with a molecular weight of about 65 kDa accumulates (Erickson *et al.*, 1985). It appears that glucocerebrosidase belongs to a family of lysosomal membrane enzymes that are not transported directly from the Golgi to the lysosome but, rather, transported as a transmembrane protein through the trans-Golgi network to the cell membrane, where it is internalized and transferred into lysosomes. Glucocerebrosides are present predominantly in mononuclear cell membranes.

A deficiency of β-glucocerebrosidase, due to a defect in the gene for the enzyme or, more rarely, for the activator protein, leads to the accumulation of its substrate, glucocerebroside, in lysosomes. This occurs particularly in the phagocytic cells of the reticuloendothelial system that are involved in the catabolism of the complex glycolipids of the membranes of leukocytes and erythrocytes. The presence of many lipid-laden vacuoles in the cytoplasm results in enlargement of the cells to 10–200 μm in diameter and displacement of the nucleus to one side. The characteristic appearance of these macrophages was first described by Gaucher, after whom they are named. The distribution of these Gaucher cells reflects the distribution of the substrate and explains the symptoms of Gaucher disease. Their accumulation in the sinusoids of the liver and in the red pulp of the spleen and the bone marrow infiltration account for the hepatosplenomegaly and skeletal defects, respectively, observed in the disease. In the more severe forms of the disease neurodegeneration occurs due to accumulation in the brain.

Gaucher disease is sub-divided into three major forms depending on the absence (type 1) or presence (types 2 and 3) of central nervous system involvement. Type 1 (adult, chronic, non-neuronopathic) is the most common form, characterized by varying degrees of hepatosplenomegaly, bone pains and/or fractures and bone marrow failure. The hallmark of type 1 disease is sparing of the central nervous system. Type 2 (infantile, acute neuronopathic) and type 3 (juvenile, sub-acute neuronopathic) are both rare; in addition to the visceral pathology characteristic of the disease, there is a neurological involvement which is overwhelming and fatal in type 2, but more slowly progressive and less predictable in type 3. The enormous variability in the phenotypic expression of the disease, even within each form, especially in types 1 and 3, and even among patients of the same ethnic descent, is due primarily to the existence of different mutations within the glucocerebrosidase gene (Grabowski *et al.*, 1990; Beutler, 1992; Horowitz and Zimran, 1994). To date, 36 mutations have been described (Table 4.1). Of these, eight mutations have been demonstrated at greater relative frequency in the general population: N370S (1226), L444P (1448), 84GG, IVS2+1, D409H (1342), R463C, recNciI (XOVR) and recTL (Beutler *et al.*, 1992; Horowitz *et al.*, 1993).

The clinical presentations among patients in the Gaucher Clinic at Shaare-Zedek Medical Center, as well as their genotypes, are described herein. The implications of genotype–phenotype correlation for prenatal diagnosis are discussed and indications for enzyme replacement therapy are suggested.

4.2 Genotypes among patients with type I Gaucher disease

4.2.1 N370S/N370S (1226/1226)

The N370S mutation is a G→A transition at nucleotide 1226 of the glucocerebrosidase cDNA (nucleotide 5841 of the active gene; Tsuji *et al.*, 1988). It is the most frequent mutation among Gaucher patients, as well as having a high incidence (73%) among Ashkenazi Jewish patients (Beutler *et al.*, 1992; Horowitz *et al.*, 1993).

Patients homozygous for the N370S mutation have type 1 Gaucher disease, this being the most common Jewish form (Figure 4.1). Nearly 90% of these patients have a mild clinical course; many are asymptomatic (Zimran *et al.*, 1989; Beutler *et al.*, 1993; Sibille *et al.*, 1993). However, some patients present moderate to severe symptomatology.

In most cases, clinical symptoms or signs tend to appear at a mean age of 23 (range 3–82 years), considerably later than in other genotypes (Table 4.2). Clinical manifestations include splenomegaly, anaemia and thrombocytopaenia (Zimran *et al.*, 1992; Zevin *et al.*, 1993). Splenectomy is required less frequently

Table 4.1. Mutation in the glucocerebrosidase gene

Mutation	nuc cDNA*	nuc gene†	Change	Description	Amino acid change	References
1023del	72	1023	C-del	deletion in leader	No change	Beutler *et al.* (1993)
84GG	84	1035	G-ins	frameshift	Premature termination	Beutler *et al.* (1991a)
IVS2+1	—	1067	G→A	splice site mutation	Aberrant mRNA	Beutler *et al.* (1992) He *et al.* (1992)
R120Q	476	3060	G→A	missense mutation	Arg^{120}→Gln	Graves *et al.* (1988)
P122S	481	3065	C→T	missense mutation	Pro^{122}→Ser	Beutler *et al.* (1993)
D140H	535	3119	G→C‡	missense mutation	Asp^{140}→His	Eyal *et al.* (1991)
K157Q	586	3170	A→C	missense mutation	Lys^{157}→Gln	Eyal *et al.* (1991)
Y212H	751	3545	T→C	missense mutation	Tyr^{212}→His	Beutler *et al.* (1993)
F2131	754	3548	T→A	missense mutation¶	Phe^{213}→Ile	Kawame and Eto (1991)
F216Y	764	4113	T→A	missense mutation	Phe^{216}→Tyr	Beutler and Gelbart (1990)
P289L	983	4332	C→T	missense mutation	Pro^{289}→Leu	He *et al.* (1993)
A309V	1043	5259	C→T	missense mutation	Ala^{309}→Val	Latham *et al.* (1991)
W312C	1053	5269	G→T	missense mutation	Trp^{312}→Cys	Latham *et al.* (1991)
T3231	1085	5302	C→T	missense mutation	Thr^{323}→Ile	He *et al.* (1993)
G325R	1090	5306	G→A	missense mutation¶	Gly^{325}→Arg	Eyal *et al.* (1990)
E326K	1093	5309	G→A‡	missense mutation	Glu^{326}→Lys	Eyal *et al.* (1991)
C342G	1141	5357	T→G	missense mutation	Cys^{342}→Gly	Eyal *et al.* (1990)
R359End	1192	5408	C→T	missense mutation	Arg^{359}→End	Beutler and Gelbart (1994)
R359Q	1193	5409	G→A	missense mutation	Arg^{359}→Gln	Kawame *et al.* (1992)
S364T	1208	5424	G→C	missense mutation	Ser^{364}→Thr	Latham *et al.* (1991)
N370S	1226	5841	A→G	missense mutation	Asn^{370}→Ser	Tsuji *et al.* (1987)
55 bp del	1263–1317	5183–5202		deletion¶		Beutler *et al.* (1993)
V394L	1297	5912	G→T	missense mutation	Val^{394}→Leu	Theophilus *et al.* (1989b)
D399N	1312	5927	G→A	missense mutation	Asp^{399}→Asn	Beutler and Gelbart (1994)

Table 4.1 Continued

Mutation	nuc cDNA*	nuc gene†	Change	Description	Amino acid change	References
D409H	1342	5957	G→C§	missense mutation¶	Asp^{409}→His	Eyal *et al.* (1990)
D409V	1343	5958	A→T	missense mutation	Asp^{409}→Val	Theophilus *et al.* (1989b)
P415R	1361	5976	C→G	missense mutation	Pro^{415}→Arg	Wigderson *et al.* (1990)
K425E	1390	6375	A→G	missense mutation	Lys^{425}→Glu	Kawame *et al.* (1992)
L444P	1448	6433	T→C§	missense mutation	Leu^{444}→Pro	Tsuji *et al.* (1987)
R463C	1504	6489	C→T	missense mutation	Arg^{463}→Cys	Hong *et al.* (1990)
IVS10–1	1505	6490	G→A	splice site mutation	Aberrant mRNA	Ohshima *et al.* (1993)
G478S	1549	6628	G→A	missense mutation	Gly^{478}→Ser	Beutler *et al.* (1993)
R496C	1603	6682	C→T	missense mutation	Arg^{496}→Cys	Kawame *et al.* (1992)
R496H	1604	6683	G→A	missense mutation	Arg^{496}→His	Beutler *et al.* (1993)
RecNciI	1448	6433	T→C	missense mutation¶	Leu^{444}→Pro	Zimran *et al.* (1990a)
	1583	6468	G→C	missense mutation¶	Ala^{456}→Pro	Latham *et al.* (1990)
	1597	6482	G→C	no mutation¶	Val^{460}→Val	Eyal *et al.* (1990)
RecTL	1342	5957	G→C	missense mutation¶	Asp^{409}→His	Latham *et al.* (1990)
	1448	6433	T→C	missense mutation¶	Leu^{444}→Pro	Eyal *et al.* (1990)
	1583	6468	G→C	missense mutation¶	Ala^{456}→Pro	Hong *et al.* (1990)
	1597	6482	G→C	no mutation¶	Val^{460}→Val	

* Nucleotide number in the cDNA.
† Nucleotide number in the gene; genomic DNA.
‡ Both changes exist on one allele.
§ Mutations exist as single point mutations and on complex alleles.
¶ Exist naturally in pseudogene.

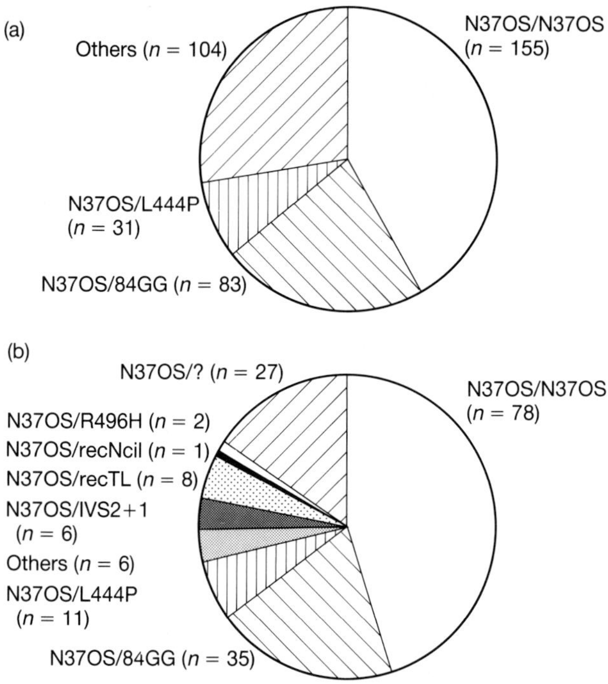

Figure 4.1. Prevalence of genotypes among type 1 Gaucher Jewish patients. (a) Cumulative data of 373 patients reported by Zimran *et al.* (1992) (38 patients), Sibille *et al.* (1993) (161 patients) and 174 patients from the Shaare-Zedek Medical Center (three most common genotypes and 'others'). (b) Detailed genotypes of 174 Jewish patients with type 1 Gaucher disease evaluated in the Shaare-Zedek Medical Center.

and, then, performed at an older age than in other genotypes (Sibille *et al.*, 1993). Bone lesions were observed among 12% of the Shaare-Zedek Medical Center's patients, including avascular necrosis (AVN) of heads of long bones and pathological fractures (Table 4.2). Similar observations have been reported elsewhere (Zimran *et al.*, 1992; Sibille *et al.*, 1993). None of the patients with this genotype present with lung involvement. It should be emphasized that all the patients summarized in Table 4.2 were referred to the clinic at Shaare-Zedek Medical Center (Jerusalem), implying a more overt presentation of Gaucher disease. However, in the general population one would suspect the potential for a subset of asymptomatic or minimally affected 'patients' with this genotype who have never elicited medical attention. Large-scale screening within Jewish populations should uncover undiagnosed homozygotes commensurate with studies of small populations (Zimran *et al.*, 1991; Beutler *et al.*, 1992).

A severity score index (SSI) was devised and calculated for each patient. The index was based on the following criteria: age at first symptoms, splenomegaly,

Table 4.2 Clinical characteristics of Gaucher patients

	Phenotype				
Genotype	Mean age at presentation (range)	Mean age at evaluation (range)	Severity score index (range)	Avascular necrosis	S/P splenectomy
N370S/N370S (*n*=73)	23.11 (3–82)	32.84 (7–85)	7.76 (1–21)	12.3%	13.7%
N370S/84GG (*n*=29)	4.9 (1–19)	24.08 (2–61)	16.09 (7–29)	51.72%	55%
N370S/IVS2+1 (*n*=7)	6.4 (2–19)	32.5 (24–46)	15.6 (9–24)	50%	100%
N370S/L444P (*n*=9)	7.37 (0–19)	23.44 (5–51)	12.37 (7–16)	22.2%	44.44%
N370S/? (*n*=22)	19.5 (2–52)	29.3 (4–67)	11.7 (2–28)	26%	34.6%

splenectomy, hepatomegaly, degree of cytopaenia, liver function tests, bone disease and other organ (e.g. lung) involvement (Zimran *et al.*, 1989).

The mean SI of patients with the N370S/N370S genotype was 7.76 (range: 1–21; Table 4.2).

4.2.2 N370S/84GG

The 84GG mutation is an insertion of an extra G at nucleotide 84 of the glucocerebrosidase cDNA (Beutler *et al.*, 1991a). This results in a shift of protein translation frame, and in premature termination 44 amino acids downstream from the first ATG. The 84GG mutation is the second most common mutation among the Ashkenazi Jewish patient population, accounting for 10–13% of the mutation alleles (Beutler *et al.*, 1992; Horowitz *et al.*, 1993; Sibille *et al.*, 1993). The N370S/84GG genotype comprises 20–22% of all genotypes in Jewish patients (Horowitz *et al.*, 1993; Sibille *et al.*, 1993). Published data include only one non-Ashkenazi patient with this genotype (Beutler *et al.*, 1991a).

Patients with this genotype generally suffer from moderate to severe symptomatic type 1 disease. The advent of clinical symptoms is at a young age (Table 4.2), in our series always before the age of 20 years. Symptoms include splenomegaly with hypersplenism, bone pains and pathological fractures and often the development of abnormal pulmonary function (Kerem and Zimran, unpublished observations). Splenectomy was performed in 55% of the patients at a mean age of 15 years; 52% of the patients with this genotype developed AVN of the hip joints and five had required total hip replacement. We are not aware of any patient with this genotype who has been diagnosed at an age over 28 (Sibille *et al.*, 1993), nor are we familiar with asymptomatic adult patients carrying the N370S/84GG genotype.

The mean SSI of the patients was 16 (range: 7–29; Table 4.2).

4.2.3 N370S/IVS2+1

The IVS2+1 is a splice site mutation (Beutler *et al.*, 1992; He and Grabowski 1992) with a G→A transition at the first nucleotide of the second intron of the active glucocerebrosidase gene, resulting in the loss of the 5′ donor splice site. Consequently, aberrant mRNAs are produced (He and Grabowski, 1992) which fail to direct synthesis of active enzyme; hence, this genotype is associated with severe phenotypic expression of the disease.

Sequence analyses of cDNAs derived from patients with this genotype revealed three discrete populations with the N370S mutation on the other allele: an exact exon 2 deletion; a deletion of exon 2 plus the first 115 bp of exon 3; and an apparently normal sequence (which is as yet unexplained, and comprised less than 10% of the clones; He and Grabowski, 1992). This mutation accounts for 1–3% of the mutant alleles among Jewish and non-Jewish Gaucher patients (Beutler *et al.*, 1992; Sibille *et al.*, 1993; Beutler and Gelbart, 1993; Horowitz *et al.*, 1993). Most type 1 patients with one IVS2+1 mutation present with N370S as the second allele. Similar to patients with the N370S/84GG genotype, the N370S/IVS2+1 genotype patients suffer from moderate to severe manifestations of the disease. They generally present at a very young age (mean 7 years, range 2–19 years; Table 4.2) with splenomegaly and hypersplenism, and develop progressive skeletal complications. In our series, AVN of the hip joint developed in three of seven patients; two patients had undergone total hip replacement (bilateral replacement in one patient) and six patients had undergone splenectomy. The mean SSI of these patients was 15.6 (range: 9–24).

4.2.4 N370S/L444P (1226/1448)

The L444P mutation is a result of a T→C transition at nucleotide 1448 of the glucocerebrosidase cDNA (Tsuji *et al.*, 1987), substituting proline for leucine and creating a new site for the restriction enzyme NciI. This mutation was found in all three types of Gaucher disease. In type 1 this mutation is found in compound heterozygotes having the genotype N370S/L444P. These patients have moderately severe symptoms. In our series there are nine patients in this category: the mean age at presentation was 7 years. Four patients had undergone splenectomy and three suffered serious skeletal complications. This genotype was found at a higher frequency (10–19%) among Gaucher patients in the USA (Sibille *et al.*, 1993; Beutler and Gelbart, 1993) than in Israel. The hypothesis that some of the patients reportedly of the N370S/L444P genotype may have been mistyped and actually were N370S/recNci or N370S/recTL (Horowitz and Zimran, 1993) is discussed below.

4.2.5 N370S/recNci (1226/XOVR)

The recNci is a designation given to a complex allele that includes three single base-pair substitutions within the coding region: L444P (T→C), R456G (G→C) and V460V (G→C; Latham *et al.*, 1991; Zimran *et al.*, 1990b; Eyal *et al.*, 1990). Sequence analysis has shown that this allele contains a fragment that is derived from the glucocerebrosidase pseudogene downstream from exon 10 (Eyal *et al.*, 1990). Patients with the N370S/recNci genotype have clinical features resembling those of patients with the N370S/L444P genotype. In our series of Israeli patients, there is only one patient with this genotype. In an American series, a few more patients have been reported with this genotype (Beutler and Gelbart, 1993).

4.2.6 N370S/recTL

recTL is another complex allele, which contains the same three point mutations as in the recNci complex allele, but with a fourth point mutation, D409H. The D409H mutation is a G→C transversion at nucleotide 1342 of the glucocerebrosidase cDNA. The D409H mutation, in homozygosity, is associated with a unique form of type 3 Gaucher disease (Zimran *et al.*, 1993). The N370S/recTL genotype, on the other hand, is associated with a rather mild phenotype. Of the seven patients in our series with this genotype, six are nearly asymptomatic, the seventh patient being a 9-year-old girl with marked splenomegaly (Zimran and Horowitz, 1994). The apparent convergence between the recTL and the recNci mutations belies the divergence in phenotypic expression in Gaucher disease. It is precisely in such instances that exact genotyping is imperative to subsequent medical care.

4.2.7 N370S/R496H (1226/1604)

The R496H mutation is a G→A transition at nucleotide 1604 of the glucocerebrosidase cDNA (Beutler *et al.*, 1992). It has as yet been identified only in patients with type 1 Gaucher disease. Interestingly, the initial patients identified with this mutation were all compound heterozygotes for 'severe' mutations, such as 84GG or IVS2+1 (Beutler *et al.*, 1992). Therefore it was assumed that, like the N370S (1226) mutation, it has a mitigating effect on phenotypic expression. In our series, four of the five patients with this mutation had the N370S/R496H genotype, and presented with mild or asymptomatic disease as predicted. The fifth, has the IVS2+1/R496H genotype. He is 1 year old, with normal growth and development but has not yet been evaluated. This child, whose father has Gaucher disease (genotype N370S/IVS2+1) and whose mother is a R496H carrier, was diagnosed *in utero*.

The homozygote state of this genotype (R496H/R496H) is probably so mild as to not require medical attention and hence has not been diagnosed (Beutler *et al.*, 1992).

4.2.8 N370S/?

Patients with the N370S/? genotype comprise about 15% of the Jewish patients (Figure 4.1) and 14% of the non-Jewish patients with Gaucher disease (Beutler and Gelbart, 1993). Since the second mutation can be one of several alleles each of which confers a unique effect on the severity of the disease, patients with this genotype present a varied clinical picture (Figures 4.2 and 4.3). It is worth noting that some of the Jewish patients with this genotype suffer from rather severe Gaucher disease, with multiple fractures and pulmonary involvement, more pronounced than that of the average patient with N370S/84GG, N370S/IVS2+1 or N370S/L444P genotypes.

4.2.9 Other genotypes

Several other genotypes exist among type 1 patients. These include: L444P/?, R496H/IVS2+1 and ?/? (i.e. two as yet unknown mutations). Among Jewish patients, they comprise only 3.4% (Figure 4.1), therefore only limited information is available about their clinical course. Among non-Jewish patients with type 1 Gaucher disease, these 'other' genotypes comprise 26% of patients, representing

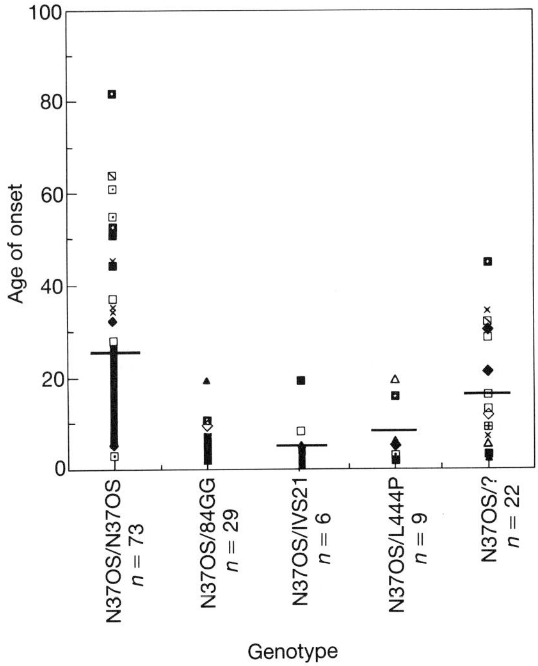

Figure 4.2. Age of onset of Gaucher disease. The age of onset of symptoms of the disease is presented according to five main genotypes (139 Jewish patients with type 1 Gaucher disease studied at the Shaare-Zedek Medical Center, Jerusalem).

a larger number of genotypes, some with rare alleles (Beutler and Gelbart, 1993).

Since Gaucher disease is the most prevalent sphingolipidosis, it is not impossible statistically that type 1 patients may suffer from other clinical disorders, including neurological diseases. The occurrence of type 1 Gaucher disease together with a neurological disease, such as myoclonic epilepsy, psychosis or Parkinson's, may be misdiagnosed as type 3 disease. Genotyping of such patients has proven useful in the differential diagnosis.

4.3 Genotypes among patients with neuronopathic Gaucher disease

4.3.1 L444P/L444P (1448/1448)

This is the most common genotype of patients with a neuronopathic form of Gaucher disease: in the homozygous state it causes type 3 disease. It is the genotype of all Gaucher patients from Norrbotten county, Sweden. In the heterozygous state of L444P/recNci, the phenotypic expression is acute neuronopathic type 2 disease (Eyal *et al.*, 1990; Grabowski *et al.*, 1990; Latham *et al.*, 1991).

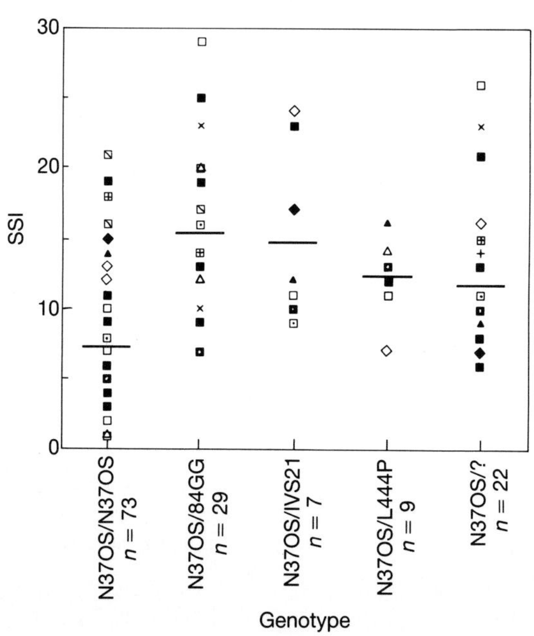

Figure 4.3. Severity score index (SSI) of Gaucher disease. The SSIs of the patients are presented according to five main genotypes (139 Jewish patients with type 1 Gaucher disease studied at the Shaare-Zedek Medical Center, Jerusalem).

Various neurologic signs, such as oculomotor apraxia, strabismus, hypertonicity and retroflexion of the head, can be found in the early stages of disease in both type 2 (during the middle of the first year of life) and type 3 (later in childhood).

There have been a few cases reported with the L444P/L444P genotype, including a few Japanese patients with severe clinical manifestations (Masuno *et al.*, 1990; Glew *et al.*, 1991), suggestive of type 1 Gaucher disease, but without neurological complications. However, the clinical course these young patients does not preclude there being a subsequent neurologic pathology, and hence reclassification as type 3. The issue of molecular diagnosis of a neuronopathic form of Gaucher disease prior to onset of neurological manifestations may be unresolved, as most of these patients are now treated routinely with enzyme replacement, thereby avoiding CNS complications (Abrahamov *et al.*, 1991). As with type 1, there is no explanation for the clinical heterogeneity observed within type 3 patients with this genotype.

In Israel, most of the L444P/L444P patients are Arab children with a rather aggressive visceral disease, but relatively less prominent neurological findings.

4.3.2 L444P/recNci

Studies based on complete sequencing of cDNAs representing both alleles from type 2 patients revealed the presence of additional point mutations on at least one allele (Eyal *et al.*, 1990; Latham *et al.*, 1990). These additional point mutations have been described above, and this complex allele designated 'recNci' (Eyal *et al.*, 1990). Patients carrying the L444P/recNci genotype present the typical clinical picture of type 2 Gaucher disease. This genotype was identified in three of eight non-Jewish patients with type 2 who have been analysed at the Tel-Aviv University; the other five genotypes were L444P/?, D409H/IVS2+1, G325R/C342G, L444P/P415R, L444P/IVS2+1 and ?/? (Horowitz *et al.*, 1993).

4.3.3 D409H/D409H (1342/1342)

The D409H mutation is a G→C transversion at nucleotide 1342 of the glucocerebrosidase cDNA. This mutation was identified originally in a cDNA from cultured fibroblasts taken from an Arab patient with oculomotor apraxia (Eyal *et al.*, 1990), and was classified as uncommon (Beutler, 1992) or rare (Grabowski *et al.*, 1990). However, recently we have reported a unique form of type 3 Gaucher disease among Arabs, predominantly from the district of Jenin, Israel, which is characterized by oculomotor apraxia and valvular sclerosis leading to severe aortic stenosis. These patients present only mild splenomegaly and no bone disease. Similar symptoms were described in a single family in Spain, but as yet the genotype of these patients has not been reported (Chabas *et al.*, 1993).

4.3.4 Other genotypes

Given the relative rarity of the neuronopathic forms, and the lack of ethnic specificity of type 2, it is not surprising that there is a larger percentage of unknown alleles among these patients. Many of the as yet unidentified alleles will probably represent rare mutations, but there are too few patients to allow determination of genotype–phenotype correlation.

It is worth noting that the neuronopathic form may also occur among Jews: we have encountered a Jewish patient with type 2 Gaucher disease (genotype L444P/IVS2+1) who died at 8 months of age (Horowitz, unpublished observations).

4.4 Implications of genotype–phenotype correlations in Gaucher disease

Biochemical diagnosis of Gaucher disease became available in the early 1970s with the introduction of an *in vitro* enzymatic assay (Beutler and Kuhl, 1970; Grabowski *et al.*, 1990). In this assay, the activity of β-glucocerebrosidase is tested in the presence of ionic detergents such as Triton X-100 or sodium deoxycholate, thus obviating the need for saposin C, the natural activator of the enzyme. This assay was found to be reliable for diagnosis of patients; however, detection of carrier status is not always as accurate due to an overlap in the *in vitro* assay activity between heterozygotes and normals (Beutler and Kuhl, 1970; Kolodny *et al.*, 1982; Matoth *et al.*, 1987; Zimran *et al.*, 1990a). Moreover, the enzymatic test cannot differentiate between the different forms of the disease, nor can it predict the severity of the disease. Mutation analysis at the DNA level overcomes these obstacles to a great extent, and provides a tool for more accurate prediction of the clinical course of Gaucher disease. It is worth mentioning that, because of the heterogeneity of clinical manifestations in some genotypes, specifically N370S/N370S, genetic counselling is sometimes difficult (Beutler, 1992; Sidransky *et al.*, 1992; Horowitz and Zimran, 1994). For example, when confronted with a case of prenatal diagnosis of a fetus homozygous for the N370S mutation, one cannot predict the future clinical course of the disease with assurance, despite the statistic that 90% of patients with this genotype are asymptomatic or only mildly affected. Other genotypes such as L444P/recNci, L444P/L444P or N370S/84GG, which are predictive of a severe form of Gaucher disease, allow a more narrow and more accurate prediction of prognosis.

The recent introduction of effective enzyme replacement therapy for Gaucher disease (Barton *et al.*, 1991; Beutler *et al.*, 1991b) adds another important dimension to molecular diagnosis. When a fetus or an individual is first diagnosed as having Gaucher disease, the evidence of a severe genotype (N370S/IVS2+1, N370S/84GG, N370S/recNci, etc.) implies the need for early enzymatic therapy. When molecular analysis reveals a mild genotype, such as N370S/N370S or N370S/R496H, treatment can be held in abeyance until such time as the signs of disease merit treatment, especially in countries where treatment is available only for symptomatic patients. Enzyme replacement therapy is unique in storage disease treatment in that it can ameliorate or even eradicate the signs of type 1 Gaucher disease. The enzyme is produced from human placenta derived from selected donors, and is modified for targeting to α-mannosyl receptors on macrophages. This treatment has proven to be both effective and safe in nearly 1000 patients worldwide, with few adverse reactions (Barton *et al.*, 1991; Beutler *et al.*, 1991b; Pastores *et al.*, 1993; Zimran *et al.*, 1993). Currently, there are two widely accepted regimens for administration of the enzyme: high dose–low frequency (60 units kg^{-1} body weight every 2 weeks) and low dose–high frequency (2.3 units kg^{-1} body weight three times a week). The major consideration is the ability of the health care schemes to cope with the high cost of the therapy. The hope for the future is that the cost of treatment will be reduced, and enzyme replacement therapy may be considered *a priori* for all Gaucher patients, as a prophylactic measure as well as a therapeutic response to severe manifestations of the disease.

References

Abrahamov A, Horowitz M, Zimran A. (1991) Early detection of the 1448/1448 genotype – associated with type 3 Gaucher disease – prior to onset of neurological manifestations. *Pediatr. Res.* **29**: 126A.

Barranger JA, Ginns EI. (1989) Glucosylceramide lipidoses: Gaucher disease. In: *The Metabolic Basis of Inherited Diseases* (eds Scriver CR, Beaudet AL, Sly WS, Valle D). McGraw-Hill, New York, pp. 1677–1699.

Barton NW, Brady RO, Damrosia JM, Di Bisceglie AJ, Dopplet SH, Hill SC, Mankin HJ, Murray GJ, Parker RI, Argoff CE, Grewal RP, Yu K *et al.* (1991) Replacement therapy for inherited enzyme deficiency: macrophage-targeted glucocerebrosidase for Gaucher's disease. *N. Engl. J. Med.* **324**: 1464–1470.

Beutler E. (1992) Gaucher disease: new molecular approaches to diagnosis and treatment. *Science* **256**: 794–799.

Beutler E, Gelbart T. (1990) Gaucher disease associated with a unique KpnI restriction site: identification of the amino acid substitution. *Ann. Hum. Genet.* **54**: 149–153.

Beutler E, Gelbart T. (1993) Gaucher disease in non-Jewish patients. *Br. J. Haematol.* **85**: 401–405.

Beutler E, Gelbart T. (1994) Two new Gaucher disease mutations. *Hum. Mutat.*, in press.

Beutler E, Kuhl W. (1970) The diagnosis of the adult of type Gaucher's disease and its carrier state by demonstration of deficiency of β-glucosidase activity in peripheral blood leukocytes. *J. Lab. Clin. Med.* **76**: 747–755.

Beutler E, Gelbart T, Kuhl W, Sorge J, West C. (1991a) Identification of the second common Jewish Gaucher disease mutation makes possible population-based screening for heterozygous state. *Proc. Natl. Acad. Sci. USA* **88**: 10544–10547.

Beutler E, Kay A, Saven A, Garver P, Thurston D, Dawson A, Rosenbloom B. (1991b) Enzyme replacement therapy for Gaucher disease. *Blood* **78**: 1183–1189.

Beutler E, Gelbart T, Kuhl W, Zimran A, West C. (1992) Mutations in Jewish patients with Gaucher disease. *Blood* **79**: 1662–1666.

Beutler E, Gelbart T, West C. (1993) Identification of six new Gaucher disease mutations. *Genomics* **15**: 203–305.

Chabas A, Cormand B, Burguera JM, Villageliu L, Grinberg D, Balcells S, Gonzalez R, Sorbino JM. (1993) Enzymatic and molecular studies in an unusual case of Gaucher's disease with cardiovascular calcifications. *Proceedings of the Second International Duodecim Symposium–Molecular Biology of Lysosomal Diseases*. Finnish Medical Society, Helsinki, p. 69A.

Eyal N, Wilder S, Horowitz M. (1990) Prevalent and rare mutations among Gaucher patients. *Gene* **96**: 277–283.

Erickson AH, Ginns GI, Barranger JA. (1985) Biosynthesis of the lysosomal enzyme glucocerebrosidase. *J. Biol. Chem.* **260**: 14319–14324.

Eyal N, Firon N, Wilder S, Kolodny EH, Horowitz M. (1991) Three unique base pair changes in a family with Gaucher disease. *Hum. Genet.* **52**: 85–88.

Glew RH, Gopalan V, Hubbell CA, Beutler E, Geil JD, Lee RE. (1991) A case of nonneurologic Gaucher's disease that biochemically resembles the neurologic types. *J. Neuropathol. Exp. Neurol.* **50**: 108–117.

Grabowski GA, Gatt S, Horowitz M. (1990) Acid-β-glucosidase: enzymology and molecular biology of Gaucher disease. *Crit. Rev. Biochem. Mol. Biol.* **25**: 385–414.

Graves PN, Grabowski GA, Eisner R, Palese P, Smith FI. (1988) Gaucher disease type 1: cloning and characterization of a cDNA encoding acid β-glucosidase from an Ashkenazi Jewish patient. *DNA* **7**: 521–528.

He GS, Grabowski GA. (1992) Gaucher disease: a G+1→A+1 IVS2 splice donor site mutation causing exon 2 skipping in the acid-β-glucosidase mRNA. *Am. J. Hum. Genet.* **51**: 810–820.

He GS, Grace ME, Grabowski GA. (1992) Gaucher disease: four rare missense mutations encoding F213I, F289Y, T323I and R463C in type I variants. *Hum. Mutat.* **1**: 423–427.

He GS, Grace ME, Grabowski GA. (1993) Four rare alleles ancoding F213I, P289L, T323I and R463C in type 1 variants. *Hum. Mutat.* **1**: 423–427.

Hong CM, Ohashi T, Yu XJ, Weller S, Barranger JA. (1990) Sequence of two alleles responsible for Gaucher disease. *DNA Cell Biol.* **9**: 233–241.

Horowitz M, Zimran A. (1994) Mutation update: mutations causing Gaucher disease. *Hum. Mutat.* **3**: 1–11.

Horowitz M, Tzuri G, Eyal N, Berebi A, Kolodny EH, Brady RO, Barton NW, Abrahamov A, Zimran A. (1993) Prevalence of nine mutations among Jewish and non-Jewish Gaucher disease patients. *Am. J. Hum. Genet.* **53**: 921–930.

Kawame H, Eto Y. (1991) A new glucocerebrosidase gene missense mutation responsible for neuronopathic Gaucher disease in Japanese patients. *Am. J. Hum. Genet.* **49**: 1378–1380.

Kawame H, Hasegawa Y, Eto Y, Maekawa Y. (1992) Rapid identification of mutations in the glucocerebrosidase gene of Gaucher disease patients by analysis of single-strand conformation polymorphism. *Am. J. Hum. Genet.* **49**: 1378–1380.

Kolodny EH, Ullman MD, Mankin HJ, Raghavan SS, Topol J, Sullivan JL. (1982) Phenotypic manifestations of Gaucher disease: clinical features in 48 biochemically verified type I patients and comment on type II patients. In: *Gaucher Disease: a Century of Delineation and Research* (eds RJ Desnick, S Gatt, GA Grabowski). Alan R. Liss, New York, pp. 33–65.

Latham TE, Theophilus BDM, Grabowski GA, Smith FI. (1991) Heterogeneity of mutations in the acid-β-glucosidase gene of Gaucher disease patients. *DNA Cell Biol.* **10**: 15–21.

Mansuno M, Shunji T, Kazuko S, Tadao O. (1990) Non-existence of a tight association between a ^{444}leucine to proline mutation and phenotypes of Gaucher disease: high frequency of a NciI polymorphism in the non-neuronopathic form. *Hum. Genet.* **84**: 203–206.

Matoth Y, Chazan S, Cnaan A, Gelertner I, Klibansky C. (1987) Frequency of carriers of chronic (type I) Gaucher disease in Ashkenazi Jews. *Am. J. Med. Genet.* **27**: 561–565.

Ohshima T, Saaki M, Matsuzaka T, Sakuragawa N. (1993) A novel splicing abnormality in a Japanese patient with Gaucher's disease. *Hum. Mol. Genet.* **2**: 1497–1498.

Pastores GM, Sibille AR, Grabowski GA. (1993) Enzyme therapy in Gaucher disease type 1: dosage efficacy and adverse effects in 33 patients treated for 6 to 24 months. *Blood* **82**: 408–416.

Sibille A, Eng CM, Kim SI, Pastores G, Grabowski GA. (1993) Phenotype–genotype correlations in Gaucher disease type I: clinical and therapeutic applications. *Am. J. Hum. Genet.* **52**: 1094–1101.

Sidransky E, Tsuji S, Martin BM, Stubblefield BK, Ginns EI. (1992) DNA mutation analysis of Gaucher patients. *Am. J. Hum. Genet.* **42**: 331–336.

Theophilus BDM, Latham TE, Grabowski GA, Smith FI. (1989a) Gaucher disease: molecular heterogeneity and phenotype–genotype correlations. *Am. J. Hum. Genet.* **45**: 212–225.

Theophilus BDM, Latham TE, Grabowski GA, Smith FI. (1989b) Comparison of RNase A, chemical cleavage, and GC clamped denaturating gradient gel electrophoresis for the detection of mutations in exon 9 of the human acid-β-glucosidase gene. *Nucl. Acids Res.* **17**: 7707–7722.

Tsuji S, Choudary PV, Martin BM, Stubblefield BK, Mayor JA, Barranger JA, Ginns EI. (1987) A mutation in the human glucocerebrosidase gene in neuronopathic Gaucher's disease. *N. Engl. J. Med.* **316**: 570–575.

Tsuji S, Martin BM, Barranger JA, Stubblefield BK, LaMarca ME, Ginns EI. (1988) Genetic heterogeneity in type 1 Gaucher disease: multiple genotypes in Ashkenazi and non-Ashkenazi individuals. *Proc. Natl Acad. Sci. USA* **85**: 2349–2352.

Wigderson M, Firon N, Horowitz Z, Wilder S, Frishberg Y, Horowitz M. (1989) Characterization of mutations in Gaucher patients by cDNA cloning. *Am. J. Hum. Genet.* **44**: 365–377.

Zevin S, Abrahamov A, Hadas-Halpern I, Kannai R, Levy-Lahad E, Horowitz M, Zimran A. (1993) Adult-type Gaucher disease in children: genetic, clinical features and enzyme replacement therapy. *Q. J. Med.* **86**: 565–573.

Zimran A, Horowitz M. (1994) RecTL: a complex allele in the glucocerebrosidase gene associated with a mild clinical course of Gaucher disease. *Am. J. Med. Genet.*, in press.

Zimran A, Sorge J, Gross E, Kubitz M, West C, Beutler E. (1989) Prediction of severity of Gaucher's disease by identification of mutations at DNA level. *Lancet* **2**: 249–352.

Zimran A, Kuhl W, Beutler E. (1990a) Detection of the 1226 (Jewish) mutation for Gaucher's disease by color PCR – a means for studying the gene frequency of the disorder. *Am. J. Clin. Pathol.* **93**: 788–791.

Zimran A, Sorge J, Gross E, Kubitz M, West C, Beutler E. (1990b) A glucocerebrosidase fusion gene in Gaucher disease: implication for the molecular anatomy, pathogenesis and diagnosis of the disorder. *J. Clin. Invest.* **85**: 219–222.

Zimran A, Gelbart T, Westwood B, Grabowski GA, Beutler E. (1991) High frequency of the Gaucher disease mutation at nucleotide 1226 among Ashkenazi Jews. *Am. J. Hum. Genet.* **49**: 885–889.

Zimran A, Kay A, Gelbart T, Garver P, Thruston D, Saven A, Beutler E. (1992) Gaucher disease: clinical, laboratory radiologic and genetic features in 53 patients. *Medicine* **71**: 337–353.

Zimran A, Abrahamov A, Gross-Zur V, Tafakji M, Rosenberg P, Hadas-Halpern I, Ferber B, Glaser Y, Horowitz M. (1993) A unique form of Gaucher disease characterized by oculomotor apraxia and valvular heart disease. *Proceedings of the Second International Duodecim Symposium – Molecular Biology of Lysosomal Diseases.* Finnish Medical Society, Helsinki, p. 107A.

5

Familial hypercholesterolaemia

Anne K. Soutar

5.1 Introduction

Familial hypercholesterolaemia (FH) is a clinical disorder characterized by the presence of a raised concentration of cholesterol in plasma leading to cholesterol deposition in the tissues, most notably as tendon xanthomata, in both the affected individual and first degree relatives (Goldstein and Brown, 1989). It is inherited as a monogenic autosomal dominant trait (Khachadurian, 1964) and, with a frequency of approximately 1 in 500 in most populations, is one of the commonest inherited disorders of metabolism. In some genetically isolated groups, the frequency can be even higher due to founder gene effects (Hobbs *et al.*, 1987; Lehrman *et al.*, 1987b; Betard *et al.*, 1992). The increase in total plasma cholesterol in FH is due almost entirely to a raised concentration of low density lipoprotein (LDL); studies of the turnover of a trace of ^{125}I-labelled LDL in FH patients have shown that this is caused by both a decrease in the fractional rate of catabolism and overproduction of plasma LDL. Since a high concentration of LDL in plasma is known to constitute a strong risk factor for the development of premature coronary heart disease (CHD), it is not surprising that FH individuals are at considerably greater risk than the general population (Slack, 1969; Stone *et al.*, 1974). Individuals who are homozygous for the defect frequently show symptoms of CHD in early childhood (Goldstein and Brown, 1989) and, until the development of radical cholesterol-lowering therapy, rarely lived beyond their second or third decade (Thompson *et al.*, 1975). Fortunately they are rare, with about 20 known cases in the UK at the present time.

The nature of the genetic defect in FH was elucidated by the early studies of Brown and Goldstein with cultured skin fibroblasts from homozygous FH patients (Brown and Goldstein, 1986). They found that cholesterol synthesis, normally subject to strict feedback regulation in cells grown with an exogenous supply of cholesterol-containing lipoproteins in the medium, was not regulated in cells from homozygous FH patients (Brown *et al.*, 1973). This was due to the absence of a high-affinity, cell-surface receptor that specifically mediated the uptake of extracellular LDL by normal cells, which they named the LDL receptor (Goldstein and Brown, 1974). Once the LDL receptor and its gene had been fully characterized (Sudhof *et al.*, 1985), it was possible to show that mutations in the structural or promoter regions of the gene were responsible for many cases of FH and more than 150 different mutant LDL receptor gene alleles have now been characterized in FH patients world-wide (Hobbs *et al.*, 1990). However, not all patients with a clinical diagnosis of FH have a mutation in their LDL receptor gene. In a few individuals, the defect lies in the gene for apoB, the ligand for the LDL receptor (Tybjaerg-Hanson *et al.*, 1990), while in others LDL receptor function is apparently normal and the defect must lie elsewhere (Harada *et al.*, 1992). Although the clinical signs resulting from other gene defects may be the same as those in FH, in this chapter the term 'familial hypercholesterolaemia' will be restricted to individuals in whom LDL receptor function is disrupted due to a mutation in the structural gene for the protein or in some gene product involved in its regulation.

5.2 Variability in the clinical symptoms of FH

As will be described in more detail below, the nature of the mutation in the LDL receptor gene determines the defect in receptor function that can be observed in cultured cells from a homozygous patient. Different gene defects can result in the production of a defective protein that allows the cell to retain up to 20% or more of normal receptor-mediated uptake of LDL or in the complete absence of any detectable LDL receptor protein or mRNA. Whether or not the fully induced activity of the receptor in skin fibroblasts in culture reflects accurately the LDL receptor activity in the relevant tissue in the whole body, namely the liver, remains open to question, but it has been observed that the clinical manifestations in homozygous FH are correlated with the residual LDL receptor activity that can be measured in the patient's cells (Sprecher *et al.*, 1985). Although the mean plasma cholesterol level in the population varies quite widely between different countries, depending mainly on dietary habits (Keys, 1970), the mean value in homozygous FH individuals in different geographical areas is remarkably similar (Table 5.1). However, there is a wide range of values in any group and the extent of the hypercholesterolaemia in a particular homozygous FH patient appears to determine the age of onset of symptoms and the age at death from CHD (Goldstein and Brown, 1989). The obvious conclusion from these observations is that the severity of the disease in homozygous FH is determined to a large extent by the nature of the mutation in the LDL receptor gene and its specific effect on receptor function. This view is supported by a recent study in which sufficient numbers of homozygous FH patients were available to allow a comparison to be made between two groups, each with one of two mutant alleles. It was observed that the group of patients with a mutation that results in a receptor protein that retains some residual function were less severely affected by CHD than those in the group with a mutation that totally abolishes receptor function in cultured cells (Moorjani *et al.*, 1993). This study was only possible because of the remarkably high incidence of true homozygous FH in the French Canadian population caused by not one but two different founder genes together with continued consanguineous marriages in isolated parts of the country. However, it should also be borne in mind that it is likely that these two groups of patients have many other gene variants in common as well as their mutant LDL receptor alleles, and that these could include genes that influence susceptibility to atherosclerosis.

In heterozygous FH, however, it is less clear what determines not only the degree of hypercholesterolaemia but also the onset and progression of overt CHD, despite the fact that it has been clear since FH was first recognized as an inherited disorder that there is wide variation in the age of onset of symptoms in affected individuals. Many reports have appeared describing the clinical features of heterozygous FH patients, some of whom succumb to fatal myocardial infarction in their prime and others who survive into old age with minimal CHD. Family studies carried out before the genetic defect had been identified indicated that variants of a single major gene, whose effects were in turn modified by several other genes, were likely to be responsible for this variation (Slack, 1969; Heiberg and Slack, 1977). A number of different studies in which heterozygous FH patients with and without CHD have been compared, have shown that males are at much greater risk than females of the same age, a distinction that is not seen in homozygous FH individuals. In the general population there is a strong correlation between cardiovascular risk, in particular the age of onset of symptoms, and the concentration of plasma cholesterol up to a value of 7.5 mmol l^{-1} (Grundy, 1986), but it is not known whether this extends into the plasma cholesterol concentration range of 7.5–14.0 mmol l^{-1} that exists in

Table 5.1. Plasma cholesterol concentration in FH in different geographical areas

Country	Mean and (range) of plasma cholesterol (mmol l^{-1})			References
	Normolipaemic*	hmz FH†	htz FH‡	
United Kingdom	6.2	19.4 (15.6–26.0)	10.4§	Thompson *et al.* (1989), Thompson (1991)
United States	6.0	18.8 (10.0–25.1)	9.5§	Sprecher *et al.* (1985), Kannel *et al.* (1971)
South Africa	Not reported	20.3 (14.7–28.0)	9.2¶	Seftel *et al.* (1980)
China	4.2	18.4 (15.6–23.6)	7.0¶	Chen *et al.* (1992), Cai *et al.* (1991)
Japan	4.5	18.8 (15.8–23.4)	9.7§	Mabuchi *et al.* (1979), Hirobe *et al.* (1982)
Lebanon	4.2‖	18.8 (12.2–31.2)	9.5¶	Khachadurian (1964)

* Men aged 30–60; † both sexes aged <25; ‡ both sexes of various ages; § clinically diagnosed htz; ¶ obligate htz; ‖ unaffected relatives.
hmz, homozygous; htz, heterozygous.

heterozygous FH patients. Although it is clear that heterozygous FH patients are at increased risk compared with the general population, as yet there is little evidence to suggest that the total plasma or LDL cholesterol concentration in heterozygous FH is correlated directly with the onset of CHD. Only a low concentration of high density lipoprotein (HDL), frequently accompanied by an increased concentration of plasma triglyceride, a low ratio of HDL to LDL cholesterol and a high concentration of plasma lipoprotein (a) (Lp(a)) emerge as strong indicators of cardiovascular risk in heterozygous FH, apart from smoking or hypertension, all of which are well-accepted risk factors for the general population (Stone *et al.*, 1974; Heiberg, 1975; Beaumont *et al.*, 1976; Streja *et al.*, 1978; Gagné *et al.*, 1979; Hirobe *et al.*, 1982; Moorjani *et al.*, 1986; Miettinen and Gylling, 1988; Seed *et al.*, 1990; Hill *et al.*, 1991; Kotze *et al.*, 1993). As more groups of heterozygous patients with defined mutations in the LDL receptor gene are identified, it is emerging that the nature of the mutation can influence the plasma cholesterol concentration (Jeenah *et al.*, 1993; Koivisto *et al.*, 1993; Kotze *et al.*, 1993; Moorjani *et al.*, 1993) but, nonetheless, currently available evidence shows that environmental or genetic factors other than the nature of the defect in the LDL receptor gene itself can have a marked influence on the phenotypic expression of FH. What is known about the way in which these factors interact with the LDL receptor defect and its effect on lipoprotein metabolism will be the subject of this chapter.

5.3 Role of the LDL receptor in plasma lipoprotein metabolism

A brief outline of the metabolism of plasma lipoproteins (Figure 5.1) will illustrate the central role played by the LDL receptor in regulating the concentration of LDL in plasma and will also indicate how and why other genes could influence the phenotype of patients with inherited defects in the receptor gene. Further details and original references can be found elsewhere (Gotto *et al.*, 1986).

Triglyceride-rich lipoproteins, very low density lipoproteins (VLDL) from the liver and chylomicrons (CM) from the gut transport lipids from their sites of synthesis and absorption to sites of utilization and storage. The particles comprise a spherical core of triglyceride surrounded by a monolayer of protein and polar lipids. The major protein component in VLDL is apoB_{100}, while that in CM is apoB_{48}, together with other smaller apoproteins including apoE, apoCI, apoCII and apoCIII, that all play a part in the metabolism of the particles. VLDL and CM are secreted into the circulation where their triglyceride core is hydrolysed by the action of the enzyme lipoprotein lipase, for which apoCII is a cofactor, to form remnant lipoproteins. These particles are also depleted of some of their surface lipid and protein components, mainly by transfer to HDL, but retain all their apoB. Although apoB_{100} in LDL is the ligand for the LDL receptor, apoB_{100} in normal VLDL is not recognized until the particle has lost much of its

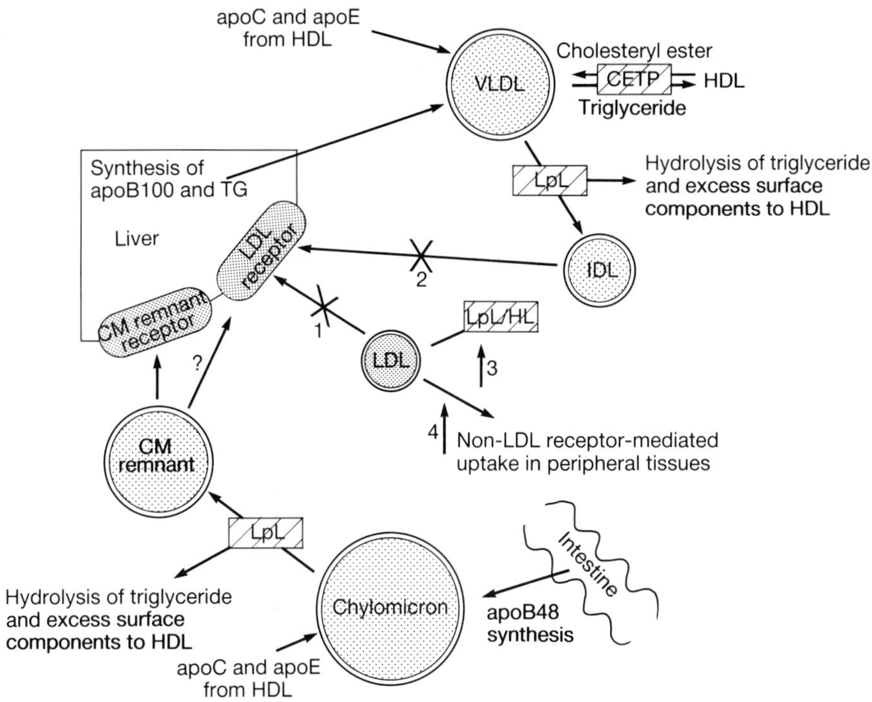

Figure 5.1 Role of the LDL receptor in the metabolism of triglyceride-rich lipoproteins in plasma. In familial hypercholesterolaemia (FH) there is reduced LDL receptor-mediated clearance of both LDL (1) and IDL (2) by the liver, the latter resulting in increased transfer of IDL to the plasma LDL pool (3), that is, overproduction of LDL. The expansion of the LDL pool in plasma results in increased clearance of LDL by non-saturable, non-specific uptake in the peripheral tissues (4), processes that may be responsible for the accelerated atherosclerosis in FH. IDL, intermediate density lipoprotein (or VLDL remnants); CM, chylomicron; LpL, lipoprotein lipase; HL, hepatic lipase; CETP, cholesteryl ester transfer protein; TG, triglyceride.

lipid and protein, particularly apoCIII, while $apoB_{48}$ in CM does not contain that part of the apoB protein that forms the receptor recognition site. Nonetheless, both VLDL and CM remnants can be recognized by the LDL receptor since they contain apoE, which has high affinity for the LDL receptor. ApoE is known to be polymorphic and the different polymorphic forms have different affinity for the LDL receptor, with the result that, in normolipaemic individuals, apoE polymorphism has a small but clearly defined effect on plasma LDL–cholesterol concentration. During normal lipoprotein metabolism, some but not all of the VLDL remnants formed are cleared by LDL receptor-mediated uptake in the liver; whether CM remnants are cleared in this way or by a separate hepatic CM remnant receptor remains unresolved, but CM are removed quantitatively at this stage. Those VLDL remnants that remain in the circulation are metabolized further and remodelled, eventually forming LDL in which the sole protein is $apoB_{100}$. Although LDL is cleared mainly by hepatic LDL receptors, some is also cleared by ill-defined non-specific mechanisms that constitute the sole pathway for clearance of LDL in patients with receptor-negative homozygous FH. The most compelling evidence for a genetically distinct CM remnant receptor comes from the observation that CM remnant clearance is apparently unimpaired in such patients (Rubinsztein *et al.*, 1990). Thus, changes or defects in LDL receptor activity influence both the input and removal of plasma LDL, explaining why measurement of LDL turnover *in vivo* shows that not only is there decreased catabolism of LDL in FH, but also overproduction (Langer *et al.*, 1972). It is also possible to identify a number of other gene products, variants of which might have an influence on the determination of lipoprotein concentration, for example apoB or apoE and all the factors that regulate their synthesis and secretion.

Patients with FH, both homozygous and heterozygous, tend to have lower than normal plasma HDL (Goldstein and Brown, 1989), but it is not clear how defects in the LDL receptor can influence HDL concentration directly. Plasma HDL serves as an acceptor for excess surface components derived from the metabolism of triglyceride-rich lipoproteins and as a reservoir of the exchangeable apoproteins in plasma. It also functions to shuttle excess cholesterol from the peripheral tissues back to the liver for disposal, a process frequently referred to as reverse cholesterol transport. Since low HDL clearly emerges as an important determinant of cardiovascular risk in heterozygous FH, it is unfortunate that the genetic and environmental factors that affect HDL concentration and metabolism are still not well understood, even in the normolipaemic population.

As well as the genetic determinants of lipoprotein metabolism, there are many other genes that have been predicted to be or been shown to be associated with increased risk of cardiovascular disease in the general population. It is likely that these also exert some influence in FH heterozygotes, although the effect may be relatively small compared with that of the increased LDL concentration.

5.4 The LDL receptor is a multifunctional protein

The extensive studies of Brown and Goldstein and their colleagues (reviewed in Brown and Goldstein, 1986), with cultured skin fibroblasts from both normal and FH subjects, showed that the LDL receptor is a cell surface glycoprotein that specifically recognizes lipoproteins that contain either apoB or apoE, mainly LDL or VLDL remnants derived from triglyceride-rich lipoproteins, as described above. The receptor protein is synthesised as a precursor whose sugar residues are modified in the Golgi apparatus during its transit to the cell membrane. The precursor and mature forms of the protein have quite distinct electrophoretic mobilities, and the maturation of the protein can be followed by immunoprecipitation of the newly synthesised protein from extracts of cells pulse-labelled with [^{35}S]methionine. After the receptor on the cell surface binds the ligand, the receptor–ligand complexes cluster in specialized areas of the cell membrane known as coated pits which invaginate to form intracellular endosomes. These then fuse with lysosomes, where the acid conditions cause the ligand and receptor to dissociate; the ligand is fully degraded by lysosomal enzymes while the receptor is recycled to the cell surface to carry out another round. Thus LDL receptor-mediated endocytosis is clearly a complex process, requiring several different protein-catalysed steps and the LDL receptor protein must contain sufficient information for all these separate functions.

Studies with cells from different FH patients confirmed the early clinical observations that the disease was genetically heterogeneous, in that biochemical techniques revealed defects in different parts of the pathway of receptor-mediated uptake of LDL (Tolleshaug *et al.*, 1983). For example, in some cells no immunodetectable LDL receptor protein was observed, while in others a protein of abnormal electrophoretic mobility was detected or the precursor form was not converted to the mature form. Measurement of the binding of labelled LDL to intact cells showed that, in some cases, an otherwise normal protein was unable to bind the ligand, while in others binding was normal but the receptor–ligand complex was not internalized or the receptor was not recycled to the cell surface after internalization. A classification system based on the nature of the cellular defect was devised (Brown and Goldstein, 1986; Table 5.2). These observations suggested that different functions of the receptor were located in different domains of the protein and could be disrupted independently by mutations involving only one or a few nucleotides, and this was confirmed once the cDNA for the receptor had been cloned (Yamamoto *et al.*, 1984) and the gene structure determined (Sudhof *et al.*, 1985).

5.5 Effect of mutations on structure and function in the LDL receptor

The LDL receptor gene spans approximately 45 kb on chromosome 19 (Francke *et al.*, 1984) and comprises 18 exons (Sudhof *et al.*, 1985). It was deduced from the cDNA sequence and from the intron/exon structure of the gene, that the LDL receptor protein consists of five distinct structural domains (Figure 5.2), several of which have extensive homology with parts of other apparently unrelated proteins. Functions have been assigned to these domains based on their biochemical properties and on the observed effect of mutations on receptor-mediated endocytosis. As more information has become available, it has emerged that the LDL receptor gene belongs to a larger gene family, with several very close relatives including the genes for the LDL receptor-related protein (LRP) (Herz *et al.*, 1988), gp330 (Raychowdhury *et al.*, 1989) and, most recently added to the list, a putative protein whose structure has been deduced from a cDNA sequence cloned from rabbit heart, which has been named the VLDL receptor (Takahashi *et al.*, 1992).

Many different mutations have now been identified in the LDL receptor genes

Table 5.2. Classification system for phenotype of mutant LDL receptor gene in cultured cells from FH patients

Class of mutant allele*		Phenotype	Location of mutations†
Class 1	Null	No LDL receptor protein or mRNA detectable	Premature termination codons in all domains
Class 2‡	Transport-defective	Precursor fails to be converted to mature protein or to be transported to cell surface	Mostly in EGF precursor-like and binding domains
Class 3	Binding-defective	Mature protein reaches cell surface, but fails to bind ligand	Clustered in binding domain
Class 4	Internalization-defective	Mature protein binds ligand on cell surface, but fails to localize in clathrin-coated pits	One, in cytoplasmic tail
Class 5	Recycling-defective	Receptor–ligand complex fails to dissociate in lysosomes; receptor fails to return to cell surface and is degraded	Clustered in EGF precursor-like domain

* From Brown and Goldstein (1986).
† Point mutations and small deletions/insertions; from Hobbs *et al.* (1992).
‡ Subdivided into 2A (total – no mature receptor protein reaches the cell surface) and 2B (partial – maturation of precursor to mature protein delayed).

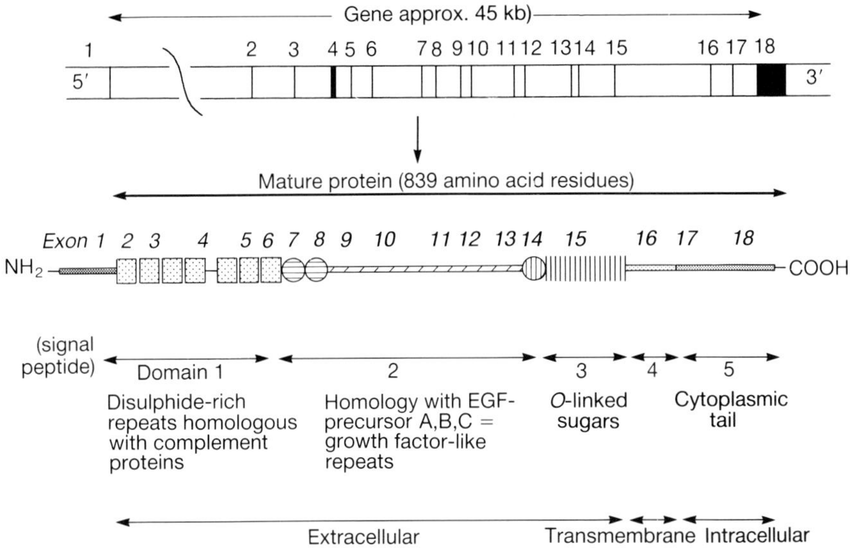

Figure 5.2 Structure of the LDL receptor and its gene. The 18 exons of the LDL receptor gene are represented by vertical lines, numbered above (Sudhof *et al.*, 1985). The protein structure, deduced from the cDNA sequence (Yamamoto *et al.*, 1984), is predicted to have five distinct structural domains, described below the diagram with the exons that code for each shown above.

of patients with FH and analysis of the defective receptor function in cells from the appropriate patient or of the mutant cDNA expressed in cultured cells has confirmed that mutations in different domains affect specific functions of the protein (Hobbs *et al.*, 1990). The mutations are varied in type and include deletions and rearrangements of segments of the gene, insertions and duplications of various sizes and nucleotide substitutions that change a single codon. No attempt will be made to describe in detail all the mutant alleles known at this time as they are listed in a recent review (Hobbs *et al.*, 1992), but they are summarized in Figure 5.3. One interesting feature of the LDL receptor gene is that many of the splice site junctions in the primary transcript are in-phase, so that a deletion encompassing one or more entire exons does not always cause a frameshift that disrupts the rest of the protein, but results in a protein in which the segment coded for by the deleted exon or exons is neatly excised, leaving the rest of the protein intact. When large groups of FH patients have been screened, the frequency of alleles in which the underlying defect is a large deletion in the LDL receptor gene is approximately 5% (Horsthemke *et al.*, 1987; Hobbs *et al.*, 1988; Langlois *et al.*, 1988; Sun *et al.*, 1992). These deletions frequently appear to have involved misalignment at the numerous Alu repeats that occur in several of the introns of the gene and in the untranslated part of the final exon. In several cases,

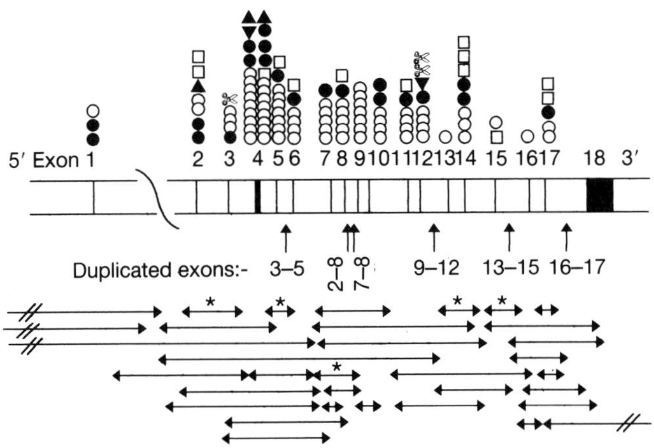

Figure 5.3 Point mutations, deletions and rearrangements in the LDL receptor genes of FH patients. The diagram of the gene shows the exons as vertical lines. Major deletions are shown below, with arrows representing the extent of the deletion. Those marked with an asterisk have been observed in more than one individual of different ethnic origin and probably represent different mutational events. Mutations involving only one or a few nucleotides and their effect on the receptor protein are shown above, as follows: ●, premature termination codon; ○, single amino acid substitution; ▼, insertion, and ▲, deletion of one or more amino acid residues; □, frameshift resulting from deletion or insertion of one or more bases; ✂, base substitution that destroys a splice site. Data from Hobbs *et al.*, 1992; Feher *et al.*, 1993; Sun *et al.*, 1994; and Webb, Sun and Soutar, unpublished observations.

unrelated patients from different parts of the world appear to have inherited a mutant allele of the receptor gene with the same deletion, but careful analysis of the deletion joint has shown that different Alu sequences must have been involved in the generation of the misalignment (Aalto-Setala *et al.*, 1989b; Rudiger *et al.*, 1991). Thus, different mutant alleles can result in an identical defect in receptor function. Deletions of only a few nucleotides have also been observed; these are either in-frame and result in the deletion of one or a few amino acid residues, or cause a frameshift that results in a protein that is normal until the position of the frameshift and then includes a number of incorrect residues, usually less that 20, before a premature stop codon is encountered. In some cases, the newly synthesised truncated protein can be detected in cells carrying such an allele, but the protein is generally very unstable and does not accumulate. In many cases, the introduction of a premature termination codon, either as the result of a frameshift or as the result of a single base change, appears to destabilize the mRNA and it is the resultant low rate of synthesis of the mutant protein that is responsible for its low concentration in the cell. Most of the mutant alleles can be assigned to one or other of the classes of defect described above, although a large number of the mutations that change a single amino acid residue anywhere in the protein affect the maturation and intracellular transport of the newly synthesised protein as well as the function specifically associated with the particular domain. In three cases, substitution of aspartic acid with glutamic acid, where the side chains of the residues differ by only a single $-CH_2$, results in a transport-defective phenotype, even though the changes, at residues 206, 283 and 321, occur in the ligand-binding domain (Hobbs *et al.*, 1992). Clearly, this process is acutely sensitive to any change in the secondary or tertiary structure of the protein.

The newly synthesised LDL receptor protein contains a 21 amino-acid-residue leader or signal sequence, coded for by the first exon, which is cleaved during translocation of the nascent polypeptide chain into the lumen of the endoplasmic reticulum. Three point mutations have been identified in this region of the gene of FH patients, and each one introduces a premature termination codon, either directly or as a result of a frameshift, that results in a null phenotype in cells from the patient (Hobbs *et al.*, 1992; Sun *et al.*, 1994).

The extracellular amino-terminal part of the mature protein comprises seven cysteine-rich 40 amino-acid-residue repeats that comprises the ligand-binding domain. The next domain of the protein is a region with extensive homology to the epidermal growth factor (EGF) precursor, which also contains three cysteine-rich repeats, two of which are adjacent to the repeats in the binding domain (Figure 5.2). Analysis of the role of these various repeats in receptor function has shown that not all are equally important for ligand recognition. One of the first observations was made in cells from an FH patient who was homozygous for an allele with a deletion encompassing exon 5 of the LDL receptor gene, that coded for a protein that lacked the sixth repeat in the binding domain but was otherwise normal. The cells were found to be unable to bind LDL but able to bind β-VLDL, for which apoE is the ligand, quite normally (Hobbs *et al.*, 1986). Further analysis of the effects of other naturally occurring mutations and of mutations introduced by site-directed mutagenesis into different repeats in the binding domain on the specificity and affinity of binding of different lipoproteins have led to the conclusion that the first repeat is not essential for binding of any ligand (van Driel *et al.*, 1987b). Of the remaining repeats, only 4 and 5 are essential for binding β-VLDL, but 2, 3, 6, 7 and repeat A in the EGF precursor domain must also be intact for normal binding of LDL (Esser and Russel, 1988; Esser *et al.*, 1988). The implications of these observations for the determination of the FH phenotype are that, in an individual with a deletion or amino acid substitution in repeats 4 or 5 of the receptor gene, there would be impaired clearance both of lipoproteins containing apoB, namely LDL, and of lipoproteins containing apoE, such as the remnants of triglyceride-rich lipoproteins that are the precursors of plasma LDL, resulting in both decreased catabolism and increased input of plasma LDL. On the other hand, mutations in the repeats flanking 4 and 5 should be less deleterious, as clearance of apoE-containing lipoproteins might be unaffected so that LDL production is relatively normal, and this has been observed in at least one case. Studies by Bilheimer and colleagues on the turnover of plasma LDL in an unusually long-lived homozygous FH patient whose parents were not obviously hypercholesterolaemic showed that, although the fractional rate of catabolism of LDL was reduced in both the patient and her parents, LDL synthesis was relatively normal (Bilheimer *et al.*, 1985; Nora *et al.*, 1985). It was later shown that this patient was homozygous for an allele of the receptor gene that was predicted to result in a single amino acid substitution in exon 6, coding for repeat 7, that would be expected to impair binding of LDL but not that of VLDL remnants or IDL (Hobbs *et al.*, 1990). It is also relevant that analyses of the mutations in the LDL receptor gene in groups of FH patients who present at lipid clinics have shown a disproportionately high frequency of point mutations in the part of exon 4 that codes for repeats 4 and 5 of the binding domain, suggesting that mutations in this region might be more deleterious for receptor function than those elsewhere (Hobbs *et al.*, 1992; Sun *et al.*, 1992; Gudnason *et al.*, 1993).

Point mutations causing single amino acid substitutions in the EGF precursor-like domain of the LDL receptor are also a frequent cause of FH, resulting for the most part in a protein that is poorly transported, if at all, to the cell surface. However, the main function of this domain has been deduced from expression *in vitro* of a cDNA from which the coding region for this domain was deleted (Davis *et al.*, 1987b), resulting in a shorter than normal receptor protein in which the other domains were essentially intact. The mutant protein reaches the cell membrane as normal, presumably because the complete absence of the EGF precursor domain is less of a problem than the presence of an incorrectly folded one and the mutant protein is not recognized as 'foreign' by the cell mechanisms responsible for modification of the precursor to the mature form and its transport through the cell. Once on the cell surface, it binds ligand with apparently normal affinity; however, the receptor–ligand complex does not dissociate in the lysosomes, with the result that the receptor cannot recycle to the cell surface after the first round of endocytosis. A similar mutation has since been identified in a Japanese homozygous FH patient with a plasma cholesterol of 13 mmol l^{-1}, in the homozygous FH range in Japan, but who had survived to 32 years of age and did not have overt CHD. Cultured cells from the patient showed severely impaired binding of LDL, but could still bind apoE-containing lipoproteins to some extent, suggesting that some ability to clear these lipoproteins may result in a less severe phenotype *in vivo* (Miyake *et al.*, 1989).

One interesting mutation in the EGF precursor domain is a single base substitution that results in substitution of the codon for the proline at position 664 of the mature protein, in growth factor-like repeat C, with that for leucine. As is usual for a receptor protein with a substitution in this region, the precursor form is processed more slowly than normal to the mature form, but the mature protein also has a lower binding affinity for LDL than the normal receptor (Knight *et al.*, 1989; Soutar *et al.*, 1989). Originally discovered in a homozygous patient of Asian Indian origin resident in the UK, it has since been found in unrelated individuals of English, Norwegian (King *et al.*, 1991) and Dutch (Defesche *et al.*, 1992) origin, as well as in other Asian Indians living in the UK (King *et al.*, 1991) or South Africa (Rubinsztein *et al.*, 1992). Analysis of the haplotype based on polymorphic sites in the LDL receptor gene showed that the mutation is likely to have occurred independently at least twice if not three times in patients with very different genetic backgrounds. The mutation is a C $\rightarrow$ T transition at a CpG dinucleotide, known to be a common cause of genetic polymorphism and inherited disease (Cooper and Youssoufian, 1988), and 16% of the known point mutations in the human LDL receptor gene have occurred at CpG dinucleotides (Hobbs *et al.*, 1992). It is of interest that, although the number of patients is small, inheritance of the same mutant allele has resulted in disease of quite different severity in these heterozygous individuals (King *et al.*, 1991), supporting the view that other genetic or environmental factors must be important in the development of CHD.

The next domain of the LDL receptor protein adjacent to the EGF precursor-like domain is a region rich in serine and threonine residues that are attachment sites for O-linked sugars. Few mutations that result in single amino acid substitutions have been observed in this domain and its function remains unclear. When expressed in heterologous cells *in vitro*, a mutant LDL receptor cDNA constructed to lack the sequence coding for all but a few residues of the O-linked sugar domain produced a receptor protein that functioned apparently normally (Davis *et al.*, 1986a). However, mutant alleles of the LDL receptor gene that have a deletion that excises the O-linked sugar domain have been found in patients with FH (Kajinami *et al.*, 1988; Koivisto *et al.*, 1993), showing that *in vivo*, if not in cultured cells *in vitro*, this domain must play some part in receptor function. The Finnish heterozygous FH patients in the single large kindred with this allele have less severe hypercholesterolaemia and higher residual LDL receptor activity in their cells than patients with a deletion that results in a null allele (Koivisto *et al.*, 1993). It is interesting that this region of the protein is probably the least conserved between species and between gene relatives in the same species. For example, it is 10 amino acid residues shorter in the rabbit than in the human LDL receptor protein (Yamamoto *et al.*, 1986), 19 residues shorter in the rabbit VLDL receptor than in the rabbit LDL receptor (Takahashi *et al.*, 1992) and totally absent from human LRP (Herz *et al.*, 1988).

The fourth domain of the LDL receptor protein is a stretch of 22 hydrophobic amino acid residues that comprise a membrane-spanning region that anchors the receptor at the cell surface. No point mutations in this domain have yet been observed in FH patients, probably because conservative changes would have little effect on function. However, truncated receptor proteins that lack this region due to premature termination codons introduced as a result of a frameshift caused by a deletion are secreted from the cell and cannot function in receptor-mediated endocytosis (Lehrman *et al.*, 1987a).

The fifth and final domain is intracellular and comprises 50 amino acid residues at the carboxy-terminal end of the receptor protein. Mutations in this domain of the gene were the first to be characterized and were found in FH patients with internalization-defective receptors in whom it had been predicted that the defect would be in the intracellular region of the protein. In some of these patients, the mutant receptor was a truncated protein that retained domains 1–4 intact, but lacked at least part of the cytoplasmic tail (Lehrman *et al.*, 1985). In cells from one FH individual, a mutation that resulted in substitution of a single amino acid residue, replacement of the tyrosine residue at position 807 with cysteine, abolished the ability of the receptor to cluster in coated pits (Davis *et al.*, 1986b). Residue-by-residue site-directed mutagenesis of this region and expression of the mutant cDNA species in cultured cells has defined four amino acid residues that are essential for directing the receptor–ligand complexes on the cell surface to cluster in coated pits (Davis *et al.*, 1987a). This internalization signal, comprising the amino acid residues NPVY in the LDL receptor, has also been identified as the consensus NPxY in a number of other cell surface receptor proteins that cluster in coated pits prior to internalization (Chen *et al.*, 1990). It also emerged from the site-directed mutagenesis experiments that receptor proteins truncated after residue 812 function apparently normally in cultured cells. However, the sequence of the carboxy-terminal 28 amino acids, residues 812–839 of the receptor protein, is highly conserved between species, suggesting that they do have a function. Expression of a mutant LDL receptor gene truncated at position 812 in transgenic mice revealed that sequences between residues 812 and 828 were necessary as a signal to direct the receptor protein to the basolateral surface of the hepatocytes (Yokode *et al.*, 1992). The cytoplasmic domain also appears to be involved in the self-association of the receptor protein (van Driel *et al.*, 1987a), which may be important for high affinity binding of ligands (Patel *et al.*, 1993). As yet, no naturally occurring mutations in the cytoplasmic domain have been found in the receptor genes of FH patients that specifically affect either basolateral sorting or self-association.

5.6 Correlation between LDL receptor gene mutations and clinical phenotype

It is clear from the description above that the extent to which receptor function is impaired, as measured in cultured cells from the patient, can vary over a wide range, with different mutations in the LDL receptor gene. Most mutations have been identified within a single kindred group, and although it is possible to discern a relationship between the nature of the mutation and the severity of the phenotype in some of these, there are rarely sufficient numbers for any statistical comparisons to be made. Where a founder effect has given rise to many FH patients with the same mutation, a number of studies of heterozygous FH patients have now revealed that different mutations can result in more or less severe hypercholesterolaemia (Table 5.3).

In some cases, it is possible to ascribe the differences to the presence of a particularly mild mutation in one group, for example in the FH Espoo patients

Table 5.3. Effect of different mutations in the LDL receptor gene on plasma cholesterol in heterozygous FH patients

Study	Mutation	Trivial name	Phenotype in cultured cells	Residual LDL-R activity in hmz cells (% of normal*)	Mean plasma cholesterol (mmol l^{-1})	Number of patients	Mean age of patients (years)
Moorjani *et al.* (1993)	Deletion of promoter and exon 1	French-Canadian-1	Null	<2	8.1	20	7.8
	trp66gly	French-Canadian-4	3 or 5	25–100	7.2†	20	8.6
Leitersdorf *et al.* (1993)	asp147his	FH Sephardic	2B	<2	10.21	15	Unknown
	Deletion of gly197	FH Lithuania	2B	<2	9.61	5	Unknown
	Stop 660	FH Lebanese	2A	<2	7.82†	21	Unknown
Koivisto *et al.* (1992, 1993)	Deletion of exons 16–18	FH Helsinki	4B	<2	11.7/9.7	66/23	47/38
	Deletion of exon 15	FH Espoo	No data	Mild‡	7.3†	10	33
	Deletion of 7 bp in exon 6	FH North Karelia	1	No data (?null)	12.1	69	47
Kotze *et al.* (1993)	asp206glu	FH Afrikaner-1	2B	5–15	9.4†	112	40
	val408met	FH Afrikaner-2	5	<2	10.8	36	40

* Data from Hobbs *et al.* (1992), based on binding, uptake and degradation of 125 LDL by cultured skin fibroblasts from patients homozygous for the mutation.
† Value significantly lower than others in same study.
‡ From studies in stimulated lymphocytes (Koivisto *et al.*, 1993).

with a deletion of exon 15 (Koivisto *et al.*, 1993), but any conclusions must be tempered by the fact that the patients with this mutation are members of a single kindred who also have a common genetic background. Nonetheless, in this group of patients there was a significant difference between the thickness of the Achilles tendon in the two groups, suggesting that the lower plasma cholesterol in the FH Espoo group was associated with a slower rate of deposition of cholesterol in tissues. Although the numbers were not great enough to show any significant difference in the incidence of CHD, it is hoped that follow-up studies will reveal a later age of onset of CHD in this group compared with other FH patients.

As a result of founder gene effects in a genetically isolated population, the frequency of FH in Afrikaners in South Africa has reached 1 in 80 and numerous heterozygous FH patients with the same mutation have been identified by population screening. This has enabled a detailed study to be made of the effect of genotype at the LDL receptor locus, as well as that of other variables, on the severity of the clinical disorder (Kotze *et al.*, 1993). Patients with an allele causing substitution of asp206 with Glu were found to have significantly lower mean plasma and LDL cholesterol concentration than another group with an allele causing substitution of val408 with met. Previous studies with cultured cells from homozygous patients with these same two mutations has shown that cells carrying the asp206glu mutation retain about 20% of normal LDL receptor activity, while cells carrying the Val408Met mutation have less than 2% residual activity (Fourie *et al.*, 1988, 1992). The lower plasma cholesterol in the group with the Asp206Glu mutation was also associated with a milder clinical phenotype, as judged by their later age of onset of CHD and lower frequency of tendon xanthomata. It is of interest, however, that within each group plasma cholesterol concentration was not itself a good indicator of CHD risk.

In other cases, it is less clear that there is a strong relationship between the mutation and the clinical phenotype in the patients, as exemplified in the study by Leitersdorf *et al.* (1993). These authors found that one group of heterozygous FH patients, with a mutation that results in an essentially null phenotype in cultured cells due to a premature termination codon at residue 660, had a lower mean plasma cholesterol than two other groups in whom different point mutations also resulted in a functionally null phenotype in cultured cells. It is possible that the differences between these groups of patients reflect their different genetic or environmental background (see Section 5.8).

5.7 Regulation of LDL receptor activity

After receptor-mediated uptake of LDL and its release from the ligand–receptor complex in the lysosomal compartment, the apoB is degraded to amino acids while the cholesteryl esters are hydrolysed to release free cholesterol. In most cultured cells, the synthesis of LDL receptors is tightly regulated by the flux of free cholesterol to maintain cholesterol homeostasis, so that cells grown in medium containing serum express few receptors. This appears to occur also in the majority of the peripheral tissues *in vivo*, with the exception of cells that utilize cholesterol for synthesis of steroid hormones where there is high expression of LDL receptor activity (Brown and Goldstein, 1986).

Quantitatively, however, the liver is the most important site of LDL receptor activity and this is reflected to some extent in cultured hepatocytes, in which LDL receptor activity is relatively resistant to down-regulation by extracellular cholesterol (Wade *et al.*, 1989). Hepatic receptor activity *in vivo* must be down-regulated to some extent, because administration of bile acid sequestrants or HMG-CoA reductase inhibitors to hyperlipoproteinaemic or normolipaemic individuals decreases their plasma cholesterol concentration by a mechanism that appears to depend upon an increase in LDL receptor synthesis in response to an increased demand for intracellular cholesterol (Bilheimer *et al.*, 1983). Treatment of heterozygous FH patients with such drugs is frequently, but not always, highly effective, presumably because the receptor protein from the normal allele is up-regulated. The response of FH patients to drug therapy, particularly with HMG-CoA reductase inhibitors, is another source of variability in the FH phenotype, but the reasons for this are not clear. Two studies describe the effect of treatment with lipid-lowering drugs on groups of patients with the same defect in the receptor gene. In one study it was concluded that patients who were heterozygous for a receptor-negative allele responded rather better to treatment with simvastatin than those with a receptor-defective allele; the mean plasma cholesterol concentrations on treatment in the two groups was the same, and the difference lay in their mean cholesterol concentrations before treatment commenced (Jeenah *et al.*, 1993). One possible explanation for this apparently paradoxical observation is that if, as has been suggested (van Driel *et al.*, 1987a; Patel *et al.*, 1993), the receptor protein functions as a dimer, then the presence of a defective protein may impede the function of the protein derived from the normal allele, while a completely absent protein cannot. In the second study, three groups of patients were compared (Leitersdorf *et al.*, 1993), each of which had a different mutation, one causing truncation of the protein at residue 660 (the Lebanese mutation), the second, deletion of residue 197 (the Lithuanian mutation) and the third, substitution of asp147 with histidine. All these mutations result in an essentially null phenotype in cultured cells (Hobbs *et al.*, 1992) and there was little difference between the mean cholesterol values before treatment. However, in the group with the Lebanese mutation, the mean decrease in plasma cholesterol after treatment with fluvastatin was 26%, while in the other two groups and in patients with undefined mutations it was 16–18% (Leitersdorf *et al.*, 1993). The reason for the greater response in one group is not understood, but clearly, genetic and environmental or cultural factors other than the residual LDL receptor activity in cells can influence the response to cholesterol-lowering drugs.

Not surprisingly, the majority of receptor-negative homozygous patients do not respond to these drugs (Uauy *et al.*, 1988) because they have no receptors, but because of the dominance of the liver in LDL receptor-mediated uptake of LDL, liver transplantation is a highly effective form of therapy for homozygous FH (Bilheimer *et al.*, 1984). However, in one receptor-negative homozygous FH patient, whose cells in culture produce no detectable LDL receptor protein under any conditions because of a premature stop codon, treatment with simvastatin resulted in a 30% decrease in plasma cholesterol (Feher *et al.*, 1993), suggesting that HMG-CoA reductase inhibitors do not only act by stimulation of LDL receptor activity.

Sterol-mediated regulation of LDL receptor activity occurs mainly at the level of transcription, and has been shown to be dependent on a 10 bp element within the 200 bp region in the 5′ untranslated region that appears to contain all the necessary information for efficient transcription of the gene. A similar sequence, termed the sterol response element or SRE-1, has been found in the promoter region of two other genes whose products are involved in cholesterol metabolism, namely those for HMG-CoA reductase and HMG-CoA synthase. When sterols are not present, this DNA element interacts with other regulatory elements in the gene to promote transcription of the receptor gene so that receptor synthesis is increased. When sterols are present, the SRE-1 is inactive, presumably because it interacts with a specific regulatory sterol-binding protein (for review, see Soutar and Knight, 1990). Thus mutations in the 5′ region of the LDL receptor gene could interfere sufficiently with normal regulation of receptor function to cause FH. Surprisingly few mutations in this region have been detected in FH patients, despite an extensive search for polymorphisms in a large group of Dutch FH patients by denaturing gradient gel electrophoresis (Top *et al.*, 1992). Three large deletions at the 5′ end that encompass the entire promoter region have been described, each of which results in a null phenotype (Hobbs *et al.*, 1987, 1988; Sun *et al.*, 1992). Only three mutations have been described that specifically affect the 200 bp promoter region, including one deletion of 3 bp and two single base substitutions; each of these lies within a binding site for Sp1 (Hobbs *et al.*, 1992), a non-specific transcription factor. Whether or not mutations in the putative regulatory DNA-binding proteins can cause FH awaits identification of the proteins concerned and characterization of their genes. It is also possible that mutations may occur in hitherto unrecognized regulatory elements more distant from the structural gene. Certainly FH patients exist in whom there are no detectable mutations in the structural gene for the LDL receptor or in its known regulatory region and in whom the defect appears to reside in LDL receptor function (Sun *et al.*, 1994).

5.8 Other genetic and environmental influences on the FH phenotype

Until recently, in all but a few rare cases, inheritance of a defective allele of the LDL receptor gene invariably appeared to result in hypercholesterolaemia, that is the defective gene was always co-dominant. One of the rare exceptions is a family from Puerto Rico in which the mother of a severely affected homozygous individual was normocholesterolaemic, despite having one copy of the mutant allele. Further inspection suggested that a second gene, not segregating with either the mutant or the normal allele of the LDL receptor gene, was having a cholesterol-lowering effect in some members of the family, both normal and FH (Hobbs *et al.*, 1989). Turnover studies have shown that those members of the family carrying this putative gene have a lower than normal synthetic rate for LDL apoB (Vega *et al.*, 1991), but none of the possible candidate genes studied to date have been shown to co-segregate with the effect. In another family, plasma cholesterol was found to be normal in a heterozygous FH individual who was also homozygous for lipoprotein lipase (LPL) deficiency, although the patient showed the hypertriglyceridaemia typically associated with LPL deficiency (Zambon *et al.*, 1993). The explanation for the dominance of the phenotype of LPL deficiency is that the metabolic defect associated with LPL occurs at an earlier point in the metabolic pathway of apoB-containing lipoprotein metabolism than that associated with the LDL receptor (see Figure 5.1). However, an additional factor may be that the particular mutation in the LDL receptor gene in the family (trp66gly) is associated with a particularly mild phenotype, both in cultured cells (Hobbs *et al.*, 1992) and in patients homozygous for the mutation (Moorjani *et al.*, 1993).

It has always been considered in Western societies that the hypercholesterolaemia in FH is relatively resistant to the modifications of diet that effectively reduce plasma cholesterol concentration in individuals with hypercholesterolaemia due to other causes. However, it is becoming clear that, in some populations, for example in China (Sun *et al.*, 1994) and in Tunisia (Slimane *et al.*, 1993), obligate heterozygous FH individuals may not be hypercholesterolaemic even when they have inherited a receptor-negative allele of the LDL receptor gene. Indeed, heterozygous FH is not recognized as a clinical disorder in China even though homozygous patients are as severely affected as those elsewhere (Cai *et al.*, 1991). This may be because these populations have a lower than normal mean plasma cholesterol and, therefore, lower coronary risk because of their low-fat diet (Chen *et al.*, 1992). The protective effect may not be due solely to diet because it has been observed that in Japan, where the general population also has a low mean concentration of cholesterol and is relatively immune to premature CHD because they consume a diet low in saturated fat and rich in fish oils, heterozygous FH individuals seem to be at the same risk as their Western counterparts (Mabuchi *et al.*, 1979).

One of the factors that appears to have a marked influence on whether a heterozygous FH individual develops premature coronary disease is the concentration of Lp(a) in the plasma (Seed *et al.*, 1990). Lp(a) concentration is under strong genetic control at the apo(a) gene locus, which is highly polymorphic. Despite an early suggestion that there was interaction between the genes for apo(a) and the LDL receptor (Utermann *et al.*, 1989) studies in which patients with defined defects in the LDL receptor gene were compared with unaffected relatives with identical Lp(a) phenotype did not support this observation (Soutar *et al.*, 1991). Studies on Lp(a) turnover have also shown that the fractional rate of catabolism in FH patients is not different from that in unaffected individuals (Knight *et al.*, 1991).

Polymorphism in the genes for the ligands of the LDL receptor is a possible source of variation in the FH phenotype. ApoE phenotype has a small but marked effect on plasma cholesterol levels in the general population, in that inheritance of an apoE4 allele tends to raise total plasma cholesterol by approximately 0.25 mmol l^{-1} and inheritance of an apoE2 allele to lower it by the same amount, compared to the population with two apoE3 alleles (Sing and Davignon, 1985). This effect has been observed in some (Dallongeville *et al.*, 1991) but not all heterozygous FH patients (Gylling *et al.*, 1991; Kotze *et al.*, 1993). Some studies have shown that FH patients with an apoE4 allele have more severe CHD and may respond less well to treatment (Eto *et al.*, 1988, 1990), but others have found the differences to be insignificant (De Knijff *et al.*, 1990; Dallongeville *et al.*, 1991). Inheritance of one defective copy of the LDL receptor gene is one of the 'secondary' genetic factors that results in expression of type III hyperlipoproteinaemia in individuals homozygous or even heterozygous for apoE2 (Emi *et al.*, 1991). An XbaI polymorphism in the gene for apoB has been shown to be in linkage disequilibrium with an unidentified genetic locus that appears to associated with either increased plasma cholesterol or increased susceptibility to CHD in some populations (for review, see Humphries, 1988). Finnish FH patients who had inherited one copy of the allele of the apoB gene with the XbaI site were found to have a mean plasma cholesterol concentration of 10.3 mmol l^{-1}, compared with mean values of 9.3 mmol l^{-1} for those without this allele and 11.2 mmol l^{-1} for those with two copies (Aalto-Setala *et al.*, 1989a). There was no difference in their response to treatment with lovastatin (Ojala *et al.*, 1991).

Co-inheritance of a defective allele of the LDL receptor and an allele of the apoB gene, $apoB_{3500}$, that is known to result in a protein with reduced affinity for the LDL receptor, has been reported in families in Germany (Rauh *et al.*, 1991) and South Africa (Rubinsztein *et al.*, 1993). Individuals heterozygous for either defect showed similar clinical symptoms normally associated with fairly severe FH, with tendon xanthoma and premature CHD. Surprisingly, the German individual who carried both defective genes was no more severely affected than others of a similar age with only one of the defects, but the Afrikaner patients had clinical symptoms intermediate between those of heterozygous and homozygous FH patients. The differences may arise because expression of the $apoB_{3500}$ gene in different individuals is not fully penetrant and in the UK approximately 10% of individuals heterozygous for this gene are normolipaemic (Myant, 1993).

Another possible source of variation in the expression of the FH phenotype or in response to treatment that has not yet been investigated is variation in the normal allele of the LDL receptor gene. A number of reports have suggested that in some (Pedersen and Berg, 1989, 1990; Schuster *et al.*, 1990; Humphries *et al.*, 1991), but not all (Klausen *et al.*, 1993), populations, variation at the LDL receptor gene locus is associated with small but significant differences in plasma cholesterol levels and it is possible that these differences could be magnified in individuals in whom one allele is defective.

5.9 Conclusions

The main conclusion that can be drawn from current knowledge is that the severity of CHD in an individual heterozygous FH patient is determined by a number of quite distinct factors. The overriding influence on the course of the disease must be the increased concentration of plasma or LDL cholesterol compared with an unaffected individual, and this is caused by the defect in LDL receptor function. Certainly in homozygous FH patients and in some groups of heterozygous patients, the nature of the mutation can be shown to correlate with the degree of the hypercholesterolaemia. Nonetheless, a few heterozygous patients with an apparently severe defect in the LDL receptor gene do not have hypercholesterolaemia, some apparently because of dietary factors and others because of genetic factors. In the more extreme cases it can be deduced that patients with more severe hypercholesterolaemia will be likely to suffer coronary disease at an earlier age, but other influences are clearly important and one of the goals of future work must be to dissect out these various factors in order that the most appropriate advice can be given to a particular family at risk from FH. This is most likely to be achieved by further comparisons of groups of patients with the same mutation, but including those with a varied genetic and environmental background, and should become possible now that relatively rapid methods for the detection of and screening for mutations in genes have become widely available.

References

Aalto-Setala K, Gylling H, Helve E, Kovanen P, Miettinen TA, Turtola H, Kontula K. (1989a) Genetic polymorphism of the apolipoprotein B gene locus influences serum LDL cholesterol level in familial hypercholesterolaemia. *Hum. Genet.* **82:** 305–307.

Aalto-Setala K, Helve E, Kovanen P, Kontula K. (1989b) Finnish type of low-density lipoprotein receptor gene mutation (FH-Helsinki) deletes exons encoding the carboxy-terminal part of the receptor protein and creates an internalisation-defective phenotype. *J. Clin. Invest.* **84:** 499–505.

Beaumont V, Jacotot B, Beaumont J-L. (1976) Ischaemic disease in men and women with familial hypercholesterolaemia and xanthomatosis. *Atherosclerosis* **24:** 441–450.

Betard C, Kessling AM, Roy M, Chamberland A, Lussier CS, Davignon J. (1992) Molecular genetic evidence for a founder effect in familial hypercholesterolemia among French Canadians. *Hum. Genet.* **88:** 529–536.

Bilheimer DW, Grundy SM, Brown MS, Goldstein JL. (1983) Mevinolin and colestipol stimulate receptor-mediated clearance of low density lipoprotein from plasma in familial hypercholesterolemia heterozygotes. *Proc. Natl Acad. Sci. USA* **80:** 4124–4128.

Bilheimer DW, Goldstein JL, Grundy SM, Starzl TE, Brown MS. (1984) Liver transplantation to provide low-density-lipoprotein receptors and lower plasma cholesterol in a child with homozygous familial hypercholesterolemia. *N. Engl. J. Med.* **311:** 1658–1664.

Bilheimer DW, East CW, Grundy SM, Nora JJ. (1985) II Clinical studies in a kindred with a kinetic LDL receptor mutation causing familial hypercholesterolemia. *Am. J. Med. Genet.* **22:** 593–598.

Brown MS, Goldstein JL. (1986) A receptor-mediated pathway for cholesterol homeostasis. *Science* **232:** 34–47.

Brown MS, Dana SE, Goldstein JL. (1973) Regulation of 3-hydroxy-3-methylglutaryl coenzyme A reductase activity in human fibroblasts by lipoproteins. *Proc. Natl Acad. Sci. USA* **70:** 2162–2166.

Cai HJ, Xie CL, Chen Q, Chen XY, Chen YH. (1991) The relationship between hepatic low-density lipoprotein receptor activity and serum cholesterol level in the human fetus. *Hepatology* **13:** 852–857.

Chen WJ, Goldstein JL, Brown MS. (1990) NPXY, a sequence often found in cytoplasmic tails, is required for coated pit-mediated internalization of the low density lipoprotein receptor. *J. Biol. Chem.* **265:** 3116–3123.

Chen Z, Peto R, Collins R, MacMahon S, Lu J, Li W. (1992) Serum cholesterol and coronary heart disease in population with low cholesterol concentrations. *Br. Med. J.* **303:** 276–282.

Cooper DN, Youssoufian H. (1988) The CpG dinucleotide and human genetic disease. *Hum. Genet.* **78:** 151–155.

Dallongeville J, Roy M, Leboeuf N, Xhignesse M, Davignon J, Lussier CS. (1991) Apolipoprotein E polymorphism association with lipoprotein profile in endogenous hypertriglyceridemia and familial hypercholesterolemia. *Arterioscler. Thromb.* **11:** 272–278.

Davis CG, Elhammer A, Russell DW, Schneider WJ, Kornfeld S, Brown MS, Goldstein JL. (1986a) Deletion of clustered O-linked carbohydrates does not impair function of low density lipoprotein receptor in transfected fibroblasts. *J. Biol. Chem.* **261:** 2828–2838.

Davis CG, Lehrman MA, Russell DW, Anderson RG, Brown MS, Goldstein JL. (1986b) The J.D. mutation in familial hypercholesterolemia: amino acid substitution in cytoplasmic domain impedes internalization of LDL receptors. *Cell* **45:** 15–24.

Davis CG, van Driel I, Russell DW, Brown MS, Goldstein JL. (1987a) The low density lipoprotein receptor. Identification of amino acids in cytoplasmic domain required for rapid endocytosis. *J. Biol. Chem.* **262:** 4075–4082.

Davis CG, Goldstein JL, Sudhof TC, Anderson RG, Russell DW, Brown MS. (1987b) Acid-dependent ligand dissociation and recycling of LDL receptor mediated by growth factor homology region. *Nature* **326:** 760–765.

De Knijff P, Stalenhoef AFH, Mol MJTM, Leuven JAG, Smit J, Erkelens DW, Schouten J, Frants RR, Havekes LM. (1990) Influence of apoE polymorphism on the response to simvastatin in patients with familial hypercholesterolemia. *Atherosclerosis* **83:** 89–97.

Defesche JC, van de Ree MA, Kastelein JJ, van Dierman DD, Janssens NW, van Doormaal JJ, Hayden MR. (1992) Detection of the Pro664-Leu mutation in the low-density lipoprotein receptor and its relation to lipoprotein(a) levels in patients with familial hypercholesterolemia of Dutch ancestry from The Netherlands and Canada. *Clin. Genet.* **42:** 273–280.

van Driel I, Davis CG, Goldstein JL, Brown MS. (1987a) Self-association of the low density lipoprotein receptor mediated by the cytoplasmic domain. *J. Biol. Chem.* **262:** 16127–16134.

van Driel I, Goldstein JL, Sudhof TC, Brown MS. (1987b) First cysteine-rich repeat in ligand-binding domain of low density lipoprotein receptor binds Ca^{2+} and monoclonal antibodies, but not lipoproteins. *J. Biol. Chem.* **262:** 17443–17449.

Emi M, Hegele RM, Hopkins PN, Wu LL, Plaetke R, Williams RR, Lalouel JM. (1991) Effects of three genetic loci in a pedigree with multiple lipoprotein phenotypes. *Arterioscler. Thromb.* **11:** 1349–1355.

Esser V, Russell DW. (1988) Transport-deficient mutations in the low density lipoprotein receptor. Alterations in the cysteine-rich and cysteine-poor regions of the protein block intracellular transport. *J. Biol. Chem.* **263:** 13276–13281.

Esser V, Limbird LE, Brown MS, Goldstein JL, Russell DW. (1988) Mutational analysis of the ligand binding domain of the low density lipoprotein receptor. *J. Biol. Chem.* **263:** 13282–13290.

Eto M, Watanabe K, Chonan N, Ishii K. (1988) Familial hyper-cholesterolaemia and apolipoprotein E4. *Atherosclerosis* **72:** 123–128.

Eto M, Sato T, Watanabe K, Iwashima Y, Makino I. (1990) Effects of probucol on plasma lipids and lipoproteins in familial hypercholesterolemic patients with and without apolipoprotein E4. *Atherosclerosis* **84:** 49–53.

Feher MD, Webb JC, Patel DD, Lant AF, Mayne PD, Knight BL, Soutar AK. (1993)

Cholesterol-lowering drug therapy in a patient with receptor-negative homozygous familial hypercholesterolaemia. *Atherosclerosis* **103:** 171–80.

Fourie AM, Coetzee GA, Gevers W, van der Westhuyzen DR. (1988) Two mutant low-density-lipoprotein receptors in Afrikaners slowly processed to surface forms exhibiting rapid degradation or functional heterogeneity. *Biochem. J.* **255:** 411–415.

Fourie AM, Coetzee GA, Gevers W, van der Westhuyzen DR. (1992) Low-density lipoprotein receptor point mutation results in expression of both active and inactive surface forms of the same mutant receptor. *Biochemistry* **31:** 12754–12759.

Francke U, Brown MS, Goldstein JL. (1984) Assignment of the human gene for the low density lipoprotein receptor to chromosome 19: synteny of a receptor, a ligand, and a genetic disease. *Proc. Natl Acad. Sci. USA* **81:** 2826–2830.

Gagné C, Moorjani S, Brun D, Toussaint M, Lupien P-J. (1979) Heterozygous familial hypercholesterolemia. Relationship between plasma lipids, lipoproteins, clinical manifestations and ischaemic heart disease in men and women. *Atherosclerosis* **34:** 13–24.

Goldstein JL, Brown MS. (1974) Binding and degradation of low density lipoproteins by cultured human fibroblasts. Comparison of cells from a normal subject and from a patient with homozygous familial hypercholesterolemia. *J. Biol. Chem.* **249:** 5153–5162.

Goldstein JL, Brown MS. (1989) Familial hypercholesterolemia. In: *The Metabolic Basis of Inherited Disease* (eds CR Scriver, AL Beaudet, WS Sly, D Valle). McGraw-Hill, New York, pp. 1215–1250.

Gotto AM, Pownall HJ, Havel RJ. (1986) Introduction to the plasma lipoproteins. In: *Methods in Enzymology, Vol. 128* (eds JP Segrest, JJ Albers). Academic Press, Orlando, pp. 3–41.

Grundy SM. (1986) Cholesterol and coronary heart disease. A new era. *J. Am. Med. Assoc.* **256:** 2849–2858.

Gudnason V, King UL, Seed M, Sun XM, Soutar AK, Humphries SE. (1993) Identification of recurrent and novel mutations in exon 4 of the LDL receptor gene in patients with familial hypercholesterolemia in the United Kingdom. *Arterioscler. Thromb.* **13:** 56–63.

Gylling H, Aalto-Setala K, Kontula K, Miettinen TA. (1991) Serum low density lipoprotein cholesterol level and cholesterol absorption efficiency are influenced by apolipoprotein B and E polymorphism and by the FH-Helsinki mutation of the low density lipoprotein receptor gene in familial hypercholesterolemia. *Arterioscler. Thromb.* **11:** 1368–1375.

Harada, SM, Tajima, S, Yokoyama, S, Miyake, Y, Kojima, S, Tsushima, M, Kawakami M, Yamamoto A. (1992) Siblings with normal LDL receptor activity and severe hypercholesterolemia. *Arterioscler. Thromb.* **12:** 1071–1078.

Heiberg A. (1975) The risk of atherosclerotic vascular disease in subjects with xanthomatosis. *Acta Med. Scand.* **198:** 249–261.

Heiberg A, Slack J. (1977) Family similarities in the age at coronary death in familial hypercholesterolemia. *Br. Med. J.* **2:** 493–495.

Herz J, Hamann U, Rogne S, Myklebost O, Gausepohl H, Stanley KK. (1988) Surface location and high affinity for calcium of a 500-kd liver membrane protein closely related to the LDL-receptor suggest a physiological role as lipoprotein receptor. *EMBO J.* **7:** 4119–4127.

Hill JS, Hayden MR, Frohlich J, Pritchard PH. (1991) Genetic and environmental factors affecting the incidence of coronary artery disease in heterozygous familial hypercholesterolemia. *Arterioscler. Thromb.* **11:** 290–297.

Hirobe K, Matsuzawa Y, Ishikawa K, Tarui S, Yamamoto A, Nambu S, Fujimoto K. (1982) Coronary artery disease in heterozygous familial hypercholesterolemia. *Atherosclerosis* **44:** 201–210.

Hobbs HH, Brown MS, Goldstein JL, Russell DW. (1986) Deletion of exon encoding cysteine-rich repeat of low density lipoprotein receptor alters its binding specificity in a subject with familial hypercholesterolemia. *J. Biol. Chem.* **261:** 13114–13120.

Hobbs HH, Brown MS, Russell DW, Davignon J, Goldstein JL. (1987) Deletion in the gene for the low-density-lipoprotein receptor in a majority of French Canadians with familial hypercholesterolemia. *N. Engl. J. Med.* **317:** 734–737.

Hobbs HH, Leitersdorf E, Goldstein JL, Brown MS, Russell DW. (1988) Multiple crm^{-}

mutations in familial hypercholesterolemia. Evidence for 13 alleles, including four deletions. *J. Clin. Invest.* **81:** 909–917.

Hobbs HH, Leitersdorf E, Leffert CC, Cryer DR, Brown MS, Goldstein JL. (1989) Evidence for a dominant gene that suppresses hypercholesterolemia in a family with defective low density lipoprotein receptors. *J. Clin. Invest.* **84:** 656–664.

Hobbs HH, Russell DW, Brown MS, Goldstein JL. (1990) The LDL receptor locus in familial hypercholesterolemia: mutational analysis of a membrane protein. *Annu. Rev. Genet.* **24:** 133–170.

Hobbs HH, Brown MS, Goldstein JL. (1992) Molecular genetics of the LDL receptor gene in familial hypercholesterolemia. *Hum. Mutat.* **1:** 445–466.

Horsthemke B, Dunning A, Humphries S. (1987) Identification of deletions in the human low density lipoprotein receptor gene. *J. Med. Genet.* **24:** 144–147.

Humphries SE. (1988) DNA polymorphisms of the apolipoprotein genes – their use in the investigation of the genetic component of hyperlipidaemia and atherosclerosis. *Atherosclerosis* **72:** 89–108.

Humphries S, Coviello DA, Masturzo P, Balestreri R, Orecchini G, Bertolini S. (1991) Variation in the low density lipoprotein receptor gene is associated with differences in plasma low density lipoprotein cholesterol levels in young and old normal individuals from Italy. *Arterioscler. Thromb.* **11:** 509–516.

Jeenah M, September W, van Roggen FG, de Villiers W, Seftel H, Marais D. (1993) Influence of specific mutations at the LDL-receptor gene locus on the response to simvastatin therapy in Afrikaner patients with heterozygous familial hypercholesterolaemia. *Atherosclerosis* **98:** 51–58.

Kajinami K, Mabuchi H, Itoh H, Michishita I, Takeda M, Wakasugi T, Koizumi J, Takeda R. (1988) New variant of low density lipoprotein receptor gene -FH Tonami. *Arteriosclerosis* **8:** 187–192.

Kannel WB, Castelli WP, Gordon T, McNamara PM. (1971) Serum cholesterol, lipoproteins and the risk of coronary heart disease. *Ann. Intern. Med.* **74:** 1–12.

Keys A. (1970) Coronary heart disease in seven countries. *Circulation* **41:** monograph no. 29.

Khachadurian AK. (1964) The inheritance of essential familial hypercholesterolemia. *Am. J. Med.* **37:** 402–407.

King UL, Gudnason V, Humphries S, Seed M, Patel D, Knight B, Soutar A. (1991) Identification of the 664 proline to leucine mutation in the low density lipoprotein receptor in four unrelated patients with familial hypercholesterolaemia in the UK. *Clin. Genet.* **40:** 17–28.

Klausen IC, Hansen PS, Gerdes LU, Rudiger N, Gregersen N, Faergeman O. (1993) A *PvuII* polymorphism of the low density lipoprotein receptor gene is not associated with plasma concentrations of low density lipoproteins including Lp(a). *Hum. Genet.* **91:** 193–195.

Knight BL, Gavigan SJ, Soutar AK, Patel DD. (1989) Defective processing and binding of low-density lipoprotein receptors in fibroblasts from a familial hypercholesterolaemic subject. *Eur. J. Biochem.* **179:** 693–698.

Knight BL, Perombelon YF, Soutar AK, Wade DP, Seed M. (1991) Catabolism of lipoprotein(a) in familial hypercholesterolaemic subjects. *Atherosclerosis* **87:** 227–237.

Koivisto PV, Koivisto UM, Miettinen TA, Kontula K. (1992) Diagnosis of heterozygous familial hypercholesterolemia. DNA analysis complements clinical examination and analysis of serum lipid levels. *Arterioscler. Thromb.* **12:** 584–592.

Koivisto PVI, Koivisto U-M, Kovanen PT, Gylling H, Miettinen TA, Kontula K. (1993) Deletion of exon 15 of the LDL receptor gene is associated with a mild form of familial hypercholesterolemia FH-Espoo. *Arterioscler. Thromb.* **13:** 1680–1688.

Kotze MJ, De Villiers WJS, Steyn K,, Kriek JA, Marais AD, Langenhoven E, Herbert JS, Graadt Van Roggen JF, Van der Westhuyzen DR, Coetzee GA. (1993) Phenotypic variation among familial hypercholesterolemics heterozygous for either one of two Afrikaner founder LDL receptor mutations. *Arterioscler. Thromb.* **13:** 1460–1468.

Langer L, Strober W, Levy RI. (1972) The metabolism of low density lipoprotein in familial type II hyperlipoproteinemia. *J. Clin. Invest.* **51:** 1528–1536.

Langlois S, Kastelein JJ, Hayden MR. (1988) Characterization of six partial deletions in the low-density-lipoprotein (LDL) receptor gene causing familial hypercholesterolemia (FH). *Am. J. Hum. Genet.* **43:** 60–68.

Lehrman MA, Goldstein JL, Brown MS, Russell DW, Schneider WJ. (1985) Internalization-defective LDL receptors produced by genes with nonsense and frameshift mutations that truncate the cytoplasmic domain. *Cell* **41:** 735–743.

Lehrman MA, Russell DW, Goldstein JL, Brown MS. (1987a) Alu–Alu recombination deletes splice acceptor sites and produces secreted low density lipoprotein receptor in a subject with familial hypercholesterolemia. *J. Biol. Chem.* **262:** 3354–3361.

Lehrman MA, Schneider WJ, Brown MS, Davis CG, Elhammer A, Russell DW, Goldstein JL. (1987b) The Lebanese allele at the low density lipoprotein receptor locus. Nonsense mutation produces truncated receptor that is retained in endoplasmic reticulum. *J. Biol. Chem.* **262:** 401–410.

Leitersdorf E, Eisenberg S, Eliav O, Friedlander Y, Berkman N, Dann EJ, Landsberger D, Sehayek E, Meiner V, Wurm M, Bard J-M, Fruchart J-C, Stein Y. (1993) Genetic determinants of responsiveness to the HMG-CoA reductase inhibitor fluvastatin in patients with molecularly defined heterozygous familial hypercholesterolemia. *Circulation* **87** (Suppl. III): 35–44.

Mabuchi H, Tatami R, Ueda K, Ueda R, Haba T, Kametani T, Watanabe A, Wakasuki T, Ito S, Koizumi J, Ohta M, Miyamoto S, Takeda R. (1979) Serum lipid and lipoprotein levels in Japanese patients with familial hypercholesterolemia. *Atherosclerosis* **32:** 435–444.

Mbewu AD, Bhatnagar D, Durrington PN, Hunt L, Ishola M, Arrol S, Mackness M, Lockley P, Miller JP. (1991) Serum lipoprotein(a) in patients heterozygous for familial hypercholesterolemia, their relatives, and unrelated control populations. *Arterioscler. Thromb.* **11:** 940–946.

Miettinen TA, Gylling H. (1988) Mortality and cholesterol metabolism in familial hypercholesterolemia. *Arteriosclerosis* **8:** 163–167.

Miyake Y, Tajima S, Funahashi T, Yamamoto A. (1989) Analysis of a recycling-impaired mutant of low density lipoprotein receptor in familial hypercholesterolemia. *J. Biol. Chem.* **264:** 16584–16590.

Moorjani S, Gagné C, Lupien PJ, Brun D. (1986) Plasma triglycerides related decrease in high-density lipoprotein cholesterol and its association with myocardial infarction in heterozygous familial hypercholesterolemia. *Metabolism* **35:** 311–316.

Moorjani S, Roy M, Torres A, Betard C, Gagné C, Lambert M, Brun D, Davignon J, Lupien P. (1993) Mutations of the low-density-lipoprotein-receptor gene, variation in plasma cholesterol, and expression of coronary heart disease in homozygous familial hypercholesterolemia. *Lancet* **341:** 1303–1306.

Myant NB. (1993) Familial defective apolipoprotein B-100: a review, including some comparisons with familial hypercholesterolaemia. *Atherosclerosis* **104:** 1–18.

Nishina PM, Johnson JP, Naggert JK, Krauss RM. (1992) Linkage of atherogenic lipoprotein phenotype to the low density lipoprotein receptor locus on the short arm of chromosome 19. *Proc. Natl Acad. Sci USA* **89:** 708–712.

Nora JJ, Lortscher RM, Spangler RD, Bilheimer DW. (1985) Familial hypercholesterolemia with 'normal' cholesterol in obligate heterozygotes. *Am. J. Med. Genet.* **22:** 585–591.

Ojala JP, Helve E, Ehnholm C, Aalto SK, Kontula KK, Tikkanen MJ. (1991) Effect of apolipoprotein E polymorphism and XbaI polymorphism of apolipoprotein B on response to lovastatin treatment in familial and non-familial hypercholesterolaemia. *J. Intern. Med.* **230:** 397–405.

Patel DD, Soutar AK, Knight BL. (1993) A mutation and an antibody that affect chemical cross-linking of low-density lipoprotein receptors on human fibroblasts. *Biochem. J.* **289:** 569–573.

Pedersen JC, Berg K. (1989) Interaction between low density lipoprotein receptor (LDLR) and apolipoprotein E (apoE) alleles contributes to normal variation in lipid level. *Clin. Genet.* **35:** 331–337.

Pedersen JC, Berg K. (1990) Gene–gene interaction between the low density lipoprotein receptor and apolipoprotein E loci affects lipid levels. *Clin. Genet.* **38:** 287–294.

Rauh G, Schuster H, Fischer J, Keller C, Wolfram G, Zollner N. (1991) Identification of a heterozygous compound individual with familial hypercholesterolemia and familial defective apolipoprotein B-100. *Klin. Wochenschr.* **69:** 320–324.

Raychowdhury R, Niles JL, McCluskey RT, Smith JA. (1989) Autoimmune target in Heymann nephritis is a glycoprotein with homology to the LDL receptor. *Science* **244:** 1163–1165.

Rubinsztein DC, Cohen JC, Berger GM, van der Westhuyzen DR, Coetzee GA, Gevers W. (1990) Chylomicron remnant clearance from the plasma is normal in familial hypercholesterolemic homozygotes with defined receptor defects. *J. Clin. Invest.* **86:** 1306–1312.

Rubinsztein DC, Coetzee GA, Marais AD, Leitersdorf E, Seftel HC, van der Westhuyzen DR. (1992) Identification and properties of the proline664-leucine mutant LDL receptor in South Africans of Indian origin. *J. Lipid Res.* **33:** 1647–1655.

Rubinsztein DC, Raal FJ, Seftel HC, Pilcher G, Coetzee GA, van der Westhuyzen DR. (1993) Characterization of six patients who are double heterozygotes for familial hypercholesterolemia and familial defective apoB-100. *Arterioscler. Thromb.* **13:** 1076–1081.

Rudiger NS, Heinsvig EM, Hansen FA, Faergeman O, Bolund L, Gregersen N. (1991) DNA deletions in the low density lipoprotein (LDL) receptor gene in Danish families with familial hypercholesterolemia. *Clin. Genet.* **39:** 451–462.

Schuster H, Humphries S, Rauh G, Held C, Keller C, Wolfram G, Zollner N. (1990) Association of DNA-haplotypes in the human LDL-receptor gene with normal serum cholesterol levels. *Clin. Genet.* **38:** 401–409.

Seed M, Hoppichler F, Reaveley D, McCarthy S, Thompson GR, Boerwinkle E, Utermann G. (1990) Relation of serum lipoprotein(a) concentration and apolipoprotein(a) phenotype to coronary heart disease in patients with familial hypercholesterolemia. *N. Engl. J. Med.* **322:** 1494–1499.

Seftel HC, Baker SG, Sandler MP, Forman FB, Joffe BI, Mendelsohn D, Jenkins T, Mieny CJ. (1980) A host of hypercholesterolaemic homozygotes in South Africa. *Br. Med. J.* **281:** 633–636.

Sing CF, Davignon J (1985) Role of the apolipoprotein E polymorphism in determining normal plasma lipid and lipoprotein variation. *Am. J. Hum. Genet.* **37:** 268–285.

Slack J. (1969) Risk of ischaemic heart disease in familial hyperlipoproteinaemic states. *Lancet* **ii:** 1380–1382.

Slimane MN, Pousse H, Maatoug F, Hammami M, Ben Farhat MH. (1993) Phenotypic expression of familial hypercholesterolaemia in Central and Southern Tunisia. *Atherosclerosis* **104:** 153–158.

Soutar AK, Knight BL. (1990) Structure and regulation of the LDL-receptor and its gene. *Br. Med. Bull.* **46:** 891–916.

Soutar AK, Knight BL, Patel DD. (1989) Identification of a point mutation in growth factor repeat C of the low density lipoprotein-receptor gene in a patient with homozygous familial hypercholesterolemia that affects ligand binding and intracellular movement of receptors. *Proc. Natl Acad. Sci. USA* **86:** 4166–4170.

Soutar AK, McCarthy SN, Seed M, Knight BL. (1991) Relationship between apolipoprotein(a) phenotype, lipoprotein(a) concentration in plasma, and low density lipoprotein receptor function in a large kindred with familial hypercholesterolemia due to the pro664→leu mutation in the LDL receptor gene. *J. Clin. Invest.* **88:** 483–492.

Sprecher DL, Hoeg JM, Schaefer EJ, Zech LA, Gregg RE, Lakatos E, Brewer HBJ. (1985) The association of LDL receptor activity, LDL cholesterol level, and clinical course in homozygous familial hypercholesterolemia. *Metabolism* **34:** 294–299.

Stone NJ, Levy RI, Fredrickson DS, Verter J. (1974) Coronary artery disease in 116 kindred with familial Type II hyperlipoproteinemia. *Circulation* **49:** 476–488.

Streja D, Steiner G, Kwiterovich PO. (1978) Plasma high density lipoproteins and ischemic heart disease. *Ann. Intern. Med.* **89:** 871–880.

Sudhof TC, Goldstein JL, Brown MS, Russell DW. (1985) The LDL receptor gene: a mosaic of exons shared with different proteins. *Science* **228:** 815–822.

Sun X-M, Webb JC, Gudnason V, Humphries S, Seed M, Thompson GR, Knight BL, Soutar

AK. (1992) Characterization of deletions in the LDL receptor gene in patients with familial hypercholesterolemia in the United Kingdom. *Arterioscler. Thromb.* **12:** 762–770.

Sun X-M., Patel DD, Webb JC, Knight BL, Fan L-M, Cai H-J, Soutar AK. (1994) Familial hypercholesterolemia in China. Identification of mutations in the LDL-receptor gene that result in a receptor-negative phenotype. *Arterioscler. Thromb.* **14:** 85–94.

Takahashi S, Kawarabayasi Y, Nakai T, Sakai J, Yamamoto T. (1992) Rabbit very low density lipoprotein receptor: a low density lipoprotein receptor-like protein with distinct ligand specificity. *Proc. Natl Acad. Sci. USA* **89:** 9252–9256.

Thompson GR. (1991) *A Handbook of Hyperlipidaemia.* Current Science Ltd, London.

Thompson GR, Lowenthal R, Myant NB. (1975) Plasma exchange in the management of homozygous familial hypercholesterolaemia. *Lancet* **i:** 1208–1211.

Thompson GR, Seed M, Niththyananthan S, McCarthy S, Thorogood M. (1989) Genotypic and phenotypic variation in familial hypercholesterolemia. *Arteriosclerosis* **9** (Suppl 1): 75–80.

Tolleshaug H, Hobgood KK, Brown MS, Goldstein JL. (1983) The LDL receptor locus in familial hypercholesterolemia: multiple mutations disrupt transport and processing of a membrane receptor. *Cell* **32:** 941–951.

Top B, Uitterlinden AG, van der Zee A, Kastelein JJ, Leuven JA, Havekes LM, Frants RR. (1992) Absence of mutations in the promoter region of the low density lipoprotein receptor gene in a large number of familial hypercholesterolaemia patients as revealed by denaturing gradient gel electrophoresis. *Hum. Genet.* **89:** 561–565.

Tybjaerg-Hanson A, Gallagher J, Vincent J, Houlston R, Talmud P, Dunning AM, Seed M, Hamsten A, Humphries SE, Myant NB. (1990) Familial defective apolipoprotein B-100: detection in the United Kingdom and Scandinavia, and clinical characteristics of ten cases. *Atherosclerosis* **80:** 235–242.

Uauy R, Vega GL, Grundy SM, Bilheimer D M. (1988) Lovastatin therapy in receptor-negative homozygous familial hypercholesterolemia: lack of effect on low-density lipoprotein concentration or turnover. *J. Pediatr.* **113:** 387–392.

Utermann G, Hoppichler F, Dieplinger H, Seed M, Thompson G, Boerwinkle E. (1989) Defects in the low density lipoprotein receptor gene affect lipoprotein (a) levels: multiplicative interaction of two gene loci associated with premature atherosclerosis. *Proc. Natl Acad. Sci. USA* **86:** 4171–4174.

Vega GL, Hobbs HH, Grundy SM. (1991) Low density lipoprotein kinetics in a family having defective low density lipoprotein receptors in which hypercholesterolemia is suppressed. *Arterioscler. Thromb.* **11:** 578–585.

Wade DP, Knight BL, Soutar AK. (1989) Regulation of low-density-lipoprotein-receptor mRNA by insulin in human hepatoma Hep G2 cells. *Eur. J. Biochem.* **181:** 727–731.

Yamamoto T, Davis CG, Brown MS, Schneider WJ, Casey ML, Goldstein JL, Russell DW. (1984) The human LDL receptor: a cysteine-rich protein with multiple Alu sequences in its mRNA. *Cell* **39:** 27–38.

Yamamoto T, Bishop RW, Brown MS, Goldstein JL, Russell DW. (1986) Deletion in cysteine-rich region of LDL receptor impedes transport to cell surface in WHHL rabbit. *Science* **232:** 1230–1237.

Yokode M, Pathak RK, Hammer RE, Brown MS, Goldstein JL, Anderson RG. (1992) Cytoplasmic sequence required for basolateral targeting of LDL receptor in livers of transgenic mice. *J. Cell. Biol.* **117:** 39–46.

Zambon A, Torres A, Bijvoet S, Gagné C, Moorjani S, Lupien PJ, Hayden MR, Brunzell JD. (1993) Prevention of raised low-density lipoprotein cholesterol in a family with familial hypercholesterolaemia and lipoprotein lipase deficiency. *Lancet* **341:** 1119–1121.

6

The molecular basis of Charcot–Marie–Tooth disease

Frank Baas, Linda J. Valentijn, Peter H.S. Meijerink and Pieter A. Bolhuis

6.1 Introduction

About a century ago, Charcot and Marie described a hereditary progressive muscular atrophy as a distinct entity (Charcot and Marie, 1886). In the same year a similar condition, the peroneal type of progressive muscular atrophy, was described by Tooth (Tooth, 1886). It is most likely that both reports concerned the same form of hereditary peripheral neuropathy, later called Charcot–Marie–Tooth (CMT) disease. The clinical features of the disease comprise distal pareses and muscle atrophy, mild sensory loss and areflexia, frequently accompanied by hollow feet (pes cavus). The disease starts in the first two decades of life and is slowly progressive, but in general does not lead to major disability.

Reports on similar neurological cases appeared in due time and a common nomenclature, 'hereditary motor and sensory neuropathy' (HMSN), was proposed (Dyck, 1984). Both names, HMSN and CMT, are currently used. Since the genetic literature uses CMT, we will use this term. HMSN is preferred by clinical neurologists.

To further define the different types of Charcot–Marie–Tooth disease, a subdivision was proposed (Dyck and Lambert, 1968a,b; Dyck, 1984). The major criteria for a subdivision of CMT into different groups are based on electrophysiological and histological examination. CMT disease can be divided into type 1 with hypertrophic nerves, signs of de- and re-myelination and reduced nerve conduction velocity (ncv), and type 2, a group with normal ncvs and a decreased number of large myelinated fibres in the nerve biopsy, also indicated as the axonal type of CMT. Other forms of CMT disease consist of CMT X, with dominant X-linked inheritance, and autosomal recessive CMT type 1. Déjerine–Sottas disease is a severe hereditary peripheral neuropathy with isolated cases or autosomal recessive inheritance (Déjerine and Sottas, 1893). See Gabreëls-Festen *et al.* (1993) for a recent overview of clinical and histopathological findings in hereditary peripheral neuropathy.

The most frequent form of CMT is type 1, with an estimated prevalence of 4–8 per 10 000 population. Linkage analysis of this group revealed genetic heterogeneity with at least two dominant loci. Most families showed linkage to chromosome 17p11.2 (CMT1A), some were linked to chromosome 1q23 (CMT1B), whereas other families were not linked to either locus. CMT X has been mapped to DXS453 at Xq between PGKP1 (Xq11.2–12) and DXS72 (Xq21.1) (Bergoffen *et al.*, 1993), although heterogeneity has been reported for CMT X (Ionasescu *et al.*, 1991). CMT 2 has been mapped to chromosome 1p (Othmane *et al.*, 1993). In this chapter we will describe the molecular basis of CMT disease type 1.

The genes for CMT1A and 1B, linked to chromosome 17p and 1q, respectively, have been identified. Identification of these genes has provided some insight into the pathogenetic mechanisms for CMT disease types 1A and 1B.

6.2 CMT1A

Linkage of CMT1A to chromosome 17 was established in 1989 (Middleton-Price *et al.*, 1989; Vance *et al.*, 1989; Raeymaekers *et al.*, 1989). Since CMT does not affect life expectancy or reproductive capacity, large families could be studied. In single families high LOD scores for location of CMT1A on chromosome 17p11.2 could be found. Precise mapping of the CMT1A locus, however, turned out to be difficult. Linkage analysis of the chromosome 17p-linked families did not yield a single locus. Only upon discovery of the fact that some CMT1A patients carried three alleles for some probes on chromosome 17p did a clear picture begin to emerge (Lupski *et al.*, 1991; Raeymaekers *et al.*, 1991). Almost all families with CMT1A carried a submicroscopic duplication of 17p11.2. This duplication contained two markers used in the linkage analysis and therefore patients could carry three alleles for these markers instead of the two alleles present in the non-affected family members. The presence of a duplication can greatly hamper linkage analysis, since false recombinants can be scored (see Figure 6.1). When the 17p duplication was recognized, linkage analysis narrowed the disease locus down to a 5–10 cM region (Lupski *et al.*, 1991; Raeymakers *et al.*, 1991, 1992).

The CMT1A-associated duplication was shown to be a direct repeat, estimated to be about 1.5 Mb large, using pulsed field gel analysis (PFG) and *in situ* hybridization (Lupski *et al.*, 1991; Hoogendijk *et al.*, 1991; Raeymaekers *et al.*, 1992; Valentijn *et al.*, 1992a). Analysis of a large series of patients showed that in almost all cases the duplication was of similar size. The presence of a *de novo* duplication in a large CMT1A family suggested that non-sister chromatid exchange was the most likely mechanism responsible for the duplication (see Figure 6.2; Raeymaekers *et al.*, 1991). Further proof that the duplication was the mutation conferring CMT1A was obtained by the analysis of isolated cases of CMT1 (Hoogendijk *et al.*, 1992). The presence of the duplication could be identified unambiguously in nine out of ten sporadic cases of CMT1, using a quantitative hybridization test. In all cases, both parents showed no signs of CMT

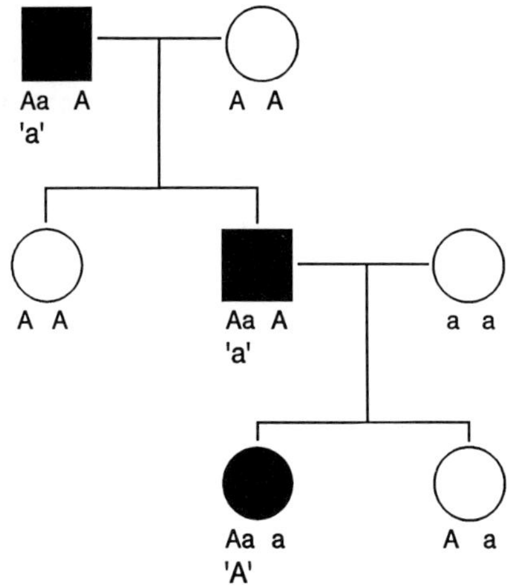

Figure 6.1. DNA duplications can lead to false recombinants in linkage analysis. In this pedigree CMT1A disease co-segregates with the duplicated allele Aa. When the duplication is not recognized as such, the genotypes will be scored incorrectly, as shown between apostrophes. In that case, the disease will show a false recombination when the Aa allele is transmitted to the affected granddaughter.

disease upon clinical and electrophysiological examination and did not carry the CMT1A-associated duplication. Therefore nine out of ten sporadic cases were *de novo* duplications. Without the genetic analysis these cases would probably have been considered recessive. It is interesting to note that, in this case, the identification of the CMT1A-associated duplication allowed genetic detection of isolated or *de novo* CMT1A cases, even without the gene responsible for the disease being known. Only a few of the cases previously considered recessive appear to be true cases of recessive inheritance.

In the case of DNA duplication, several mechanisms for a disease can be envisaged. The most straightforward explanation is interruption of a gene at the duplication border. The product of the fusion gene should then be considered responsible for the disease. Another possibility is that the altered gene dose, due to the duplication, is responsible for the phenotype. A third possibility is that the duplication alters chromatin structure and thereby affects the expression of gene(s) located in the duplication. The first option, interruption of a gene at the duplication border, requires that all duplications have their breakpoint in a defined region, that is, at the gene of interest.

As mentioned before, most CMT1A duplications seemed to be of similar size, compatible with any of the three above-mentioned alternatives. The identification of CMT1A-like abnormalities in two patients with cytogenetically visible duplications or trisomies 17p suggested that altered gene dose was the mechanism responsible for CMT1A disease (Chance *et al.*, 1992; Lupski *et al.*, 1992). These

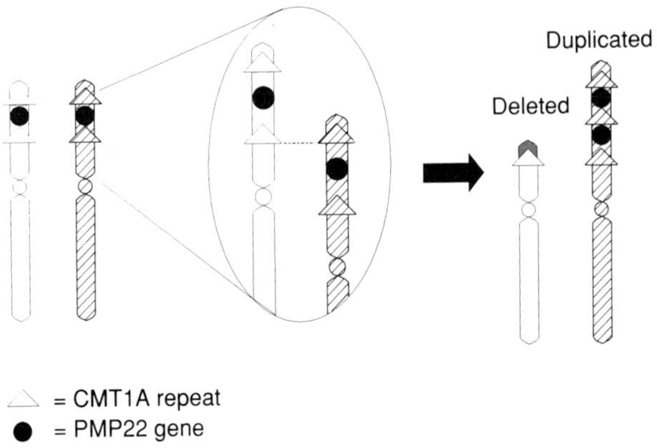

Figure 6.2. Unequal non-sister chromatid exchange during meiosis results in a duplicated and deleted chromosome. Due to misalignment of two non-sister chromosomes at the low-copy CMT1A repeat, a cross-over event will result in a duplicated and deleted chromosome.

patients showed reduced ncvs. The fact that the reduction of ncv was only found after identification of the CMT1A-associated duplication is due to the other profound clinical abnormalities associated with the partial trisomy 17p.

In view of the constant size of the CMT1A-associated duplication, the gene responsible for CMT1A could be located anywhere in the 1.5 Mb region. The search for the CMT1A gene was greatly facilitated by the identification of the mutation in the mouse mutant Trembler (Low, 1976a,b). This mutant was suggested to be a potential model for CMT1 disease (Vance, 1991). The animals develop a peripheral neuropathy with hypomyelination and the disease is inherited in an autosomal dominant way. In addition, the Tr locus was mapped to mouse chromosome 11, which is syntenic to human chromosome 17. Two allelic mutants of Trembler (Tr) were described, Tr and Tr^J. Genetic analysis of the Tr and Tr^J mice showed point mutations in the gene for the peripheral myelin protein-22, PMP22, which was mapped at the Tr locus (Suter *et al.*, 1992a,b). PMP22 encoded a novel Schwann cell-specific membrane protein of unknown function.

In view of the synteny of human chromosome 17 with mouse chromosome 11, the gene for PMP22 was a potential candidate for CMT1A. Shortly after the identification of the Tr mutations, the human homologue of PMP22 was mapped in the CMT1A duplication (Matsunami *et al.*, 1992; Patel *et al.*, 1992; Timmerman *et al.*, 1992; Valentijn *et al.*, 1992a). The PMP22 gene, which was not interrupted by the duplication, was considered the prime candidate for CMT1A. However, all evidence that PMP22 was in fact the CMT1A gene was still circumstantial. Final proof of involvement of alterations of PMP22 in CMT1A disease was obtained by the analysis of the PMP22 gene in CMT patients lacking the CMT1A-associated duplication. The identification of a mutation in the

PMP22 gene of a large CMT1A family that did not carry the 1.5 Mb duplication was first evidence that PMP22 was indeed the gene responsible for CMT1A (Valentijn *et al.*, 1992b). This PMP22 mutation was proved to be identical to the Tr^J mutation. Identification of a *de novo* PMP22 mutation in a case of *de novo* CMT1 provided further evidence that PMP22 was the CMT1A gene (Roa *et al.*, 1993). In this case the mutation was different from the Tr or Tr^J mutation. All three PMP22 mutations identified in CMT1A are located in the predicted transmembrane domains (see Table 6.1 and Figure 6.3). Thus CMT1A is caused by either the presence of an additional copy of, or a mutation in, the PMP22 gene. How gene dosage or mutation can confer the same phenotype will be discussed later.

Table 6.1. Mutations in PMP22 and P_0 resulting in peripheral neuropathies in man and mice

Gene	Disease	Mutation	References
PMP22	CMT1A	Duplication	Lupski *et al.*(1991, 1992) Raeymaekers *et al.* (1991)
	CMT1A	leu16pro	Valentijn *et al.* (1992b)
	CMT1A	ser79cys	Roa *et al.* (1993)
	Tr^J	leu16pro	Suter *et al.* (1992a)
	Tr	gly150asp	Suter *et al.* (1992b)
	HNPP	Deletion	Chance *et al.* (1993)
	HNPP	Frameshift	Nicholson *et al.* (1994)
P_0	CMT1B	del ser34	Kulkens *et al.* (1993)
	CMT1B	arg98his	Hayasaka *et al.* (1993a)
	CMT1B	ile30met	Hayasaka *et al.* (1993b)
	CMT1B	asp90glu	Hayasaka *et al.* (1993c)
	CMT1B	lys96glu	Hayasaka *et al.* (1993c)

6.2.1 Mechanism of the CMT1A duplication

A detailed analysis of the genotypes of *de novo* CMT1A duplications suggested that the duplication originated as a result of an unequal cross-over event between the two parental chromosomes during meiosis (Figure 6.2). The majority of the CMT1A duplications are of similar size, which suggests that the mechanism for unequal non-sister chromatid exchange in CMT1A uses a specific site. The identification of a low-copy repeat (CMT1A-REP) flanking the 1.5 Mb large duplication supports this hypothesis (Pentao *et al.*, 1992). Unequal crossing-over due to misalignment of this site during meiosis is most probably the mechanism that generates the CMT1A duplication. Thus far, only two cases of a shorter duplication have been described (Ionasescu *et al.*, 1993; Valentijn *et al.*, 1993). Both duplications contain the PMP22 gene and thus support increased gene dose

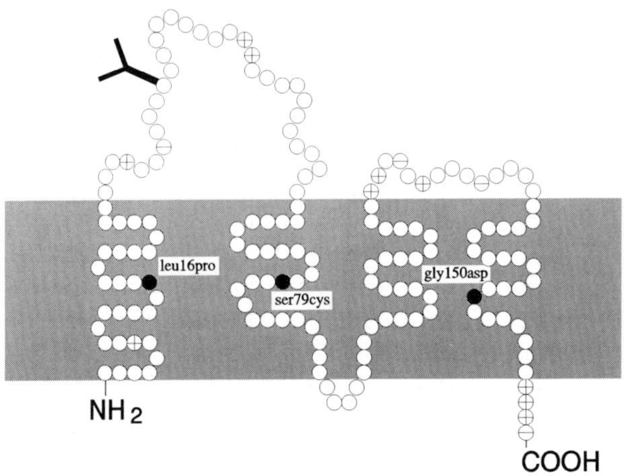

Figure 6.3. PMP22 mutations in CMT1A and Trembler. Schematic representation of PMP22, deduced from the predicted amino acid sequence. The positions of the PMP22 mutations are indicated (see also Table 6.1).

as pathogenetic mechanism for CMT1A. One duplication has been mapped in detail (Valentijn *et al.*, 1993). It is only 460 kb in size and its proximal breakpoint seems to coincide with the CMT1A-REP region. The distal breakpoint is located just beyond the PMP22 gene. The CMT1 phenotype of this family is not markedly different from the phenotype observed in patients with the common 1.5 Mb CMT1A duplication. This excludes the possibility that genes located in the 1 Mb region distal to PMP22 in the CMT1A duplication have a major effect on the CMT1A phenotype. However, in view of the large inter- and intra-pedigree variation of the CMT1A phenotypes, minor modifying effects cannot be excluded.

The mechanism of duplication by unequal non-sister chromatid exchange implies that, besides the duplicated chromosome, a deleted chromosome is generated as a reciprocal event (Figure 6.2). The deleted chromosome was shown to be associated with another form of peripheral neuropathy, hereditary neuropathy with liability to pressure palsies (HNPP; Chance *et al.*, 1993). HNPP is characterized by an apparent susceptibility of the peripheral nerves to trauma, resulting in transient palsies and sensory loss. Ncvs are reduced during an episode of palsies and nerve biopsies show sausage-like focal thickening of the myelin sheath (tomacula). Thus, not only the presence of three copies, but also a single copy of the 'CMT1A' region, results in a disease of the peripheral nervous system (PNS). In one family with HNPP without the deletion of the 1.5 Mb region, a 2 bp deletion in the PMP22 gene was identified (Nicholson *et al.*, 1994). This deletion most probably acts as a null mutation, since it creates a stop codon early in the protein coding sequence. Therefore, PMP22 is also a very likely candidate for HNPP.

6.3 CMT1B

Charcot–Marie–Tooth disease type 1B is located on chromosome 1q in the region of the Duffy locus (Guiloff *et al.*, 1982; Bird *et al.*, 1983). CMT1B is indistinguishable from type 1A on clinical and histopathological grounds. Linkage analysis showed that this form mapped close to the Fc-γ-RII receptor gene and it has been speculated that this gene is a candidate for CMT1B (Lebo *et al.*, 1991a, b). The gene for protein zero (P_0), the major component of peripheral myelin, was recently mapped to this region (Oakey *et al.*, 1992). In view of its involvement in myelination, P_0 was a good candidate gene. Sequence analysis of P_0 in CMT1B families subsequently showed mutations in the coding region of P_0, giving strong evidence that this gene is responsible for CMT1B (Hayasaka, 1993a,b,c; Kulkens *et al.*, 1993; Table 6.1).

6.4 Myelination

In the following section we will discuss the process of myelin formation and how alterations of PMP22 or P_0 might affect this process. The mechanism of myelination is not completely resolved yet. The current opinion is that Schwann cells wrap myelin around the axon by 'tucking under'. In this model, the axon and Schwann cell body remain in the same position, while the tip of the innermost membrane loop of the Schwann cell slides between the axon and the already-surrounding myelin. This innermost turn of the myelin sheath does not compact, in contrast to other layers of the Schwann cell membrane. Compaction of myelin occurs by removal of cytoplasm forming the major dense line, followed by close attachment of the outer surfaces of the membrane forming the intraperiod line.

Based on abundance, myelin proteins are divided into major and minor proteins. The major proteins are P_0, proteolipid protein (PLP) and myelin basic protein (MBP). The minor proteins are myelin-associated glycoprotein (MAG), 2′,3′-cyclic nucleotide 3′-phosphodiesterase (CNP), a transport protein (P2) and the oligodendrocyte-myelin glycoprotein (OMGP). Despite the similarities in structural organization of PNS and CNS myelin, their protein composition is different. The major proteins in peripheral myelin are P_0 and MBP, whereas CNS myelin contains PLP and MBP. MBP is a highly soluble charged protein, existing in different forms by alternative splicing, and is only detected in the major dense line of myelin, a region of compacted cytoplasm between two layers of the cell membrane. PLP and P_0 are membrane proteins. P_0 is the most abundant protein of the PNS. P_0 is a member of the immunoglobulin superfamily (Amzel and Poljak, 1979; Lemke and Axel, 1985). This transmembrane protein is located mainly at the intraperiod line representing the compacted interaction between the two outer surfaces of the cell membrane wrapped around the axon. Homophilic interactions of P_0 play an important role in the compaction of PNS myelin. These interactions are mediated by the immunoglobulin-like extracellular domain of P_0 (D'Urso *et al.*, 1990; Filbin *et al.*, 1990). This function is demonstrated clearly in transgenic mice, homozygous for a non-functional P_0 allele. These mice have major abnormalities in their PNS myelin (Giese *et al.*, 1992). In view of its role in compaction of myelin, alterations of P_0 can be envisaged to result in a disease of the PNS. Thus far, four different mutations of P_0 have been identified in CMT1 patients lacking the chromosome 17p duplication. All mutations are dominant and located in the extracellular Ig-like domain (see Table 6.1). The dominant effect of P_0 mutations can be explained by the homophilic interactions of P_0. Expression of both a normal and an altered copy of the P_0 gene could lead to the formation of abnormal homo- and heterodimers. These abnormal homo- and heterodimers could be non-functional or even toxic, affecting compaction of myelin. Also, the possibility that processing of P_0 is affected by the mutations cannot be excluded.

In view of the effect of alterations of P_0 mutations on the formation of peripheral myelin, it is important to discuss the effect of mutations of PLP, which can be regarded as the CNS counterpart of P_0. Mutations of PLP result in Pelizaeus–Merzbacher disease (PMD), a severe X-chromosomal disease with defective myelination of the CNS. PMD shows some interesting similarities with CMT disease. Several point mutations of PLP have been identified in man, mouse, rat and dog (Dautigny *et al.*, 1986; Nave *et al.*, 1986; Hudson *et al.*, 1989; Nadon *et al.*, 1990). Interestingly, one case of PMD was described with a duplication of Xq21–q22 (Cremers *et al.*, 1987). The PLP gene was not altered, and the two normal copies of the PLP gene are supposed to be responsible for PMD. In this respect the gene dose mechanism, as suggested for PMP22, might also be applied here. Another interesting feature of PLP mutations is exemplified in the murine plp mutants Jimpy and Rumpshaker. Both have severe defects in myelination of the CNS, though in Jimpy the number of oligodendrocytes is strongly reduced, whereas in Rumpshaker no such reduction is found (Schneider *et al.*, 1992). In these animal models, it seems as if two functions of PLP are separated by the two mutations. It can be concluded that PLP has two functions, one as an important factor for glial cell development and the other being a structural component of CNS myelin.

6.4.1 How do alterations of PMP22 affect the peripheral nervous system?

Expression of myelin genes is under strict control. Regeneration of a peripheral nerve after crush injury provided a good model to study the mechanism of myelination. After injury, the axons distal to the site of transection degenerate, Schwann cells lose axonal contact and dedifferentiate, myelin-specific gene expression stops and existing myelin is phagocytozed. Subsequently, the dedifferentiated Schwann cells start to express genes usually expressed in their pre-myelinating stage. After regeneration of the axon, Schwann cells make contact with the regenerated axon and start to differentiate again. Pre-myelinating gene expression ceases and myelin-specific gene expression occurs. This course of events suggests that Schwann cells need axonal contact to maintain their differentiated myelinating state. The signals provided by the axons and the putative receptors on the Schwann cells are not identified yet.

PMP22 is a small 22 kDa membrane protein of 160 amino acids, expressed at relatively high levels in the PNS. Expression of its gene is down-regulated in Schwann cells after trauma of the peripheral nerve (Spreyer *et al.*, 1991; Welcher *et al.*, 1991). Its expression followed that of the major myelin genes. Sequence analysis showed that PMP22 was similar to the growth arrest-specific gene 3 (gas-3) (Schneider *et al.*, 1988; Manfioletti *et al.*, 1990). Gas-3 expression increased in fibroblasts upon growth arrest, induced by either serum starvation or contact inhibition. Analysis of the tissue-specific expression of gas-3/PMP22 showed that the highest levels of mRNA were present in peripheral nerve, estimated to be 0.1% of the total mRNA population. The levels of PMP22 mRNA in other tissues were at least 10-fold lower. In view of the fact that the PMP22 gene is not interrupted by the duplication, CMT1A could either be due to the altered gene dose or the duplication could affect expression of the gene. Analysis of the regulation of PMP22 expression in fibroblasts of CMT1A patients revealed no major differences when compared with normal controls (Valentijn *et al.*, 1992a). Therefore, the altered gene dose is considered to be the cause of CMT1A.

The predicted amino acid sequence of PMP22 suggests that it is a membrane protein with four transmembrane domains and a potential glycosylation site (Figure 6.3). N-linked glycosylation of the protein was demonstrated (Manfioletti *et al.*, 1990). A search for conserved motifs has not yielded any additional ones yet. In view of the high expression levels in Schwann cells, PMP22 is considered a major component of peripheral myelin, although it has not been identified as such in protein analysis of PNS myelin (see above). At present, no good quantitative data are available for the PMP22 protein level. The protein does not show on Coomassie brilliant blue- or silver-stained protein gels of PNS myelin, which suggests that its expression level might be much lower than that of the other components of myelin, such as P_0, MBP and MAG

The regulation of PMP22 during the *in vitro* growth of fibroblasts is compatible with a role in either allowing entrance to or maintenance of a cell in a quiescent state. However, the observed increase of PMP22 mRNA after serum starvation is slow and therefore a role for PMP22 in maintenance of a cell in G_0 seems more likely. The regulation of PMP22 in the peripheral nerve after injury is also compatible with this model. The Schwann cells of the nerve distal to the lesion will dedifferentiate upon loss of axonal contact and the PMP22 mRNA levels go down. After re-innervation, Schwann cells will myelinate the newly formed axons and the expression of PMP22 follows expression of the other myelin genes like MBP (Snipes *et al.*, 1992). Whether the presumed regulatory function of PMP22 is the only function of PMP22 in the Schwann cell is unknown. In view of the fact that CMT1 disease can be caused by mutations in either PMP22 or the structural protein P_0, a possible structural role cannot be ruled out. The expression pattern during development, the very high mRNA levels of PMP22 in Schwann cells and its localization in the membrane are compatible with such a role. This could be a function in addition to the possible regulatory function. A precedent for such a dual role of a myelin protein is exemplified in the two murine mutants Jimpy and Rumpshaker (Schneider *et al.*, 1992).

In view of the gene dosage effect of PMP22 on the formation of PNS myelin, it could either be a component of a multi-subunit complex or have a regulatory function that is tightly controlled. In the first case, either excess (CMT1A duplication), shortage (HNPP deletion) or a mutation altering stability or structure of PMP22 will alter the subunit ratio and/or complex formation. In the second case, the gene dose or mutations have to affect proliferation of Schwann cells. A role for PMP22 in transducing signals from the axon to the Schwann cell is also conceivable. In view of the highly co-regulated expresssion of PMP22 and the other myelin genes, the first hypothesis, that PMP22 is part of a multi-subunit complex, seems appealing, although the second possibility that PMP22 has a regulatory role cannot be excluded. Precise localization of PMP22 in the Schwann cell by electron microscopy and detailed analyses of PMP22 expression during regeneration might provide more insight.

References

Amzel LM, Poljak RJ. (1979) Three dimensional structure of immunoglobulins. *Annu. Rev. Biochem.* **48:** 961–997.

Bergoffen J, Trofatter J, Pericak-Vance MA, Haines JL, Chance PF, Fischbeck KH. (1993) Linkage localization of X-linked Charcot–Marie–Tooth disease. *Am. J. Hum. Genet.* **52:** 312–318.

Bird TD, Ott J, Giblett ER. (1982) Evidence for linkage of Charcot–Marie–Tooth neuropathy to the Duffy locus on chromosome 1. *Am. J. Hum. Genet.* **34:** 388–394.

Bird TD, Ott J, Giblett ER, Chance PF, Sumi SM, Kraft GH. (1983) Genetic linkage evidence for heterogeneity in Charcot–Marie–Tooth neuropathy (HMSN type I). *Ann. Neurol.* **14:** 679–684.

Chance PF, Bird TD, Matsunami N, Lensch BS, Brothman AR, Feldman GM. (1992) Trisomy 17p associated with Charcot–Marie–Tooth neuropathy type 1A phenotype: evidence for gene dosage as mechanism for CMT1A. *Neurology* **42:** 2295–2299.

Chance PF, Alderson MK, Leppig KA, Lensch MW, Matsunami N, Smith B, Swanson PD, Odelberg SJ, Disteche CM, Bird TD. (1993) DNA deletion associated with hereditary neuropathy with liability to pressure palsies. *Cell* **72:** 143–151.

Charcot JM, Marie P. (1886) Sur une forme particuliere d'atrophie musculaire progressive, souvent familiale, debutant par les pieds et les jambes et atteignant plus tard les mains. *Rev. Med.* **6:** 97–138.

Cremers FPM, Pfeiffer RA, van de Pol TJR, Hofker MH, Kruse TA, Wieringa B, Ropers HH. (1987) An interstitial duplication of the X chromosome in a male allows physical fine mapping of probes from the Xq13–q22 region. *Hum. Genet.* **77:** 23–27.

D'Urso D, Brophy P, Staugaitis S, Gillespie C, Frey A, Stempak J, Colman D. (1990) Protein zero of peripheral nerve myelin: biosynthesis, membrane insertion, and evidence for homotypic interaction. *Neuron* **4:** 449–460.

Dautigny A, Mattei M-G, Morello D, Alliel PM, Pham-Dinh D, Amar L, Arnaud D, Simon D, Mattei J-F, Guenet J-L, Avner P. (1986) The structural gene coding for myelin-associated proteolipid protein is mutated in jimpy mice. *Nature* **321:** 867–869.

Déjerine J, Sottas J. (1893) Sur la nevrite interstitielle hypertrophique et progressive de l'enfance. *Comp. Rend. Soc. Biol.* **45:** 63–96.

Dyck PJ, Lambert EH. (1968a) Lower motor primary sensory neuron diseases with peroneal muscular atrophy. I. Neurologic, genetic, and electrophysiologic findings in hereditary polyneuronopathies. *Arch. Neurol.* **18:** 603–618.

Dyck PJ, Lambert EH. (1968b) Lower motor and primary sensory neuron disease with peroneal muscular atrophy. II. Neurologic, genetic, and electrophysiologic findings in various neuronal degenerations. *Arch. Neurol.* **18:** 619–625.

Dyck PJ. (1984) Inherited neuronal degeneration and atrophy affecting peripheral motor, sensory and autonomic neurons. In: *Peripheral Neuropathy* (eds PJ Dyck, PK Thomas, EH Lambert, R Bunge). WB Saunders, Philadelphia, pp. 1600–1642.

Filbin MT, Walsh FS, Trapp BD, Pizzey JA, Tennekoon GI. (1990) Role of myelin P_0 protein as a homophilic adhesion molecule. *Nature* **344:** 871–872.

Gabreëls-Festen AAWM, Gabreëls FJM, Jennekens FGI. (1993) Hereditary motor and sensory neuropathies, present status of types I, II and III. *Clin. Neurol. Neurosurg.* **95:** 93–107.

Giese KP, Martini R, Lemke G, Soriano P, Schachner M. (1992) Mouse P_0 gene disruption leads to hypomyelination, abnormal expression of recognition molecules and degeneration of myelin and axons. *Cell* **71:** 565–576.

Guiloff RJ, Thomas PK, Contreras M, Armitage S, Schwarz G, Sedgwick EM. (1982) Evidence for linkage of type I hereditary motor and sensory neuropathy to the Duffy locus on chromosome 1. *Ann. Hum. Genet.* **46:** 25–27.

Guiloff RJ, Thomas PK, Contreras M, Armitage S, Schwarz G, Sedgwick EM. (1982) Linkage of autosomal dominant type I hereditary motor and sensory neuropathy to the Duffy locus on chromosome 1. *J. Neurol. Neurosurg. Psychiatr.* **45:** 669–674.

Hayasaka K, Ohnishi A, Takada G, Fukushima, Murai Y. (1993a) Mutation of the myelin P_0 gene in Charcot–Marie–Tooth neuropathy type 1. *Biochem. Biophys. Res. Commun.* **194:** 1317–1322.

Hayasaka K, Takada G, Ionasescu VV. (1993b) Mutation of the myelin P_0 gene in Charcot–Marie–Tooth neuropathy type 1B. *Hum. Mol. Genet.* **2:** 1369–1372.

Hayasaka K, Himoro M, Sato W, Takada G, Uyemura K, Shimizi N, Bird TD, Conneally PM, Chance PF. (1993c) Charcot–Marie–Tooth neuropathy type 1B is associated with mutations of the myelin P_0 gene. *Nature Genetics* **5:** 31–34.

Hoogendijk JE, Hensels GW, Zorn I, Valentijn L, Janssen EAM, de Visser M, Barker DF, Ongerboer de Visser BW, Baas F, Bolhuis PA. (1991) The duplication in Charcot–Marie–Tooth disease type 1a spans at least 1100 kb on chromosome 17p11.2. *Hum. Genet.* **88:** 215–218.

Hoogendijk JE, Hensels GW, Gabreels-Festen AAWM, Gabreels FJM, Janssen EAM, De Jonghe P, Martin J-J, Van Broeckhoven C, Valentijn LJ, Baas F, de Visser M, Bolhuis PA. (1992) *De-novo* mutation in hereditary motor and sensory neuropathy type I. *Lancet* **339:** 1081–1082.

Hudson LD, Puckett C, Berndt J, Chan J, Gencic S. (1989) Mutation of the proteolipid protein gene PLP in a human X chromosome-linked myelin disorder. *Proc. Natl Acad. Sci. USA* **86:** 8128– 8131.

Ionasescu VV, Trofatter J, Haines JL, Summers AM, Ionasescu R, Searby C. (1991) Heterogeneity in X-linked recessive Charcot–Marie–Tooth neuropathy. *Am. J. Hum. Genet.* **48:** 1075–1083.

Ionasescu VV, Ionasescu R, Searby C, Barker DF. (1993) Charcot–Marie–Tooth neuropathy type 1A with both duplication and non-duplication. *Hum. Mol. Genet.* **2:** 405–410.

Kulkens T, Bolhuis PA, Wolterman RA, Kemp S, Nijenhuis S te, Valentijn LJ, Hensels GW, Jennekens FGI, Visser M de, Hoogendijk JE, Baas F. (1993) Deletion of the serine 34 codon from the major peripheral myelin protein P_0 gene in Charcot–Marie–Tooth disease type 1B. *Nature Genetics* **5:** 35–39.

Lebo RV, Lynch ED, Wiegant J, Moore K, Trounstine M, van der Ploeg M. (1991a) Multicolor fluorescence *in situ* hybridization and pulsed field electrophoresis dissect CMT1B gene region. *Hum. Genet.* **88:** 13–20.

Lebo RV, Chance PF, Dyck PJ, Redila-Flores MT, Lynch ED, Golbus MS, Bird TD, King MC, Anderson LA, Hall J, Wiegant J, Jiang Z, Dazin PF, Punnett HH, Schonberg SA,

Moore K, Shull MM, Gendler, S, Hurko O, Lovelace RE, Latov N, Trofatter J, Conneally PM. (1991b) Chromosome 1 Charcot–Marie–Tooth disease (CMT1B) locus in the Fc- gamma receptor gene region. *Hum. Genet.* **88:** 1–12.

Lemke G, Axel R. (1985) Isolation and sequence of a cDNA encoding the major structural protein of peripheral myelin. *Cell* **40:** 501–508.

Low PA. (1976a) Hereditary hypertrophic neuropathy in the trembler mouse. Part 1. Histopathological studies: light microscopy. *J. Neurol. Sci.* **30:** 327–341.

Low PA. (1976b) Hereditary hypertrophic neuropathy in the trembler mouse. Part 2. Histopathological studies: electron microscopy. *J. Neurol. Sci.* **30:** 343–368.

Lupski JR, Montes de Oca-Luna R, Slaugenhaupt S, Pentao L, Guzzetta V, Trask BJ, Saucedo-Cardenas O, Barker DF, Killian JM, Garcia CA, Chakravarti A, Patel PI. (1991) DNA duplication associated with Charcot–Marie–Tooth disease type 1A. *Cell* **66:** 219–232.

Lupski JR, Wise CA, Kuwano A, Pentao L, Parke JT, Glaze DG, Ledbetter DH, Greenberg F, Patel PI. (1992) Gene dosage is a mechanism for Charcot–Marie–Tooth disease type 1A. *Nature Genetics* **1:** 29–33.

Manfioletti G, Ruaro ME, Del Sal G, Philipson L, Schneider C. (1990) A growth arrest-specific (gas) gene codes for a membrane protein. *Mol. Cell. Biol.* **10:** 2924–2930.

Matsunami N, Smith B, Ballard L, Lensch MW, Robertson M, Albertsen H, Hanemann CO, Muller HW, Bird TD, White R, Chance PF. (1992) Peripheral myelin protein-22 gene maps in the duplication in chromosome 17p11.2 associated with Charcot–Marie–Tooth 1A. *Nature Genetics* **1:** 176–179.

Middleton-Price HR, Harding AE, Berciano J, Pastor JM, Huson SM, Malcolm S. (1989) Absence of linkage of hereditary motor and sensory neuropathy type I to chromosome 1 markers. *Genomics* **4:** 192–197.

Middleton-Price HR, Harding AE, Monteiro C, Berciano J, Malcolm S. (1990) Linkage of hereditary motor and sensory neuropathy type I to the pericentromeric region of chromosome 17. *Am. J. Hum. Genet.* **46:** 92–94.

Nadon NL, Duncan ID, Hudson LD. (1990) A point mutation in the proteolipid protein gene of the 'shaking pup' interrupts oligodendrocyte development. *Development* **110:** 529–537.

Nave K-A, Lai C, Bloom FE, Milner RJ. (1986) Jimpy mutant mouse: a 74-base deletion in the mRNA for myelin proteolipid protein and evidence for a primary defect in RNA splicing. *Proc. Natl Acad. Sci. USA* **83:** 9264–9268.

Nicholson GA, Valentijn LJ, Cherryson AK, Kennerson ML, Bragg TL, DeKroon RM, Ross DA, Pollard JD, McLeod JG, Bolhuis PA, Baas F. (1994) A frameshift mutation in the PMP22 gene in hereditary neuropathy with liability to pressure palsies. *Nature Genetics*, in press.

Oakey RJ, Watson ML, Seldin MF. (1992) Construction of a physical map on mouse and human chromosome 1: comparison of 13 Mb of mouse and 11 Mb of human DNA. *Hum. Mol. Genet.* **1:** 613–620.

Othmane KB, Middleton LT, Loprest LJ, Wilkinson KM, Lennon F, Rozear MP, Stajich JM, Gaskell PC, Roses AD, Pericak-Vance MA, Vance JM. (1993) Localization of the gene (CMT2A) for autosomal dominant Charcot–Marie–Tooth disease type 2 to chromosome 1p and evidence of genetic heterogeneity. *Genomics* **17:** 370–375.

Patel PI, Franco B, Garcia C, Slaugenhaupt SA, Nakamura Y, Ledbetter DH, Chakravarti A, Lupski JR. (1990) Genetic mapping of autosomal dominant Charcot–Marie–Tooth disease in a large French–Acadian kindred: identification of new linked markers on chromosome 17. *Am. J. Hum. Genet.* **46:** 801–809.

Patel PI, Roa BB, Welcher AA, Schoener-Scott R, Trask BJ, Pentao L, Snipes GJ, Garcia CA, Francke U, Shooter EM, Lupski JR, Suter U. (1992) The gene for the peripheral myelin protein PMP-22 is a candidate for Charcot–Marie–Tooth disease type 1A. *Nature Genetics* **1:** 159–165.

Pentao L, Wise CA, Chinault AC, Patel PI, Lupski JR. (1992) Charcot–Marie–Tooth type 1A duplication appears to arise from recombination at repeat sequences flanking the 1.5 Mb monomer unit. *Nature Genetics* **2:** 292–300.

Raeymaekers P, Timmerman V, De Jonghe P, Swerts L, Gheuens J, Martin J-J, Muylle L, De

Winter G, Vandenberghe A, Van Broeckhoven C. (1989) Localization of the mutation in an extended family with Charcot–Marie–Tooth neuropathy (HMSN I). *Am. J. Hum. Genet.* **45:** 953–958.

Raeymaekers P, Timmerman V, Nelis E, De Jonghe P, Hoogendijk JE, Baas F, Barker DF, Martin JJ, De Visser M, Bolhuis PA, Van Broeckhoven C and the HMSN Collaborative Research Group: (1991) Duplication in chromosome 17p11.2 in Charcot–Marie–Tooth neuropathy type 1a (CMT 1a). *Neuromusc. Dis.* **1:** 93–97.

Raeymaekers P, Timmerman V, Nelis E, Van Hul W, De Jonghe P, Martin J-J, Van Broeckhoven C and the HMSN Collaborative Research Group: (1992) Estimation of the size of the chromosome 17p11.2 duplication in Charcot–Marie–Tooth neuropathy type 1a (CMT1a). *J. Med. Genet* **29:** 5–11.

Roa BB, Garcia CA, Suter U, Kulpa DA, Wise CA, Mueller J, Welcher AA, Snipes GJ, Shooter EM, Patel PI, Lupski LJR. (1993) Charcot–Marie–Tooth disease type 1A. Association with a spontaneous point mutation in the PMP22 gene. *N. Engl. J. Med.* **329:** 96–101.

Schneider C, King RM, Philipson L. (1988) Genes specifically expressed at growth arrest of mammalian cells. *Cell* **54:** 787–793.

Schneider A, Montague P, Griffiths I, Fanarraga M, Kennedy P, Brophy P, Nave K-A. (1992) Uncoupling of hypomyelination and glial cell death by a mutation in the proteolipid protein gene. *Nature* **358:** 758–761.

Snipes GJ, Suter U, Welcher AA, Shooter EM. (1992) Characterization of a novel peripheral nervous system myelin protein (PMP-22/SR13). *J. Cell. Biol.* **117:** 225–238.

Spreyer P, Kuhn G, Hanemann CO, Gillen C, Schaal H, Kuhn R, Lemke G, Muller HW. (1991) Axon-regulated expression of a Schwann cell transcript that is homologous to a 'growth arrest-specific' gene. *EMBO J.* **10:** 3661–3668.

Suter U, Moskow JJ, Welcher AA, Snipes GJ, Kosaras B, Sidman RL, Buchberg AM, Shooter EM. (1992a) A leucine-to-proline mutation in the putative first transmembrane domain of the 22-kDa peripheral myelin protein in the trembler-J mouse. *Proc. Natl Acad. Sci. USA* **89:** 4382–4386.

Suter U, Welcher AA, Ozcelik T, Snipes GJ, Kosaras B, Francke U, Billings-Gagliardi S, Sidman RL, Shooter EM. (1992b) Trembler mouse carries a point mutation in a myelin gene. *Nature* **356:** 241–244.

Timmerman V, Nelis E, Van Hul W, Nieuwenhuijsen BW, Chen KL, Wang S, Othman KB, Cullen B, Leach RJ, Hanemann CO, De Jonghe P, Raeymaekers P, van Ommen G-J B, Martin J-J, Muller HW, Vance JM, Fischbeck KH, Van Broeckhoven C. (1992) The peripheral myelin protein gene PMP-22 is contained within the Charcot–Marie–Tooth disease type 1A duplication. *Nature Genetics* **1:** 171–175.

Tooth HH. (1886) *The Peroneal Type of Progressive Muscular Atrophy*. HK Lewis, London.

Valentijn LJ, Bolhuis PA, Zorn I, Hoogendijk JE, van den Bosch N, Hensels GW, Stanton VP Jr, Housman DE, Fischbeck KH, Ross DA, Nicholson GA, Meershoek EJ, Dauwerse HG, van Ommen G-J B, Baas F. (1992a) The peripheral myelin gene PMP-22/GAS-3 is duplicated in Charcot–Marie–Tooth disease type 1A. *Nature Genetics* **1:** 166–170.

Valentijn LJ, Baas F, Wolterman RA, Hoogendijk JE, van den Bosch NHA, Zorn I, Gabreels-Festen AAWM, de Visser M, Bolhuis PA. (1992b) Identical point mutations of PMP-22 in Trembler-J mouse and Charcot–Marie–Tooth disease type 1A. *Nature Genetics* **2:** 288–291.

Valentijn LJ, Baas F, Zorn I, Hensels GW, de Visser M, Bolhuis PA. (1993) Alternatively sized duplication in Charcot–Marie–Tooth disease type 1A. *Hum. Mol. Genet.* **2:** 2143–2146.

Vance JM, Nicholson G, Yamaoka LH, Stajich J, Stewart CS, Speer C, Hung W-Y, Roses AD, Barker D, Pericak-Vance MA. (1989) Linkage of Charcot–Marie–Tooth neuropathy type 1a to chromosome 17. *Exp. Neurol* **104:** 186–189.

Vance JM. (1991) Hereditary motor and sensory neuropathies. *J. Med. Genet.* **28:** 1–5.

Welcher AA, Suter U, De Leon M, Snipes GJ, Shooter EM. (1991) A myelin protein is encoded by the homologue of a growth arrest-specific gene. *Proc. Natl Acad. Sci. USA* **88:** 7195–7199.

7

The genetics of Wilms' tumour

John K. Cowell

7.1 Introduction

For many genes involved in biochemical/developmental pathways, damage of the coding sequence will give rise to the absence or altered expression of the protein product which, in turn, gives rise to a particular phenotype. Although the same is also true for some cancers, such as retinoblastoma (Hogg *et al.*, 1992, 1993; Onadim *et al.*, 1993), in the majority of cases involving complex organs, such as the kidneys, it is unlikely that inactivation of a single gene will give rise to the malignant phenotype. When dealing with multistep carcinogenesis it is not always possible to relate the occurrence of specific mutations in a particular gene to a given phenotype. In this chapter the relationship between the heterogeneity associated with the children's kidney cancer, Wilms' tumours, and abnormalities involving different genes and chromosome regions will be discussed.

7.2 Wilms' biology

Wilms' tumour (WT) is a malignancy of the developing kidney affecting approximately 1:10 000 children (Matsunaga, 1981). Although tumours occur essentially sporadically, there is a low incidence (1%) of familial cases, with an apparently autosomal dominant mode of inheritance and incomplete penetrance (Knudson and Strong, 1972). Based on their epidemiological study of age of onset and frequency of bilateral tumours in the population, Knudson and Strong (1972) suggested that WT conformed to their popularized two-hit hypothesis, where only two mutations are required to initiate tumorigenesis. By this theory, hereditary cases would carry a constitutional mutation so only one additional mutational event is required in the appropriate cell type to initiate tumorigenesis. These tumours therefore have an earlier age of onset than sporadic cases, which must acquire two random mutations (presumably in the same gene) for the tumour to develop and the lag time between two mutations accounts for the later onset of tumours in these patients.

An important clue to the location of the WT predisposition gene came from the observation that individuals with aniridia (congenital absence of irises) were at significantly increased risk of the development of Wilms' tumours (Miller *et al.*, 1964). This observation was extended by Riccardi and co-workers who showed that Wilms'–aniridia (WA) patients often also have abnormal gonadal development (G) and mental retardation (R) – the WAGR syndrome (Riccardi *et al.*, 1978; Francke *et al.*, 1979). The vast majority of WAGR patients carry a constitutional heterozygous chromosome deletion involving chromosome region 11p13 (Figure 7.1), suggesting that genes critical for the normal development of the iris, kidney and gonads are located there. Less than 1% of all WT patients, however, carry 11p deletions and, of those that do, only 50% develop tumours (Narahara *et al.*, 1984). Presumably the second critical mutation is not acquired in these cases, thus demonstrating the recessive nature of the WT predisposition gene. This has led to this class of genes being called 'tumour suppressor genes' or 'recessive oncogenes'.

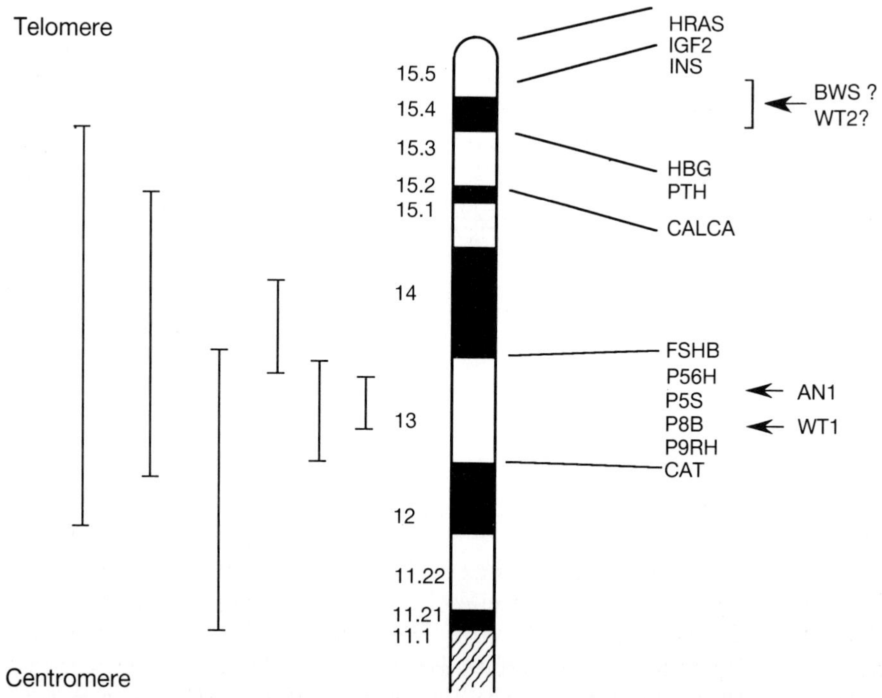

Figure 7.1. Diagrammatic representation of the short arm of chromosome 11. The extent of selected deletions from patients with WAGR syndrome are shown by the bars on the left and the location of genetic loci used in LOH and deletion studies are shown on the right, together with the sites of the WT1 and WT2 genes.

7.3 Loss of heterozygosity

Individuals who were constitutionally heterozygous for DNA markers at particular 11p loci (Figure 7.1) were shown to be apparently homozygous in their tumours. This loss of heterozygosity (LOH) has generally been accepted as a mechanism for 'exposing' mutations in recessive oncogenes. LOH studies for 11p markers involving large numbers of tumours demonstrated allele loss in only 15–20% of sporadic unilateral WT (Mannens *et al.*, 1988; Wadey *et al.*, 1990; Coppes *et al.*, 1992a). In the majority of these cases, although LOH was seen at 11p13, allele loss extended to include the distal region of 11p15. In some tumours LOH occurred exclusively in 11p15, suggesting that there may be another Wilms' tumour suppressor gene at that site. The general mechanisms leading to LOH were described by Cavenee *et al.* (1983), and those most commonly found for 11p in WT were deletions, non-disjunction of the whole chromosome and mitotic recombination. In our own study, all tumours retained two copies of 11p (Solis *et al.*, 1988) and, in the majority of cases, LOH was due to mitotic recombination (Wadey *et al.*, 1990). Thus, LOH studies did not support the idea that the WT gene in 11p13 (WT1) was frequently involved in Wilms' tumorigenesis.

Although WT is often considered to be a hereditary cancer (Knudson and Strong, 1972), there are very few examples of families with affected individuals in more than one generation. Despite the apparently low penetrance (40–60%) of the gene, in the few families which have been described, no evidence for linkage to chromosome 11 markers could be found either at p13 or p15 (Grundy *et al.*, 1988; Huff *et al.*, 1988; Schwartz *et al.*, 1991). It appears, therefore, that the gene for familial WT lies elsewhere. The question, therefore, is where might the other critical genes be? Accepting that LOH might answer this question, Maw *et al.* (1992) undertook a systematic survey of sporadic tumours using probes from most of the chromosome arms and found allele loss on 16q in 20% of cases. Despite this, Huff *et al.* (1992) were unable to demonstrate linkage to chromosome 16 markers.

7.4 Isolation of WT1

Although there seem to be several genes implicated in Wilms' tumorigenesis, the 11p13 deletions provided the basis for gene cloning strategies for one of them and early studies involved the fine structure analysis of small deletions (Figure 7.1) to define the smallest region of overlap in 11p13 which must contain the WT gene (Porteous *et al.*, 1987; Lewis *et al.*, 1988; Cowell *et al.*, 1989). DNA markers were isolated that lay within the smallest deletions (Figure 7.1) and these contributed to the eventual isolation of a candidate gene, WT1 (Call *et al.*, 1990; Gessler *et al.*, 1990). The WT1 gene encodes a 3.2 kb transcript occupying approximately 50 kb of genomic chromosome DNA. It contains 10 exons ranging in size from 180 to 240 bp. The last four exons encode, individually, four zinc finger motifs of the Cys2-His2 type which are characteristic of DNA-binding proteins which regulate transcription (see Figure 7.2). The first six exons code for a proline–glutamine-rich (PGR) region which is also typically associated with transcription regulators. The structure of this gene alone, therefore, strongly supports a role for WT1 in the orchestration of the normal development of the kidney. Two alternative splice sites were recognized within the WT1 transcript which gives rise to four distinct mRNAs. The first splice introduces 17 amino acids just proximal to the zinc fingers and the second splice results in the insertion of three amino acids (+KTS) between the 3rd and 4th zinc finger. The function of these different mRNAs is not known but the second splice site interrupts the zinc finger domain and alters the DNA binding specificity of the WT1 protein (Bickmore *et al.*, 1992). WT1 is expressed predominantly in those cells undergoing the transition from mesenchyme to epithelium, i.e. during the differentiation of the metanephric blastema to the nephrons, the formation of the mesothelium from the mesenchymal lining of the coelom and the production of the sex cords from the mesenchyme of the primitive gonad. Expression was also seen in the glomerular epithelium of the related mesonephros. In tumours, expression was found in the mesonephric glomeruli and cells approximating to these structures. Tumours which were predominantly blastemal, or showed epithelial differentiation, had higher levels of expression compared with those which were predominantly mesenchymal (Pritchard-Jones and Fleming, 1991). Thus, WT1 expression is only found in the malignant counterparts of the cell types showing its expression during normal development. The presence of WT1 expression in the genital ridge, fetal gonad and mesoderm is consistent with the fact that both sporadic and syndrome-associated WT are accompanied by an increased frequency of abnormalities involving the genitourinary (GU) system (Breslow and Beckwith, 1982). Thus, WT1 expression has pleiotrophic effects on the development of different tissues. In turn this may explain the genital abnormalities in WAGR patients. In one WAGR patient, Andersen *et al.*, (1978) reported the coincidence of gonadoblastoma rather than WT. The expression of WT1 in gonadal tissue was not surprising since these two tissues are derived from embryologically adjacent tissue in the developing embryo (van Heyningen and Hastie, 1992).

7.5 Mutations of WT1

Although WT1 lay within chromosome region 11p13 this, in itself, was not proof of its role in tumorigenesis. The identification of intragenic deletions, however, is often an indication of the authenticity of a candidate tumour suppressor gene. The first reports of mutations in WT were large deletions encompassing the WT1 gene (Call *et al.*, 1990, Gessler *et al.*, 1990). Any other gene within that deletion, however, would also be a candidate. A low resolution search for structural rearrangements of WT1 in Wilms' tumours using Southern blot analysis met with limited success. In our own study, only 8% showed abnormal banding patterns (Cowell *et al.*, 1991), one of which was a homozygous deletion of exon 10 which eliminated the 4th zinc finger. Tadokoro *et al.* (1992) only found one homozygous deletion in 42 tumours using Southern blotting. This 8 kb deletion removed exons 6 and 7 and generated a frameshift, resulting in a downstream stop codon.

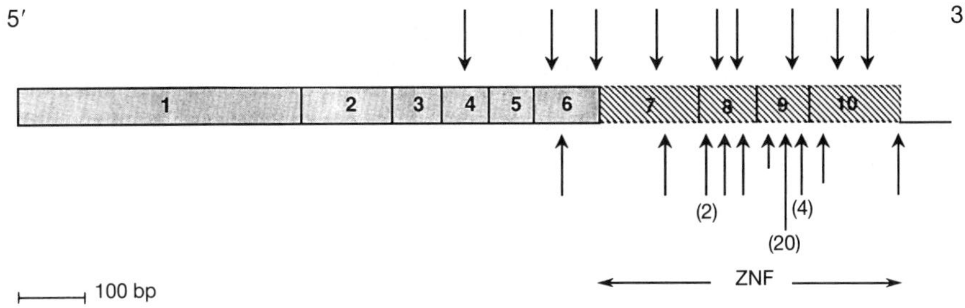

Figure 7.2. Diagrammatic representation of the WT1 gene indicating the relative size of the 10 exons. The last four exons (striped) encode the four zinc fingers (ZNF). The arrows above the gene indicate the approximate position of mutations found in sporadic tumours. The arrows below the line indicate the location of mutations found in constitutional cells from patients with Denys–Drash syndrome. The numbers in brackets indicate the frequency of the same mutations in different patients.

The intragenic 11 kb germinal deletion reported by Huff *et al.* (1991), which removed exon 6, was the only one from 53 different tumours they studied and was from a bilaterally affected patient. This mutation created a frameshift generating a stop codon downstream. Interestingly, the second event in each of the bilateral tumours in this patient was different, although the effect was the same. In the left WT the wild-type allele was lost due to somatic recombination. In the right tumour the normal chromosome was lost and the one carrying the mutation was duplicated. Pelletier *et al.* (1991a) described small, intragenic deletions in two individuals with WT and GU abnormalities. Thus, all of these mutations resulted in the generation of premature stop codons and were either homozygous or hemizygous in the tumours. It was assumed that the majority of mutations in WT1 would be more subtle and could not be detected using this relatively crude procedure. Large surveys of WT1 in Wilms' tumours, however, failed to confirm this. Curiously, there are examples of heterozygous mutations in tumours. It is possible that a mutant protein could interfere with the function of the normal protein in a dominant-negative fashion, as suggested for p53. Another possibility is that a heterozygous mutation may interact with a mutation in another gene elsewhere in the genome. Haber *et al.* (1990), for example, reported a heterozygous deletion in WT1 which resulted in the loss of the third zinc finger. Although the tumour was heterozygous for the WT1 mutation it was homozygous for chromosome 11. This suggests that another recessive mutation has been exposed on chromosome 11 and that the WT1 mutation occurred as a later event. The majority of mutations, so far, affect the zinc finger region of the gene (Figure 7.2) which may reflect that these regions have been the most extensively studied or, more likely, that disruption of the zinc finger function is more effective in destroying WT1 function.

7.6 Analysis of tumours from WAGR patients

The disappointingly few tumours showing deletions/mutations in WT1 cast some doubt about its central role in tumorigenesis, although it may be that its function is influenced by other genes (see below). This posed the question, does the remaining allele in the tumour cells from WAGR patients undergo mutational inactivation during tumorigenesis? In our own survey of three such tumours (Baird *et al.*, 1992a; Santos *et al.*, 1993) all three showed inactivating mutations in the remaining allele. One was a C→T point mutation and the others were small duplications, all of which led to the production of a premature stop codon downstream. Brown *et al.* (1992) reported a large deletion in the tumour from a WAGR patient which also generates a stop codon. Presumably, therefore, inactivation of the WT1 gene is an important event in these tumours, at least, although this is not the general observation in sporadic tumours.

7.7 Mode of action of WT1

The WT1 zinc finger motif was shown to have approximately 50% sequence homology with that of the early growth response (EGR) family of transcriptional activators (Rauscher *et al.*, 1990). The products from these genes are required to initiate changes in the expression of specific target genes required for entry into the cell cycle or for the initiation of the differentiated phenotype. The +KTS version of the WT1 gene also recognizes the consensus GCGGGGGCG binding site in the promoter region of genes which are recognized by EGR genes but the –KTS does not (Rauscher *et al.*, 1990). Madden *et al.* (1991) showed that, in transfection assays, WT1 can suppress the activity of promoters which contain this motif, including IGF2 (Drummond *et al.*, 1992). Furthermore, the mutation in the zinc finger region described by Haber *et al.* (1990), destroys the DNA binding capacity of WT1 (Rauscher *et al.*, 1990). It is the proline–glutamine– rich region in the 5′ part of WT1 (exons 1–6) which is responsible for repressing gene activity (Madden *et al.*, 1993), the zinc finger motif serving only to bind DNA.

The protein produced by alternative splicing resulting in the inclusion of 17 extra amino acids immediately 5′ to the zinc finger region still binds to DNA but does not repress gene expression. Morris *et al.* (1991) developed antibodies to WT1 and showed that the full length protein is 429 amino acids long but does not appear to form protein–protein complexes. Although the exact function of WT1 is not known, it is the only member of the EGR1 family of proteins which can suppress gene activity and one suggestion is that it does so as an antagonist, possibly by occupying the EGR1 site (Rauscher *et al.*, 1990; Morris *et al.*, 1991). By inhibiting cell proliferation, a tissue-specific programme of gene expression might be initiated which results in differentiation. Although the target gene(s) for WT1 are not known, over-expression of growth factors such as IGF2 (Scott *et al.*, 1985) and platelet-derived growth factor-A (PDGF-A) (Frazier *et al.*, 1987) may be important in tumorigenesis. Gashler *et al.* (1992) showed that the PDGF-A promoter had multiple binding sites for WT1 and, in transient tranfection assays,

WT1 could reduce the levels of expression of PDGF. The same was shown to be true for the IGF2 gene (Drummond *et al.*, 1992).

7.8 Denys–Drash syndrome

The observation that patients with 11p13 deletions also have abnormal GU development suggested that a gene important in the development of the gonads was located in this region. Denys–Drash syndrome (DDS) is a rare condition defined by the presence of a characteristic nephrology, leading to progressive renal failure, and GU abnormalities often manifesting as male pseudohermaphroditism (Denys *et al.*, 1967; Drash *et al.*, 1970). DDS patients are also predisposed to the development of WT. Another feature of this syndrome is that phenotypically female patients usually have a 46XY karyotype and ambiguous genitalia. The nephropathy (mesangial sclerosis) may be associated with either genital abnormalities or WT, some have all three phenotypes (Jadresic *et al.*, 1990). Those DDS patients who develop WT usually have bilateral tumours, with an earlier age of onset than sporadic tumours, suggesting a genetic predisposition. The observation that one DDS patient also carried a constitutional 11p13 deletion (Jadresic *et al.*, 1991) suggested that WT1 may be involved. Furthermore, in our study, one tumour from a DDS patient showed LOH for 11p (Wadey *et al.*, 1990). In this tumour we found a homozygous mutation in WT1 (Baird *et al.*, 1992a). Pelletier and colleagues (1991b) showed that all DDS patients, in fact, carry constitutional mutations in WT1 which mostly involved two particular nucleotides and resulted in the generation of missense mutations in exons 8 and 9. Since then, mutations have been found in other exons (Baird *et al.*, 1992b; Ogawa *et al.*, 1993a) but the majority (Figure 7.2) are still in exons 8 and 9 (Coppes *et al.*, 1993). The most common mutation, which occurs in over 50% of cases, converts an arginine amino acid to a tryptophan at position 394. A mutation which prevents splicing at one of the alternative splice sites in intron 9 has also been reported in DDS patients. An interesting observation by Coppes *et al.* (1992b) was that a mutation in exon 9 giving rise to DDS in a small child was also present in the unaffected father. This mutation resulted in an Arg→Trp substitution which had, in previous cases, resulted in the DDS phenotype. There is a possibility that the father was a mosaic for the mutation but, if not, this observation raises the possibility that WT1 mutations may not always be completely penetrant. Thus, DDS patients predominantly carry missense mutations which is in contrast to those seen in WT. There has been much speculation as to how the amino acid changes seen in DDS cause the abnormal development of the kidney and gonads. The majority of mutations affect the zinc finger DNA-binding domain of the WT1 protein which presumably affects its ability to regulate the expression of other genes. During development it is the cells which express WT1 at the highest levels (glomerular epithelial cells) which are the precursors of the aberrant structures seen in the nephropathy, demonstrating the central role of WT1 in the differentiation in these cells.

It is clear that the WT1 mutations seen in DDS patients have a more profound effect than simply reducing gene expression as a result of inactivation. In WAGR patients, for example, the associated developmental abnormalities are generally less severe than those seen in DDS. It has been suggested (Pelletier *et al.*, 1991b) that, as a result of mutation, the major disruption of the genital system is due to a gain of gene function which has a more profound effect than loss of function. Whether this gain of function allows the WT1 protein to bind to new sites, and affect transcription of other genes in the developmental pathway, is still not clear. WAGR patients clearly do not develop nephropathy and the associated GU abnormalities are usually cryptoorchidism (undescended testes) or hypospadias (misplaced urethra). Thus, the renal system seems more tolerant to reduced WT1 expression during embryogenesis than does the genital system. Loss of function, however, may play some role in the development of WT, since the majority of tumours from DDS patients undergo LOH.

Some patients with XY gonadal dysgenesis do not develop WT but rather have an increased risk for the development of gonadoblastoma. These patients (Frasier *et al.*, 1964) form the Frasier syndrome (FS). Patients with FS do not show mutations in the zinc finger region of WT1 as seen in DDS patients (Poulat *et al.*, 1993). WT1, therefore, does not seem to be important in the development of gonadoblastoma and the molecular distinction between FS and DDS provides a means of separating the two syndromes.

Similarly, the expression of WT1 in cells in the developing gonads correlates well with the disordered differentiation in DDS patients. DDS patients have both Mullerian and Wolffian structures, implying that WT1 may play a key role in primary sex determination. The signals for the regression of the Mullerian ducts and development of the Wolffian structures are not present during development in DDS patients with 46XY karyotypes and their phenotypes range from streak gonads to the absence of Mullerian and Wolffian structures. Because individuals carrying the same mutation in WT1 have different phenotypes, this possibly suggests that the gene product interacts with other differentiation factors which can modify the phenotype.

7.9 Beckwith–Wiedemann syndrome

In 1963 Beckwith and Wiedemann independently described a rare congenital disorder, characterized by an excess growth of tissues and organs (organomegaly), disporportionate growth on the left or right sides of the body (hemihypertrophy), and a predisposition to the development of intra-abdominal tumours (Beckwith, 1963; Wiedemann, 1964). The most common tumour was WT but adrenocortical carcinoma, embryonal rhabdomyosarcoma, hepatoblastoma and pancreatoblastoma were also common (Wiedemann, 1983). Interestingly, sporadic cases of these tumours not associated with Beckwith–Wiedemann syndrome (BWS) showed LOH for 11p markers, implicating the same genes responsible for WT in tumorigenesis in these tissues. In rhabdomyosarcoma the critical region for LOH was 11p15.5 (Scrable *et al.*, 1987). Although the majority of BWS cases are sporadic, up to 15% show evidence of an inherited predisposition with apparently autosomal dominant expression, low penetrance and variable expressivity (Best and Hoekstra, 1981). Genetic linkage analysis demonstrated that the BWS locus is in 11p15 (Koufos *et al.*, 1989; Ping *et al.*, 1989). On rare occasions (Waziri *et al.*, 1983; Turleau *et al.*, 1984), BWS patients carried constitutional abnormalities involving the short arm of chromosome 11 which were frequently duplications of the 11p15 region which contains the growth-promoting genes for insulin and IGF2 (Figure 7.1). However, as more reports emerged, it became clear that constitutional chromosome translocations were also found in BWS patients, the breakpoints of which were always in 11p15. It appears, therefore, that an, as yet unidentified, gene in 11p15 is important for the BWS phenotype, but whether this is the same gene involved in the development of a variety of children's cancers is not clear. That two different genes were present in 11p15 was suggested by Mannens *et al.* (personal communication) who used *in situ* hybridization to analyse the 11p15 region and showed that the breakpoints associated with the translocations from different BWS patients were in different positions.

A role for a gene in 11p15 in WT development was also suggested by Henry *et al.* (1989) from a study of tumours from WAGR patients carrying 11p13 deletions. In this study, LOH for 11p15 was also seen in these tumours. Thus, despite the frequent mutation of the remaining WT1 allele in the tumours from these patients they can also show LOH for 11p15 loci. Because these tumours do not show homozygous deletions, the LOH seen at 11p15 is not due to non-disjunction of chromosome 11 but rather LOH has occurred independently at 11p15, as a result of mitotic recombination. This observation suggests that positive selection for recessive mutations in a gene in 11p15 is important for Wilms' tumorigenesis. The possibility that this 11p15 gene may be more important for sporadic Wilms' tumorigenesis was suggested by our study demonstrating the absence of WT1 mutations in tumours showing LOH for 11p13 (Cowell *et al.*, 1993), because LOH extended to 11p15 in these cases. The location of a potential gene in 11p15 (termed WT2 in Figure 7.1) was further defined by Koi *et al.* (1993) who showed that a 6–7 Mbp region of 11p15, when transferred into a rhabdomyosarcoma cell line, suppressed malignancy. This region of the chromosome did not include the IGF2 gene.

7.10 Origins of WT

An appreciation of how the kidney develops is important to our understanding of how disruption of specific genes can give rise to tumorigenesis. Invasion of the nephrogenic blastema by the ureteric bud (which is an outgrowth of the Wolffian duct), after about 6 weeks of normal development, results in an inductive interaction between these two structures which causes the condensation of the blastemal cells around the advancing nephrogenic tube, resulting in their differentiation into nephrons (Figure 7.3). As the nephrogenic tube advances and invades the mesonephros, a wave of differentiation proceeds through the renal lobes, which are the organizational unit of the metanephric kidney. This wave of differentiation continues until week 36 of development. Beckwith and colleagues

(Beckwith *et al.*, 1990) used this information to devise a characterization of WT based on the location of the tumour and identified two distinct subtypes; perilobular nephrogenic rests (PLNR) and intralobular nephrogenic rests (ILNR). The nephrogenic rest refers to cells which have not followed the normal differentiation pattern in the kidney and these are considered the precursor lesion of WT. The PLNR are restricted to the periphery of the renal lobe and are therefore considered to have arisen later in kidney development, whereas ILNR can be found anywhere in the renal lobe and are considered early events.

ILNR and PLNR also differ in some important epidemiological characteristics. The age of diagnosis of patients with ILNR is earlier than those with PLNR, suggesting that these, if any, arise in patients with a genetic predisposition. In keeping with this suggestion, ILNR are common in children with DDS and WAGR syndromes but BWS is more often associated with PLNR. In bilateral tumours, which arise at the same time in the patient, PLNR are more common whereas ILNR are more common in metachronous bilateral WT. These observations suggest that, because of the strong association of ILNR with aniridia, DDS and WAGR, the WT1 gene is linked with the pathogenesis of ILNR whereas another WT gene may be related to BWS. It is interesting that NR are not found more frequently in patients with 11p13 deletions, so clearly a second event is important and may even involve the sequential interaction of other genes. An apparently premalignant stage of WT is described as nephroblastomatosis, which is characterized by the persistence of foci in stem cells in the kidney. Wilms'

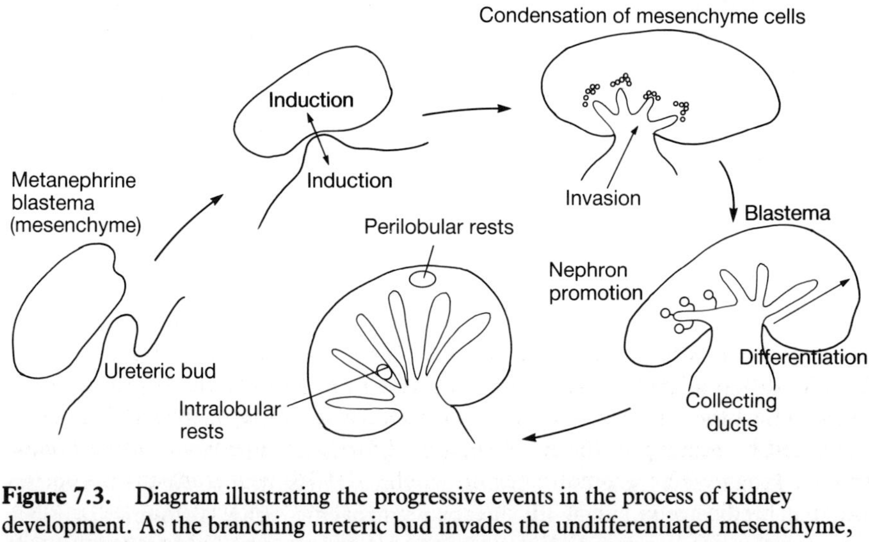

Figure 7.3. Diagram illustrating the progressive events in the process of kidney development. As the branching ureteric bud invades the undifferentiated mesenchyme, nephrons are induced to form, which in turn induce the branching ureter to form collecting ducts. Tumorigenic events which occur early during this process leave undifferentiated cells deep (intralobular) in the kidney whereas later events are more peripheral (perilobular).

tumour occurs more frequently in kidneys showing evidence of nephroblastomatosis and is found coincidentally in 90% of unilateral tumours and 100% of bilateral tumours (Beckwith *et al.*, 1990). The transition from nephroblastomatosis to WT is not a common event however, since approximately 1% of fetal biopsies show nephroblastomatosis but only 1:10 000 children develop tumours.

7.11 Genomic imprinting

Gene function can be modified through mechanisms other than mutation. Genomic imprinting, which is described as a gamete-specific modification causing differential expression of the two alleles of a gene, can also be responsible (Sapienza, 1989). It had been noticed that there are regions in certain parts of the mouse genome which are silenced selectively and, if transgenes are introduced into these regions, they too become inactivated. In some cases this inactivation is inherited through the maternal germline, whereas it is through the paternal germline in others. These genes, which could be transmitted in an inactive form, showed evidence of imprinting (see Chapter 1). It is assumed that this inactivation results from DNA methylation which is reconstituted faithfully when transmitted through the female germline so it is a reversible phenomenon. In tumours showing LOH for 11p markers it was always the maternal chromosome which was lost (Pal *et al.*, 1990). In WAGR patients the deletion chromosome is always derived from the paternal germline (Huff *et al.*, 1991). Furthermore, in those cases of BWS where 11p15 duplications are involved, these are always paternal. There does not seem to be any evidence that this part of chromosome 11 is highly methylated in the tissues studied but in the mouse the equivalent chromosome region to 11p15.5 also shows this phenomenon (De Chiara *et al.*, 1990, 1991). This is suggested to reflect tissue-specific imprinting of the 11p15.5 region. The assumption is that imprinting inactivates one allele and LOH removes the functioning allele, although there are complications with this simplistic interpretation of facts (Feinberg, 1993).

7.12 Interactions with IGF2

The insulin-like growth factors (IGFs) are the most potent mitogenic factors for kidney cells in culture. High levels of IGF2 have been demonstrated in the developing embryonic kidney (Scott *et al.*, 1985) but only low levels are seen in the fully differentiated tissue. High IGF2 levels are also seen in WT which may simply reflect the undifferentiated, embryonic nature. Most of the IGF2 function is initiated by binding to the IGF1 receptor (IGF1R) which is a transmembrane tyrosine kinase. The promoter region of the IGF1R gene contains numerous potential binding sites for WT1. Werner *et al.* (1993) showed that the mRNA levels for the WT1 and IGF1R genes were inversely correlated, suggesting that WT1 may be a negative regulator of IGF1R. The suggestion is that, by mutating WT1, increased expression of IGF1R results which, in turn, promotes cell growth. By activating WT1 expression during normal kidney differentiation, the growth potential of the cells will be reduced and differentiation ensues. How this mechanism might lead to tumorigenesis is not clear, since the WT1 gene is apparently normal in the vast majority of WT (Coppes *et al.*, 1993; Little *et al.*, 1992) so IGF1R regulation should also be normal.

Two genes, H19 (whose function is not known) and IGF2, which map to the 11p15 region in humans (Figure 7.1), undergo reciprocal imprinting in mice, with maternal expression of H19 and paternal expression of IGF2. The same was shown to be true for the homologous human genes (Ogawa *et al.*, 1993b; Rainer *et al.*, 1993). When WTs which had not undergone LOH at 11p15 were analysed, 70% showed biallelic expression of one or both genes. This observation suggested that relaxation of imprinting is a new epigenetic mutational mechanism and, therefore, that loss of imprinting is equivalent to the uniparental disomy seen in BWS where two doses of the IGF2 gene are expressed. Because transgenic mice lacking a functional paternal IGF2 gene are growth retarded, IGF2 has been proposed as a candidate for the BWS gene. The IGF2 gene, however, is unlikely to be the tumour suppressor gene located in 11p15, since it maps outside the region which can suppress malignancy (Koi *et al.*, 1993) and is some distance from the breakpoints in the constitutional translocations predisposing some patients to BWS.

7.13 Transgenic mice

The importance of WT1 in normal development was demonstrated by generating mice deficient for WT1 function (Kreidberg *et al.*, 1993). Mice with constitutional homozygous mutations died *in utero* after 13–15 days. Heterozygous mice, on the other hand, develop normally and do not develop tumours later in life. In the double mutant, although the Wolffian ducts were present, the ureteric bud did not develop and, even though the metanephric blastema formed, there was evidence of increased apoptosis although the surviving cells appeared normal. In *in vitro* assays, however, the blastema could not be induced to differentiate by the spinal cord which is usually the most potent inducer of differentiation. Thus, the kidneys fail to develop because of the absence of the ureteric bud. It appears, therefore, that ureteric outgrowth is dependent on a signal from the blastema and that WT1 is required for the generation of that signal. As suggested by the observations in DDS patients, the gonads failed to develop in mutant mice. Since death by renal failure usually only occurs at birth, it seemed unlikely that this was the cause of death in the homozygous mutant mice. Post-mortem analysis showed that there was also a failure of the heart to develop properly as well as the diaphragm and pleural tissues, and this is the most likely cause of death. These observations demonstrate that WT1 is important in maintaining the integrity of the mesothelium. The multiple developmental abnormalities seen in these mice reflect the tissues where WT1 is expressed most prominently.

7.14 Conclusions

The cloning of the WT1 gene represented an early success for the positional cloning strategy of isolating genes which lie in defined chromosome regions. The fact that development of tumours does not always occur in patients carrying mutations in WT1 suggests that this gene may not, after all, play a fundamental role in tumorigenesis. Rather, it appears that this gene is more likely involved in the control of fundamental events in the embryonic development of a series of tissues and organs including the kidney and gonads. This suggestion was confirmed in transgenic mouse studies. However, as a result of the careful analysis of phenotypic abnormalities associated with WT predisposition, in combination with cytogenetic and genetic analysis, the presence of other genes important in tumour predisposition has been discovered and a few possible locations for these genes suggested. It will be interesting in the future to see how all of these genes interact to produce a rapidly growing pool of tumour precursor cells which ultimately develop into fully transformed cells.

References

Andersen S, Geertingen P, Larsen HW, Mikkelson M, Parbing A, Vestermark S, Warburg M. (1978) Aniridia, cataract and gonadoblastoma in a mentally retarded girl with deletion of chromosome 11. *Ophthalmologica* **76:** 171–177.

Baird PN, Groves N, Haber DA, Housman DE, Cowell JK. (1992a) Identification of mutations in the WT1 gene in tumours from patients with the WAGR syndrome. *Oncogene* **7:** 2141–2149.

Baird PN, Santos A, Groves N, Jadresic L, Cowell JK. (1992b) Constitutional mutations in the WT1 gene in patients with Denys–Drash syndrome. *Hum. Mol. Genet.* **1:** 301–305.

Beckwith JP. (1963) Extreme cytomegaly of the adrenal fetal cortex, omphalocele hyperplasia of kidneys and pancreas, and Leydig-cell hyperplasia: another syndrome? *Western Soc. Pediatr. Res.* (Nov. 11th).

Beckwith JB, Kiviat NB, Bonadio JF. (1990) Nephrogenic rests, nephroblastomatosis, and the pathogenesis of Wilms' tumor. *Pediatr. Pathol.* **10:** 1–36.

Best LG, Hoekstra RE. (1981) Wiedemann–Beckwith syndrome: autosomal-dominant inheritance in a family. *Am. J. Med. Genet.* **9:** 291–299.

Bickmore WA, Oghene K, Little MH, Seawright A, van Heyningen V, Hastie ND. (1992) Modulation of DNA binding specificity by alternative splicing of the Wilms' tumour WT1 gene transcript. *Science* **257:** 235–237.

Breslow NE, Beckwith JB. (1982) Epidemiological features of Wilms' tumour: results of the national Wilms' tumour study. *J. Natl Cancer Inst.* **68:** 429.

Brown KW, Watson JE, Poirier V, Mott MG, Berry PJ, Maitland NJ. (1992) Inactivation of the remaining allele of the WT1 gene in a Wilms' tumour from a WAGR patient. *Oncogene* **7:** 763–768.

Call KM, Glaser T, Ito CY, Buckler AJ, Pelletier J, Haber DA, Rose EA, Kral A, Yeger H, Lewis WH, Jones C, Housman DE. (1990) Isolation and characterization of a zinc finger polypeptide gene at the human chromosome 11 Wilms' tumour locus. *Cell* **60:** 509–520.

Cavenee WK, Dryja TP, Phillips RA, Benedict WF, Godbout R, Gallie BL, Murphree AL, Strong LC, White RL. (1983) Expression of recessive alleles by chromosomal mechanisms in retinoblastoma. *Nature* **305:** 779–784.

Coppes MJ, Bonetta L, Huang A, Hoban P, Chilton-MacNeill S, Campbell CE, Weksberg R,

Yeger H, Reeve AE, Williams BRG. (1992a) Loss of heterozygosity mapping in Wilms' tumour indicates the involvement of three distinct regions and a limited role for nondisjunction or mitotic recombination. *Genes Chrom. Cancer* **5:** 326–334.

Coppes MJ, Liefers GJ, Higuchi M, Zinn AB, Balfe JW, Williams BRG. (1992b) Inherited WT1 mutation in Denys–Drash syndrome. *Cancer Res.* **52:** 6125–6128.

Coppes MJ, Liefers GJ, Paul P, Yeger H, Williams BRG. (1993) Homozygous somatic WT1 point mutations in sporadic unilateral Wilms' tumor. *Proc. Natl Acad. Sci. USA* **90:** 1416–1419.

Cowell JK, Wadey RB, Buckle B, Pritchard J. (1989) The aniridia–Wilms' tumour association: molecular and genetic analysis of chromosome deletions on the short arm of chromosome 11. *Hum. Genet.* **82:** 123–126.

Cowell JK, Wadey RB, Haber DA, Call KM, Housman DE, Pritchard J. (1991) Structural rearrangements of the WT1 gene in Wilms' tumour cells. *Oncogene* **6:** 595–599.

Cowell JK, Groves N, Baird PN. (1993) Loss of heterozygosity at 11p13 in Wilms' tumour does not necessarily involve mutations in the WT1 gene. *Br. J. Cancer* **67:** 1259–1261.

De Chiara TM, Efstratiadis A, Robertson E. (1990) A growth-deficiency phenotype in heterozygous mice carrying an insulin-like growth factor II gene disrupted by targeting. *Nature* **345:** 78–80.

De Chiara TM, Robertson EJ, Efstratiadis A. (1991) Parental imprinting of the mouse insulin-like growth factor II gene. *Cell* **64:** 849–859.

Denys P, Malvaux P, Van den Berghe H, Tanghe W, Proesmans W. (1967) Association d'un syundrome anatomo-pathologique de pseudohermaphrodisme masculin, d'une tumeur de Wilms', d'une nephropathie parenchymateuse et d'un mosaicism XX/XY. *Arch. Fr. Pediatr.* **24:** 729–739.

Drash A, Sherman F, Hartman WH, Blizzard RM. (1970) A syndrome of pseudohermaphroditism, Wilms' tumour, hypertension, and degenerative renal disease. *J. Pediatr.* **76:** 585–593.

Drummond IA, Madden SL, Rohwer-Nutter P, Bell GI, Sukhatme VP, Rauscher FJ. (1992) Repression of the insulin-like growth factor II gene by the Wilms' tumor suppressor WT1. *Science* **257:** 674–678.

Feinberg AP. (1993) Genomic imprinting and gene inactivation in cancer. *Nature Genetics* **4:** 110–113.

Francke U, Holmes LB, Atkins L, Riccardi VM. (1979) Aniridia–Wilms' tumour association; evidence for specific deletion of 11p13. *Cytogenet. Cell Genet.* **24:** 185–192.

Frasier SD, Bashmore RA, Mosier HD. (1964) Gonadoblastoma associated with pure gonadal dysgenesis in monozygotic twins. *J. Paediatr.* **64:** 740–745.

Gashler AL, Bonthron DT, Madden SL, Rauscher FJ, Collins T, Sukhatme VP. (1992) Human platelet derived growth factor A chain is transcriptionally repressed by the Wilms' tumour suppressor gene WT1. *Proc. Natl Acad. Sci. USA* **89:** 10984–10988.

Gessler M, Poustka A, Cavenee W, Neve RL, Orkin SH, Bruns GAP. (1990) Homozygous deletion in Wilms' tumours of a zinc-finger gene identified by chromosome jumping. *Nature* **343:** 774–778.

Grundy P, Koufos A, Morgan K, Li FP, Meadows AT, Cavenee WK. (1988) Familial predisposition to Wilms' tumour does not map to the short arm of chromosome 11. *Nature* **336:** 375–376.

Haber DA, Buckler AJ, Glaser T, Call KM, Pelletier J, Sohn RL, Douglass EC, Housman DE. (1990) An internal deletion within an 11p13 zinc finger gene contributes to the development of Wilms' tumour. *Cell* **61:** 1257–1269.

Henry I, Grandjouan S, Coullin P, Barichard F, Huerre-Jeanpierre C, Glaser T, Philips T, Lenoir G, Chaussain JL, Junien C. (1989) Tumor-specific loss of 11p15.5 alleles in del11p13 Wilms' tumor and in familial adrenocortical carcinoma. *Proc. Natl Acad. Sci. USA* **86:** 3247–3251.

van Heyningen V, Hastie ND. (1992) Wilms' tumour: reconciling genetics and biology. *Trends Genet.* **8:** 16–21.

Hogg A, Onadim Z, Baird PN, Cowell JK. (1992) Detection of heterozygous mutations in the RB1 gene in retinoblastoma patients using single-strand conformation polymorphism analysis and polymerase chain reaction sequencing. *Oncogene* **7:** 1445–1451.

Hogg A, Bia B, Onadim Z, Cowell JK. (1993) Molecular mechanisms of oncogenic mutations in tumours from patients with bilateral and unilateral retinoblastoma. *Proc. Natl Acad. Sci. USA* **90:** 7351–7355.

Huff V, Compton DA, Chao L-Y, Strong LC, Geiser CF, Saunders GF. (1988) Lack of linkage of familial Wilms' tumour to chromosomal band 11p13. *Nature* **336:** 377–378.

Huff V, Miwa H, Haber DA, Call KM, Housman D, Strong LC, Saunders GF. (1991) Evidence for WT1 as a Wilms' tumour (WT) gene: intragenic germinal deletion in bilateral WT. *Am. J. Hum. Genet.* **48:** 997–1003.

Huff V, Reeve AE, Leppert M, Strong LC, Douglass EC, Geiser CF, Li FP, Meadows A, Callen DF, Lenoir G, Saunders GF. (1992) Nonlinkage of 16q markers to familial predisposition to Wilms' tumor. *Cancer Res.* **52:** 6117–6120.

Jadresic L, Leake J, Gordon I, Dillon MJ, Grant DB, Pritchard J, Risdon RA, Barratt TM. (1990) Clinicopathologic review of twelve children with nephropathy, Wilms' tumor, and genital abnormalities (Drash syndrome). *J. Pediatr.* **117:** 717–725.

Jadresic L, Wadey RB, Buckle B, Barratt TM, Mitchell CD, Cowell JK. (1991) Molecular analysis of chromosome region 11p13 in patients with Drash syndrome. *Hum. Genet.* **86:** 497–501.

Knudson AG, Strong LC. (1972) Mutation and cancer: a model for Wilms' tumour of the kidney. *J. Natl Cancer Inst.* **48:** 313–324.

Koi M, Johnson LA, Kalikin LM, Little PFR, Nakamura Y, Feinberg AP. (1993) Tumor cell growth arrest caused by subchromosomal transferable DNA fragments from chromosome 11. *Science* **260:** 361–364.

Koufos A, Grundy P, Morgan K, Aleck KA, Hadro R, Lampkin BC, Kalbakji A, Cavenee WK. (1989) Familial Wiedemann–Beckwith syndrome and a second Wilms' tumour locus both map to 11p15.5. *Am. J. Hum. Genet.* **44:** 711–719.

Kreidberg JA, Sariola H, S, Loring JM, Maeda M, Pelletier J, Housman D, Jaenisch R. (1993) WT-1 is required for early kidney development. *Cell* **74:** 679–691.

Lewis WH, Yeger H, Bonetta L, Chan HSL, Kang J, Junien C, Cowell JK, Jones C, Dafoe LA. (1988) Homozygous deletion of a DNA marker from chromosome 11 in sporadic Wilms' tumour. *Genomics* **3:** 25–31.

Little MH, Prosser J, Condie A, Smith PJ, van Heyningen V, Hastie ND. (1992) Zinc finger point mutations within the WT1 gene in Wilms' tumor patients. *Proc. Natl Acad. Sci. USA* **89:** 4791–4795.

Madden SL, Cook DM, Morris JF, Gashler A, Sukhatme VP, Rauscher FJ. (1991) Transcriptional repression mediated by the WT1 Wilms' tumour gene product. *Science* **253:** 1550–1553.

Madden SL, Cook DM, Rauscher FJ. (1993) A structure–function analysis of transcriptional repression mediated by the WT1, Wilms' tumor suppressor protein. *Oncogene* **8:** 1713–1720.

Mannens M, Slater RM, Heytig C, Bliek J, De Kraker J, Coad N, De Pagter-Holthuizen P, Pearson PL. (1988) Molecular nature of genetic changes resulting in loss of heterozygosity of chromosome 11 in Wilms' tumours. *Hum. Genet.* **81:** 41–48.

Matsunaga E. (1981) Genetics of Wilms' tumour. *Hum. Genet.* **57:** 231–246.

Maw MA, Grundy PE, Millow LJ, Eccles MR, Dunn RS, Smith PJ, Feinberg AP, Law DJ, Paterson MC, Telzerow PE, Callen DF, Thompson AD, Richard RI, Reeve AE. (1992) A third Wilms' tumour locus on chromosome 16q. *Cancer Res.* **52:** 3094–3098.

Miller RW, Fraumeni JR, Manning MD. (1964) Association of Wilms' tumour with aniridia, hemihypertrophy, and other congenital malformations. *N. Engl. J. Med.* **270:** 922–927.

Morris JF, Madden SL, Tournay OE, Cook DM, Sukhatme VP, Rauscher FJ. (1991) Characterization of the zinc protein encoded by the WT1 Wilms' tumor locus. *Oncogene* **6:** 2339–2348.

Narahara K, Kikkawa K, Kimira S, Kimoto H, Ogata M, Kasai M, Matsuoka K. (1984) Regional mapping of catalase and Wilms' tumour, aniridia, genitourinary abnormalities, and mental retardation triad loci to the chromosome segment 11p1305–p1306. *Hum. Genet.* **66:** 181–185.

Ogawa O, Eccles MR, Yun K, Mueller RF, Holdaway MDD, Reeve AE. (1993a) A novel insertional mutation at the third zinc finger coding region of the WT1 gene in Denys–Drash syndrome. *Hum. Mol. Genet.* **2:** 203–204.

Ogawa O, Eccles MR, Szeto J, McNoe LA, Yun K, Maw MA, Smith PJ, Reeve AE. (1993b) Relaxation of insulin-like growth factor II gene imprinting implicated in Wilms' tumour. *Nature* **362:** 749–751.

Onadim Z, Hogg A, Cowell JK. (1993) Mechanisms of oncogenesis in patients with familial retinoblastoma. *Br. J. Cancer* **68:** 958–964.

Pal N, Wadey RB, Buckle B, Yeomans E, Pritchard J, Cowell JK. (1990) Preferential loss of maternal alleles in sporadic Wilms' tumour. *Oncogene* **5:** 1665–1668.

Pelletier J, Bruening W, Li FP, Haber DA, Glaser T, Housman DE. (1991a) WT1 mutations contribute to abnormal genital system development and hereditary Wilms' tumour. *Nature* **353:** 431–434.

Pelletier J, Bruening W, Kashtan CE, Mauer SM, Manivel JC, Striegel JE, Houghtin DC, Junien C, Habib R, Fouser L, Fine RN, Silverman BL, Haber DA, Housman D. (1991b) Germline mutations in the Wilms' tumour suppressor gene are associated with abnormal urogenital development in Denys–Drash syndrome. *Cell* **67:** 437–447.

Ping AJ, Reeve AE, Law DJ, Young MR, Boehnke M, Feinberg AP. (1989) Genetic linkage of Beckwith–Wiedemann syndrome to 11p15. *Am. J. Hum. Genet.* **44:** 720–723.

Porteous DJ, Bickmore W, Christie S, Boyd PA, Cranston G, Fletcher JM, Gosden JR, Rout D, Seawright A, Simola KJO, van Heyningen V, Hastie ND. (1987) HRAS-1 selected chromosome transfer generates markers that colocalise aniridia- and genitourinary dysplasia-associated translocation breakpoints and the Wilms' tumour gene within band 11p13. *Proc. Natl Acad. Sci. USA* **84:** 5355–5359.

Poulat F, Morin D, Konig A, Brun P, Giltay J, Sultan C, Dumas R, Gessler M, Berta P. (1993) Distinct molecular origins for Denys–Drash and Frasier syndromes. *Hum. Genet.* **91:** 285–286.

Pritchard-Jones K, Fleming S. (1991) Cell types expressing the Wilms' tumour gene (WT1) in Wilms' tumours: implications for tumour histogenesis. *Oncogene* **6:** 2211–2220.

Rainer S, Johnson LA, Dobry CJ, Ping AJ, Grundy PE, Feinberg AP. (1993) Relaxation of imprinted genes in human cancer. *Nature* **362:** 747–749.

Rauscher FJ, Morris JF, Tournay OE, Cook DM, Curran T. (1990) Binding of the Wilms' tumour locus zinc finger protein to the EGR-1 consensus sequence. *Science* **250:** 1259–1262.

Riccardi VM, Sujansky E, Smith AC, Francke U. (1978) Chromosome imbalance in the aniridia–Wilms' tumour association: 11p interstitial deletion. *Pediatrics* **61:** 604–610.

Santos A, Osorio-Almeida L, Baird PN, Silva JM, Boavida MG, Cowell JK. (1993) Insertional inactivation of the WT1 gene in tumour cells from a patient with WAGR syndrome. *Hum. Genet.* **92:** 83–86.

Sapienza C. (1989) Genome imprinting and dominance modification. *Ann. N.Y. Acad. Sci.* **564:** 24–38.

Schwartz CE, Haber DA, Stanton VP, Strong LC, Skolnick MH, Housman DE. (1991) Familial predisposition to Wilms' tumour does not segregate with the WT1 gene. *Genomics* **10:** 927–930.

Scott J, Cowell JK, Robertson M, Priestley EL, Wadey MR, Hopkins B, Pritchard J, Bell G, Rall IL, Graham BC, Knott J. (1985) Insulin-like growth factor-II gene expression in Wilms' tumour and embryonic tissues. *Nature* **317:** 260–262.

Scrable HJ, Witte DP, Lampkin BC, Cavenee WK. (1987) Chromosomal localisation of the human rhabdomyosarcoma locus by mitotic recombination mapping. *Nature* **329:** 645–647.

Solis V, Pritchard J, Cowell JK. (1988) Cytogenetics of Wilms' tumours. *Cancer Genet. Cytogenet.* **34:** 223–234.

Tadokoro K, Fujii H, Ohshima A, Kakizawa Y, Shimizu K, Sakai A, Sumiyoshi K, Inoue T, Hayashi Y, Yamada M. (1992) Intragenic homozygous deletion of the WT1 gene in Wilms' tumour. *Oncogene* **7:** 1215–1221.

Turleau C, De Grouchy J, Chavin-Colin F, Martelli H, Voyer M, Charlas R. (1984) Trisomy

11p15 and Beckwith–Wiedemann syndrome: a report of two cases. *Hum. Genet.* **67**: 219–221.

Wadey RB, Pal NP, Buckle B, Yeomans E, Pritchard J, Cowell JK. (1990) Loss of heterozygosity in Wilms' tumour involves two distinct regions of chromosome 11. *Oncogene* **5**: 901–907.

Waziri M, Patil SR, Hanson JW, Bartley JA. (1983) Abnormalities of chromosome 11 in patients with features of Beckwith–Wiedemann syndrome. *J. Pediatr.* **102**: 873–876.

Werner H, Re GG, Drummon IA, Sukhatme VP, Rauscher FJ, Sens DA, Garvin AJ, LeRoith D, Roberts CTJ. (1993) Increased expression of the insulin-like growth factor I receptor gene, IGF1R in Wilms' tumour is correlated with modulation of IGF1R promoter activity by the WT1 Wilms' tumour gene product. *Proc. Natl Acad. Sci. USA* **90**: 5828–5832.

Wiedemann HR. (1964) Complexe malformatif familial avec hernie ombilicle et macroglossie: un syndrome nouveau? *J. Genet. Hum.* **13**: 223–232.

Wiedemann HR. (1983) Tumours and hemihypertrophy associated with Wiedemann–Beckwith syndrome. *Eur. J. Pediatr.* **141**: 129.

8

How a dynamic mutation manifests in myotonic dystrophy

Catherine L. Winchester and Keith J. Johnson

8.1 Introduction

Myotonic dystrophy (DM) is the commonest form of adult muscular dystrophy, characterized by myotonia and progressive muscle weakness, with an incidence of 1 in 8000 in most populations (Harper, 1989). It is a heterogeneous disorder affecting a wide range of systems. In addition to skeletal muscle problems, patients often suffer from cataracts, cardiac conduction defects, mental retardation, premature balding and testicular atrophy. Clinical diagnosis therefore not only includes neurological examination but also the detection of myotonia electromyographically (EMG) and slit lamp investigations to identify lens opacities. Biochemical, electrophysiological and histological changes have also been demonstrated in DM patients. Although there does not appear to be one underlying biochemical defect, whole body resistance to insulin has been observed (Moxley *et al.*, 1984). Electrophysiological studies have implicated that the fundamental defect is a change in membrane function. Alterations in ion exchanges and conductivity, such as reduced resting potential accompanied by an increased intracellular sodium concentration (Rudel and Lehman-Horn, 1985), runs of action potentials and altered sodium and potassium exchanges (Hull and Roses, 1976) have been observed in erythrocytes of DM patients. Reduced phosphorylation of membrane proteins has also been demonstrated in preparations of these cells from patients (Roses and Appel, 1973). The variety of membrane anomalies suggests that the defect is an alteration in a fundamental structural membrane component but as yet no protein has been implicated completely.

The disease can be broadly classified into three clinical groups; minimally affected or late onset, classical adult onset and severe congenital onset. People with the mildest form of DM often go undiagnosed, as the disease occurs late in life, usually with cataracts as the only visible sign and minimal muscle involvement. The classical form of DM usually develops in early adult life and is characterized by progressive muscle stiffness and weakness. Congenital DM (CDM) is the most severe form of the disease, inherited almost solely from affected mothers. It presents in newborn babies who suffer from respiratory distress and mental impairment but who do not show the classic symptoms of myotonia and muscle weakness. If these children survive the neonatal period then these symptoms present in early adulthood or late adolescence.

Although these clinical distinctions exist, variation in severity of symptoms and age of onset occurs within, as well as between, families and the picture is complicated by incomplete penetrance and variable expressivity. Anticipation, an increase in severity of symptoms and decrease in age of onset, has also been observed as the disease is transmitted from generation to generation (Howeler *et al.*, 1989).

Despite the clinical variability of DM, it has been shown by family studies to be the result of a single gene defect which is inherited in an autosomal dominant manner.

8.2 Identification of the myotonic dystrophy gene locus and the underlying mutation

In 1992, the myotonic dystrophy gene locus and the underlying mutation were finally identified (Aslanidis *et al.*, 1992; Brook *et al.*, 1992; Buxton *et al.*, 1992; Fu *et al.*, 1992; Harley *et al.*, 1992; Mahadevan *et al.*, 1992). Cloning the gene was made difficult initially by the lack of any visible cytogenetic abnormality or knowledge of a defined biochemical defect. However DM was one of the first diseases which segregates in a Mendelian fashion to be linked to protein markers. Linkage was reported between DM and the Lutheran (Lu) blood group antigen, the ABH secretor (Se) blood group and the complement factor C3 (Mohr, 1954; Renwick *et al.*, 1971; Harper *et al.*, 1972; Eiberg *et al.*, 1983). These markers and the DM locus were assigned to chromosome 19 when the gene coding for C3 was cloned and mapped to it (Whitehead *et al.*, 1982). This information was later confirmed by the discovery of linkage between the DM locus and restriction fragment length polymorphisms (RFLPs) detected by C3 cDNA probes (Davies *et al.*, 1983).

Since then a detailed genetic map of chromosome 19 consisting of genes and anonymous DNA fragments that detect RFLPs has been built up (Figure 8.1). This enabled the 1 Mb critical region defining the DM locus with flanking markers to be identified and mapped (Shaw *et al.*, 1986; Johnson *et al.*, 1988, 1989; Brunner *et al.*, 1989; Korneluk *et al.*, 1989; Harley *et al.*, 1991a) and subsequently the gene cloned. One of the strongest pieces of evidence supporting this localization was the detection of linkage disequilibrium between the disease locus and markers from this region in European and Japanese populations (Harley *et al.*, 1991b; Yamagata *et al.*, 1992) as well as the more isolated French-Canadian and Finnish populations (MacKenzie *et al.*, 1989; Nokelainen

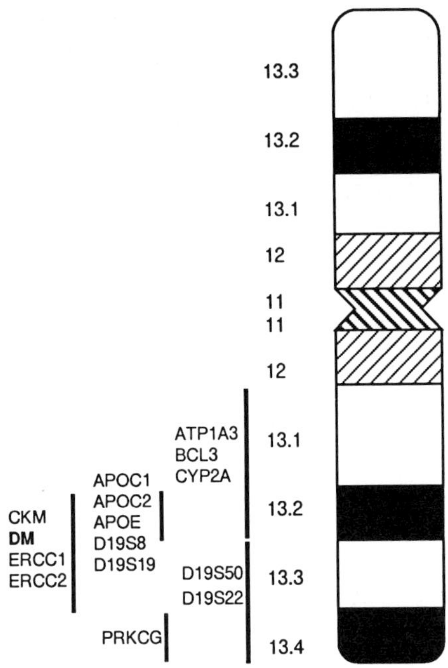

Figure 8.1. An ideogram of human chromosome 19 showing the DM critical region at 19q13 defined by the polymorphic markers used to map the disease gene (Shaw *et al.*, 1986; Johnson *et al.*, 1988, 1989; Brunner *et al.*, 1989; Korneluk *et al.*, 1989; Harley *et al.*, 1991a).

et al., 1990). The finding in the Europeans was surprising given that the disease is dominant and causes a significant decrease in reproductive fitness, and this was only explained with the discovery that a dynamic mutation underlies this disorder.

One of the expressed DNA clones from within this critical region detected unstable DNA fragments in the genomic DNA of patients that showed length polymorphisms when digested with a variety of restriction enzymes (Aslanidis *et al.*, 1992; Buxton *et al.*, 1992; Harley *et al.*, 1992). This cDNA has differing laboratory names and has been termed 45H9 (referring to a 700 bp truncated fragment isolated from a human fetal brain cDNA library; Aslanidis *et al.*, 1992), cDNA 25 (a 4.5 kb genomic clone identified from a human fetal brain cDNA library contaminated with genomic clones; Buxton *et al.*, 1992) and pBB0.7 (a subclone of a strongly conserved human genomic clone spanning the DM critical region; Harley *et al.*, 1992). All three probes detected the same two allele EcoRI insertion/deletion polymorphism in unaffected individuals and a larger variable DM patient-specific allele. The EcoRI polymorphism is not due to a restriction site alteration but to the insertion/deletion of consecutive Alu repeats (Mahadevan *et al.*, 1993a) 5 kb distal to the unstable region (Shelbourne *et al.*, 1992) (Figure 8.2). Alleles of 8.6 kb and 9.8 kb (Buxton *et al.*, 1992) are identified when Southern blots of EcoRI-digested control genomic DNA are probed and larger bands, up to 5 kb longer than the larger normal allele, are detected on Southern blots of EcoRI-digested genomic DNA of DM patients (Figure 8.3). The 8.6 kb

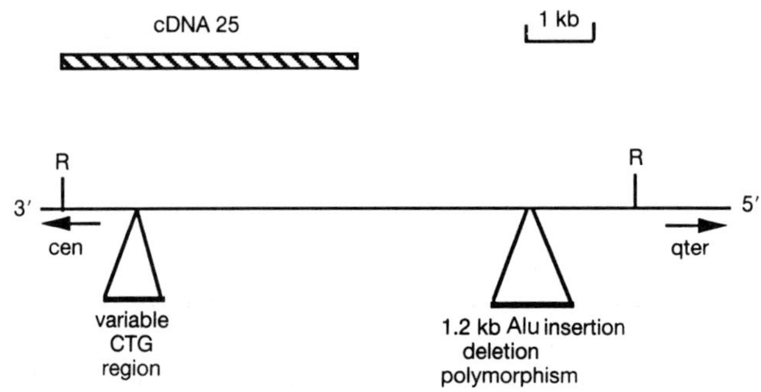

Figure 8.2. A diagram of the DM gene showing its orientation with respect to the centromere and telomere of chromosome 19 and the relative positions of the CTG repeat region and the Alu insertion/deletion region within the gene. The EcoRI restriction sites (R), and the probe, cDNA 25, with which the DM-specific polymorphisms can be detected, are also shown.

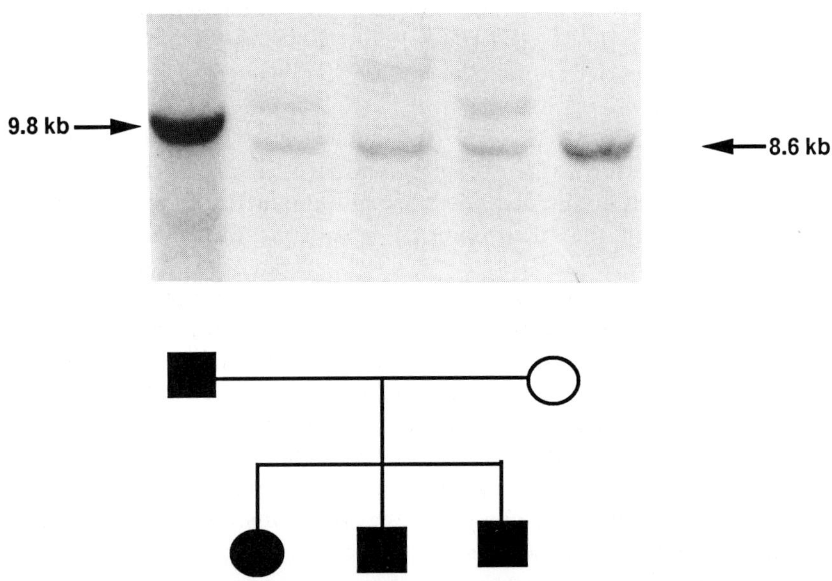

Figure 8.3. A Southern blot of a DM family's DNA digested with the enzyme EcoRI and probed with the radiolabelled DM-specific probe, cDNA 25. The pedigree is shown beneath. The unaffected mother is homozygous for the normal 8.6 kb deletion allele which she transmits to her children. The affected father has a normal 9.8 kb insertion allele and a small expansion, shown as a diffuse band on the gel. The minimally expanded allele is transmitted with further expansions to his more severely affected children. The affected children are heterozygous and show one normal 8.6 kb allele, inherited from their mother, and a larger DM-specific allele, inherited from their father.

allele contains two Alu repeats, whilst the 9.8 kb allele and the enlarged DM alleles are associated with five Alu repeats. The larger DM-specific fragments vary in length between and within DM families and are shown to increase in size in successive generations, correlating with an earlier age of disease onset and an increase in the severity of symptoms. The discovery of this unstable DNA at the DM locus provided a genetic explanation for the clinical phenomenon of anticipation which presents in DM (Howeler *et al.*, 1989; Harley *et al.*, 1993). Sequence analysis of genomic clones spanning the expanding DM region and the use of the polymerase chain reaction (PCR) to investigate variations in the allele sizes between normal and affected individuals, revealed that the mutation causing the DNA instability is a trinucleotide (CTG) repeat which is highly polymorphic in the normal population (Brook *et al.*, 1992; Davies *et al.*, 1992) and which increases dramatically in length in DM patients (Brook *et al.*, 1992; Fu *et al.*, 1992; Mahadevan *et al.*, 1992). Genomic clones containing the CTG repeat were used to probe DNA from a number of different animal species to check for sequence conservation, and to probe cDNA libraries to show that the trinucleotide repeat amplifying in DM is within a gene and is transcribed. Database comparisons of cDNA sequences showed that the CTG repeat is situated in the 3′ untranslated region of a gene encoding a cAMP-dependent protein kinase, approximately 500 bp upstream of the polyadenylation signal.

Therefore the extensive search climaxed in the discovery of the DM locus at 19q13.3 and the dynamic expansion of a trinucleotide (CTG) repeat as the mutation underlying the disease. This exciting discovery revealed DM to be the third disease connected with this type of mutation, following the discovery of the genes and the expansion mutations underlying spinal and bulbar muscular atrophy (SBMA), also known as Kennedy's disease (La Spada *et al.*, 1991), and fragile X syndrome (FraXA; Verkerk *et al.*, 1991). It has been postulated that this novel type of mutation may be a new common class of human mutation resulting in inherited disorders. Earlier this year, an expansion mutation was shown to be responsible for Huntington's disease (Huntington's Disease Collaborative Research Group, 1993), spinocerebellar ataxia type 1 (SCA-1; Orr *et al.*, 1993) and a second fragile site near FraXA which also underlies a mental retardation syndrome, FraXE (Sutherland and Baker, 1992; Flynn *et al.*, 1993; Knight *et al.*, 1993). As more disease genes are identified, it remains to be seen whether this is a common mechanism but it is likely that dynamic mutations will prove to be a major class of human genetic disorders.

8.3 Structure of the DM gene and putative gene product

Comparisons of cDNA sequences showed that the gene responsible for the DNA instability involved in DM encoded a protein kinase. Protein kinases have important regulatory roles in all cellular processes and a mutation in this class of gene could explain the variable phenotype observed in DM and the previously reported changes in phosphorylation of membrane proteins observed in erythrocytes of patients (Roses and Appel, 1973).

The DM gene is about 13 kb in length with 15 exons (Mahadevan *et al.*, 1993b) (Figure 8.4) and is transcribed in the orientation telomere to centromere (Brook *et al.*, 1992). Characterization of the full length gene enabled the identification of the complete coding sequence and the prediction of a mature mRNA transcript (Mahadevan *et al.*, 1993b). This led to the prediction of a putative amino acid sequence comprising 624 amino acids, and so, despite no protein product being isolated, a function for the DM gene could be investigated.

Exons 1–8 comprise the N terminus of the predicted protein, with the majority of this domain, exons 2–8, showing homology to a family of serine–threonine protein kinases. The initial part of the N terminus does not show homology to any known protein. The intermediate region of the putative protein, exons 9–12, shows homology to the coiled coil domains of filament proteins. This region of the gene is also characterized by the Alu repeat polymorphism preceding exon 9. The C terminus of the predicted protein is hydrophobic and is thought to act as a membrane anchor. It is here in the last exon, exon 15, that the CTG repeat is situated, downstream from the translation stop signal.

The DM gene is expressed predominantly in tissues in which DM manifests, such as the heart and skeletal muscle, with low levels of expression in the brain cortex, lung, pancreas and testis. No gene expression has been observed in adult and fetal liver or fetal stomach, small bowel and cerebellum (Hofmann-Radvanyi *et al.*, 1993).

A second distinct gene has been located in the DM region that partially overlaps the DM gene at the 5′ end (Jansen *et al.*, 1992). It is transcribed telomere to centromere resulting in a 3 kb transcript that has no homology to known genes or proteins. There is, however, a mouse homologue (Jansen *et al.*, 1992). To date,

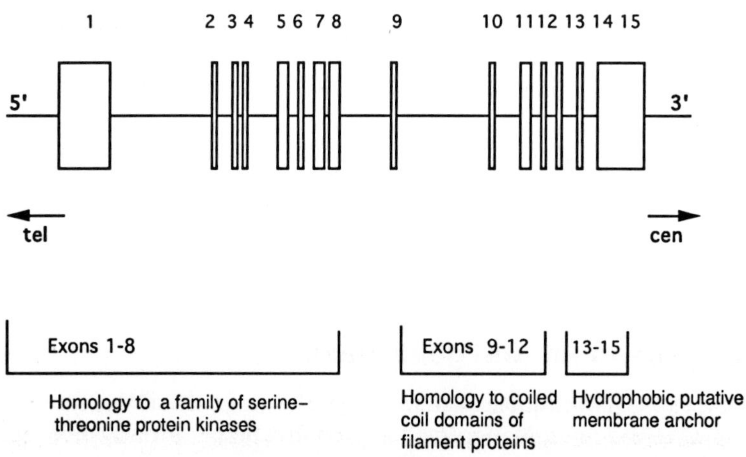

Figure 8.4. A schematic representation of the human DM gene with the predicted domains of the putative serine–threonine protein kinase. Data compiled from Brook *et al.* (1992) and Mahadevan *et al.* (1993b).

the role of this gene in the manifestation of DM is not known.

8.4 How does a triplet repeat mutation cause a dominant disease?

The presence of the expansion mutation in the DM gene causes the development of a dominant debilitating disease. However, the unstable CTG repeat is not translated and therefore the mutation is not incorporated into the functional protein. Recent research has focused on determining at which stage the mutation takes effect. It is possible that the CTG amplification disrupts gene transcription, mRNA processing and/or gene translation, resulting in abnormal levels of protein or different functional forms.

Experiments have been performed to quantify gene transcription and to determine the levels of DM mRNA in affected and unaffected tissues. Lack of transcription of enlarged DM genes in congenitally affected children has been reported (Hofmann-Radvanyi *et al.*, 1993). However, an increase in mRNA levels in tissues of a congenitally affected child has also been reported (Sabourin *et al.*, 1993), whilst in adult tissues reduced levels of mRNA from alleles with relatively large CTG expansions have been observed (Fu *et al.*, 1993). The seemingly contradictory nature of these investigations may be technical as different methods were applied to different patient types and as yet we know very little about the transcription patterns of the two unaffected alleles from people, although co-equal expression from two normal alleles has been observed (Fu *et al.*, 1993). If mRNA levels are altered it is possible that this is due to a change in the rate of transcription of the DM allele or a change in the stability of the DM allele mRNA transcript. The CTG repeat is situated in the 3′ untranslated region, a major site involved in transcription stability (Mullner and Kuhn, 1988; Mullner *et al.*, 1989), and any change in mRNA levels could lead to changes in protein levels. Since the putative gene product is a protein kinase, it is possible that it interacts with one or more gene products leading to the disruption of one or more cellular processes.

Other proposed alterations include splicing, imprinting and disruption of chromatin structure. Although alternatively spliced mRNA transcripts that encode protein isoforms that differ at their C and N termini have been isolated from heart and brain of patients (Jansen *et al.*, 1992; Fu *et al.*, 1993), it is not known whether they would have a functional difference.

8.5 Significance of the CTG repeat

Amplification of the unstable region in the 3′ end of the DM gene using the polymerase chain reaction (PCR), has enabled some investigation into the relationship between genotype, the CTG expansion mutation, and phenotype, the manifestation of the disease. Despite the highly variable phenotypic expression of the DM gene, there is an approximate correlation with the number of CTG repeats, the severity of symptoms and the age of onset. However, the tissue from which the DNA is extracted is important because of the somatic heterogeneity of the repeat observed in different tissues (discussed later). In general, the greater the amplification the more severe the disease and the earlier symptoms develop (Brook *et al.*, 1992; Fu *et al.*, 1992; Mahavedan *et al.*, 1992). The number of CTG units is polymorphic in the normal population with a range of 5–37 repeats (Brunner *et al.*, 1992) (Figure 8.5). A trimodal distribution is observed in European populations, with the most frequently occurring allele being 5 (Imbert *et al.*, 1993). The second mode consists of three major alleles of copy number 11, 12 and 13, and one minor allele of 14. The final mode has no peak frequency but represents alleles of 19 and above.

Despite the variation observed in DNA stability and the size of the amplification, all DM patients and mutation carriers have the same Alu insertion allele. The stably inherited 5 repeat allele and the heterogeneous 19 and above repeat alleles are also associated with this allele, whilst the 11, 12, 13 and 14 repeat alleles are associated with the deletion allele (Imbert *et al.*, 1993). Certain normal alleles are associated with the insertion allele, five copies of the CTG repeat and 16 copies upwards, and virtually all cases of DM are linked to the insertion allele (Figure 8.6). This suggests that the DM mutation arose from a limited pool of founder chromosomes. Other evidence includes the lack of sporadic cases and the ethnic distribution of myotonic dystrophy. The amplification mutation found on chromosome 19 in the European population has also been shown to be the cause

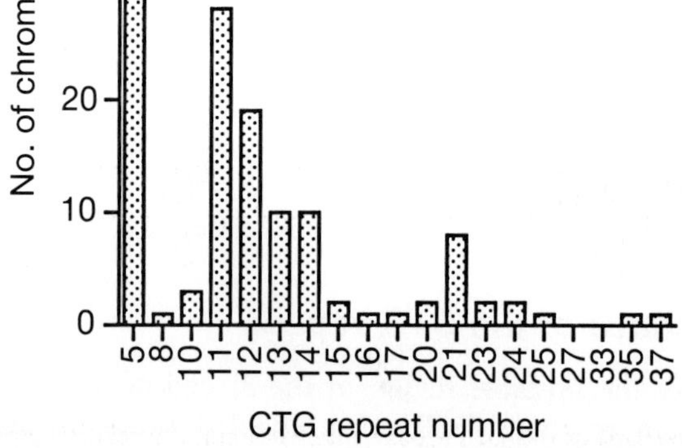

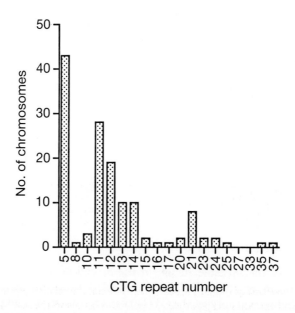

Figure 8.5. Frequency distribution of the DM CTG repeat in European populations. Graph compiled from data from Davies *et al.* (1992), Brunner *et al.* (1992) and Imbert *et al.* (1993).

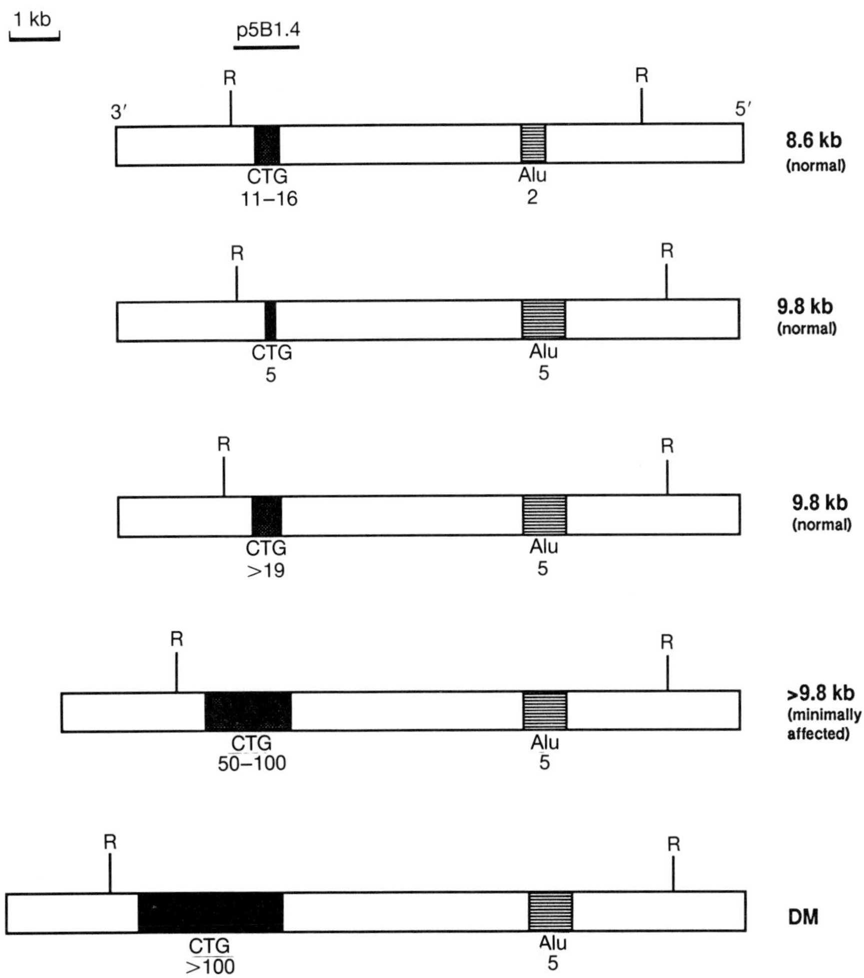

Figure 8.6. Diagrams of the DM kinase gene showing the relative positions of the CTG repeat region and the Alu insertion/deletion polymorphism and the association between the size of repeat, Alu number and the allele size detected by EcoRI digestion (R indicates restriction sites) of genomic DNA and hybridization with the radiolabeled DM-specific probe p5B1.4 (Shelbourne *et al.*, 1992).

of myotonic dystrophy in Japanese patients, although the frequencies of normal alleles are different (Davies *et al.*, 1992; Yamagata *et al.*, 1992). This observation, and the very low prevalence of DM in African, Cantonese, Oceanians and Thai populations, is consistent with the migration of northern Eurasian groups from Africa (Ashizawa and Epstein, 1991), implying that the mutation occurred after this geographical divergence but before the separation of European and Japanese populations.

8.6 Stability of the CTG repeat and effects on phenotypic variability

Apart from the recent report of two sisters who are homozygous for the DM mutation (Cobo *et al.*, 1993a), DM patients are found to be heterozygous, with one allele in the normal range and one allele that has amplified. Minimally affected people have amplified alleles of at least 50 repeats, with an upper limit approaching 100 copies. More severely affected people show large expansions of the trinucleotide repeat, up to several kb in length, with congenital cases usually showing the greatest increase, correlating with the most severe symptoms (Table 8.1). However, there are no definite repeat size boundaries for the three clinical groups and there are overlaps between these groups. These include people with relatively large repeats who have remained asymptomatic (Brunner *et al.*, 1992), DM patients who have repeat lengths in the normal range (Mahadevan *et al.*, 1992) and other individuals with repeat alleles usually in the minimally affected range who have been severely affected (Lavedan *et al.*, 1993b). Although overlap between the classes exists, correlation of repeat length and phenotype tends to be consistent within families but can be weak between families. These associations have important implications for diagnosis and genetic counselling and, therefore, family history as well as repeat length needs to be considered. The choice of tissue from which the DNA is extracted is also critical for accurate predictive testing in view of the wide range of expansion sizes reported from different CDM fetal tissues (Lavedan *et al.*, 1993b).

Stability of the CTG region during transmission is unclear, although in general the amplification mirrors the phenomenon of clinical anticipation. However, amplification does not always occur and when it does it is not necessarily to the same extent, even in offspring that inherit the same DM allele. In general, for repeat alleles with copy number exceeding 500, the degree of expansion seen on transmission is greater from females than males (Lavedan *et al.*, 1993b). This is explained largely by the fact that males with expansions of more than 500 repeats tend to pass on contractions to their offspring if they are fertile. These factors have important consequences for subsequent generations and need to be considered in genetic counselling.

The 5 repeat allele and the 11, 12, 13 and 14 repeat alleles appear to be inherited stably whilst the low frequency and heterogeneity of the repeat alleles above 19 indicate that CTG units of this length are unstable. Small expansions of

Table 8.1. Correlation of disease symptoms with CTG repeat expansion

CTG repeat number (genotype)	Disease symptoms (phenotype)
5–37 repeats	Unaffected
50–100 repeats	Asymptomatic gene carriers or minimally affected with mild symptoms such as cataracts and slight muscle problems late in life
100–4000 repeats	Classical adult onset with myotonia and muscle weakness. Severe symptoms include premature balding, testicular atrophy, cardiac conduction defects and mental retardation. Congenitally affected patients generally show the most severe symptoms

alleles in this group would result in repeat sizes that fall within the minimally affected range, as there is a very small difference in the two classes. These initial mutations are thought to lead to minor DNA instability, as the minimally expanded alleles can be inherited stably over a few generations with occasional expansions but no change in phenotype. This has been demonstrated in one of the CEPH families (Weber and Wong, 1993). However, at some stage, subsequent amplifications of alleles in the 50–100 range cross a threshold level of instability and eventually mutations occur that result in massive expansions of the CTG repeat and dramatic changes in phenotype. All DM patients originate from family members with minimal expansions whose repeat length has caused a cascade of instability and resulted in DNA expansions on transmission, correlating with the progression of DM from minimally affected to congenitally affected in just a few generations. Expanded disease-associated alleles generally increase in length from generation to generation, with an increase in disease severity. However, recent reports of mild penetrance and reduction in the length of the CTG repeat have shown that this is not always the case.

The presence of semi-stable minimal DM mutations suggests that potential predisposing alleles for DM may be much more prevalent in the population than previously thought. There may be an undiagnosed pool of people who carry and transmit these alleles with little or no phenotypic appearance. Minimally affected people usually are not diagnosed as isolated cases but are only detected when descendents are affected due to CTG amplification. In general, the larger the minimally expanded allele the less stable it is at meiosis, resulting in a larger disease-associated allele. In fact, seemingly sporadic cases have been shown to be due to this (Shelbourne *et al.*, 1993).

In the majority of cases, transmission of the expanded DM gene has led to an increase in the copy number of the CTG trinucleotide repeat, consequently accompanied by earlier manifestation of the disease and increased severity in symptoms. However, recent studies have shown that *contraction* of the repeat on transmission also occurs but at a much lower frequency. With the exception of four maternally derived cases (Cobo *et al.*, 1993b; Redman *et al.*, 1993), the vast majority of reductions have arisen from the inheritance of the paternal DM allele (Shelbourne *et al.*, 1992; Abeliovich *et al.*, 1993; Brunner *et al.*, 1993; Harley *et al.*, 1993; Hunter *et al.*, 1993; Lavedan *et al.*, 1993a; Mulley *et al.*, 1993; O'Hoy *et al.*, 1993). In some cases, the repeat number has decreased to such an extent that it falls within the normal range and the offspring are asymptomatic, whilst in other cases the copy number is reduced but it is still higher than normal and offspring have symptoms of at least equal severity to their affected parent. In some cases anticipation still occurs despite an apparent decrease in repeat number on transmission. This finding is almost certainly due to the fact that the repeat lengths have until now been determined from lymphocytes and recent evidence strongly suggests that the repeat length in patient muscle samples is much larger than that in their lymphocyte DNA (Anvret *et al.*, 1993). Follow-up of these CTG contraction patients is extremely important for determining DNA stability and it will be interesting to see if further reductions occur in future generations, or whether the smaller repeat size favours transmission with subsequent amplification.

Although there is no difference in the rate of transmission of the mutation, it appears that males with large amplifications are less likely than affected females to have offspring with larger repeats (Mulley *et al.*, 1993).This may be the result of testicular atrophy, commonly incurred by DM affected males, causing reduced fertility, or may be due to selection of spermatozoa with smaller repeats. It has also been proposed that the DM alleles of affected men are restricted from amplification when they exceed 2 kb in length (Lavedan *et al.*, 1993a). The contraction data show that males with repeats of less than 2 kb generally transmit amplified alleles to their offspring but expansion is restricted when their alleles exceed 2 kb. This limit does not seem to apply to female transmissions, although an affected mother bearing an amplification of 7 kb (Cobo *et al.*, 1993b), one of the largest reported so far, passed on a 4 kb allele to her congenitally affected son. The occurrence of reductions could explain the incomplete penetrance and the persistence of DM alleles in the population. The mechanism of reduction is not known but gene conversion, replication slippage, unequal sister strand exchange, direct deletion of the mutation and germline heterogeneity have been suggested. In one case haplotype, analysis has identified discontinuous gene conversion as the cause of the contraction (O'Hoy *et al.*, 1993).

The instability of the CTG repeat extends further than meiotic instability to mitotic instability. Variable sized fragments are observed as smeared, diffuse hybridization signals rather than discrete allele-specific bands on Southern blots when patient DNA is investigated with a DM gene-specific probe (Aslanidis *et al.*, 1993; Buxton *et al.*, 1992; Harley *et al.*, 1992; Abeliovich *et al.*, 1993; Lavedan *et al.*, 1993a,b). Somatic mosaicism has important implications in prenatal and presymptomatic diagnosis, as DNA analysis of lymphocytes may not represent the repeat length in the affected tissues. Larger CTG expansions have been observed in skeletal muscle than in lymphocytes (Anvret *et al.*, 1993;

Thornton *et al.*, 1994). It may therefore be more accurate to analyse muscle DNA as this is the primarily affected tissue. However, for prenatal diagnosis, chorionic villus samples may be sufficient as they show very little of the somatic heterogeneity that is observed in lymphocyte preparations from older patients (Myring *et al.*, 1992). It is possible that differences in repeat length in different tissues may contribute to the variability in symptom type and severity. Somatic instability is also thought to be an explanation for the occurrence of CTG repeat contractions. Offspring with smaller repeat lengths could arise if different sized repeat alleles existed in the gametes.

8.7 Variability in phenotype and inheritance of congenital DM

Congenital DM is the most severe form of DM and is generally associated with the largest expansions of the CTG repeat. Newly born infants suffering from this form of the disease inherit repeat lengths that are much larger than those of their transmitting parents and other DM sufferers, with the number of CTG repeats ranging from as few as 700 to over 4000. However, despite the importance of repeat size in the manifestation of the disease, it is not the only factor.

CDM is almost exclusively maternally inherited from classically affected women with large expansions, who are showing symptoms of DM during their pregnancy and at parturition. The clinical status of the transmitting mother is thought to be a contributing factor, as neurologically normal women who have mild symptoms, such as minor lens opacities, have not been shown to give birth to congenitally affected offspring. However there have been a few reports of asymptomatic women giving birth to congenitally affected children (Howeler and Busch, 1990; Goodship *et al.*, 1992; Redman *et al.*, 1993). It is estimated that the risk of a classically affected woman having a congenitally affected child is about 10% (Koch *et al.*, 1991). Family studies have shown that the chance of a congenitally affected child being born to a classically affected woman is greater if she has already had one congenital child, with the probability estimated to increase to about 50% (Harper, 1975). This means that all future pregnancies in which the fetus inherits the DM allele from a mother of a CDM child have virtually a 100% risk of also being congenitally affected. Later-born children with child onset DM also have worse symptoms than their older siblings (Andrews and Wilson, 1992).

DNA expansion alone is not enough to explain the congenital form of DM or the almost exclusive maternal inheritance, as there is overlap between the DM classes and their associated repeat lengths. Genomic imprinting and the inheritance of mitochondrial DNA have been proposed and subsequently disproved as mechanisms for maternal transmission of CDM. Genomic imprinting involves differential methylation of maternal and paternal alleles resulting in differential expression. It was suggested that if a normal paternal allele could be inactivated by methylation then the expanded maternal allele would be expressed, resulting in the severe form of the disease. However, no methylation differences between paternally and maternally derived alleles have been identified (Jansen *et al.*, 1993; Shaw *et al.*, 1993). Various mitochondrial abnormalities have been suggested as explanations for the inheritance of CDM, however, no convincing differences have been reported (Thyagarajan *et al.*, 1993). Another sex-related difference thought to contribute to CDM is the effect of altered metabolic factors, resulting from the multisystemic nature of DM, crossing the placenta and harming the developing embryo *in utero* (Harper and Dyken, 1972). Investigations concentrated on sex-related explanations because of the lack of paternally transmitted cases, although this non-occurrence may be explained by the reduced fitness of severely affected males and particularly those males bearing large repeats. However, due to a recent report of a paternally derived case of CDM (Nakagawa *et al.*, 1994), other explanations need to be explored.

8.8 Future prospects

Direct detection of the DM mutation has revolutionized genetic testing for this disorder and has enabled successful and accurate prenatal and presymptomatic diagnosis. However, there is as yet no clear understanding of the cellular mechanisms underlying the apparent meiotic and mitotic instability of the repeat, although it is almost certain that the expansions are due to problems arising during DNA replication. One possibility for future therapy in DM would be to develop the ability to intervene in the pathway leading to expansion in such a way that repair mechanisms could reduce the number of repeats again. In fact, if we can understand the process by which apparent contraction of the repeats occurs spontaneously on transmission in about 10% of male meioses, we may be able to effect an overall reduction in repeat number by this approach. The aetiology of the pathophysiological changes associated with repeat expansion is also unclear and there are as yet no models from studies of other organisms that can explain this.

In the present absence of a clear understanding of the aetiology and physiology of the disease, the prospects for well designed therapy are relatively limited. It is becoming clear that the somatic heterogeneity seen as differences in expansion sizes between lymphocyte and muscle biopsies explains some of the problems of genotype–phenotype correlation. The muscle expansions are nearly always significantly larger than those detected in lymphocytes of classically affected adult onset patients. The sample sizes on which this assumption is based are, however, still relatively small, and clarification of this question remains to be achieved by much larger surveys that are currently underway.

Current work into the development of a mouse genetic model of DM will enable many of these questions to be answered. The availability of a DM mouse will give us insights into the mechanisms controlling the stability of the repeat in this and other diseases caused by dynamic mutations. One of the key issues to be addressed in such a model system will be the relative contribution of alterations in the level of expression of the kinase gene to the phenotype and the possible involvement of alterations in expression levels of other genes in this disorder.

In conclusion, the discovery of dynamic mutations as a major class of human genetic disorders has immediately explained, at a molecular level, some puzzling genetic phenomena like variable penetrance and expressivity, anticipation and sex differences in inheritance patterns. However, because this is such a new and novel concept in human genetics there remain a number of difficult and exciting challenges to be met before this discovery translates into improved treatments for patients.

Acknowledgements

Work at Charing Cross and Westminster Medical School is supported by the Muscular Dystrophy Group of Great Britain, the Wellcome Trust, Medical Research Council and the University of London Central Research Fund. CLW gratefully acknowledges the Livingstone Trust for the provision of a research fellowship. We thank Nessa Carey for helpful discussions and encouragement.

References

Abeliovich D, Lerer I, Pashut-Lavon I, Shmueli E, Raas-Rothschild A, Frydman M. (1993) Negative expansion of the myotonic dystrophy unstable sequence. *Am. J. Hum. Genet.* **52:** 1175–1181.

Andrews PI, Wilson J. (1992) Relative disease severity in siblings with myotonic dystrophy. *J. Child Neurol.* **7:** 161–167.

Anvret M, Ahlberg G, Grandell U, Hedberg B, Johnson K, Edstrom L. (1993) Larger expansions of the CTG repeat in muscle than in lymphocytes from patients with myotonic dystrophy. *Hum. Mol. Genet.* **2:** 1397–1400.

Ashizawa T, Epstein HF. (1991) Ethnic distribution of myotonic dystrophy gene. *Lancet* **338:** 642–643.

Aslanidis C, Jansen G, Amemiya C, Shutler G, Mahadevan M, Tsilfidis C, Chen C, Alleman J, Wormskamp NGM, Vooijs M, Buxton J, Johnson K, Smeets HJM, Lennon G, Carrano AV, Korneluk RG, Wieringa B, de Jong PJ. (1992) Cloning of the essential myotonic dystrophy region and mapping of the putative defect. *Nature* **355:** 548–551.

Brook JD, McCurrach ME, Harley HG, Buckler AJ, Church D, Aburatani H, Hunter K, Stanton VP, Thirion J-P, Hudson T, Sohn R, Zemelman B, Snell RG, Rundle SA, Crow S, Davies J, Shelbourne P, Buxton J, Jones C, Juvonen V, Johnson K, Harper PS, Shaw DJ, Housman DE. (1992) Molecular basis of myotonic dystrophy: expansion of a trinucleotide (CTG) repeat at the 3′ end of a transcript encoding a protein kinase family member. *Cell* **68:** 799–808.

Brunner H, Smeets H, Lambermon HMM, Coerwinkel-Driessen N, van Oost BA, Wieringa B, Ropers H-H. (1989) A multipoint linkage map around the locus for myotonic dystrophy on chromosome 19. *Genomics* **5:** 589–595.

Brunner HG, Nillesen W, van Oost BA, Jansen G, Wieringa B, Ropers H-H, Smeets HJM. (1992) Presymptomatic diagnosis of myotonic dystrophy. *Am. J. Med. Genet.* **29:** 780–784.

Brunner HG, Jansen G, Nillesen W, Nelen MR, de Die CEM, Howeler CJ, van Oost BA, Wieringa B, Ropers H-H, Smeets HJM. (1993) Brief report: reverse mutation in myotonic dystrophy. *N. Engl. J. Med.* **328:** 476–480.

Buxton J, Shelbourne P, Davies J, Jones C, Van Tongeren T, Aslanidis C, de Jong P, Jansen G, Anvret M, Riley B, Williamson R, Johnson K. (1992) Detection of an unstable fragment of DNA specific to individuals with myotonic dystrophy. *Nature* **355:** 547–548.

Cobo A, Martinez JM, Martorell L, Baiget M, Johnson K. (1993a) Molecular diagnosis of homozygous myotonic dystrophy in two asymptomatic sisters. *Hum. Mol. Genet.* **2:** 711–715.

Cobo AM, Baiget M, Lopez de Munain A, Pozza JJ, Emparanza JI, Johnson K. (1993b) Sex-related differences in intergenerational expansion of myotonic dystrophy gene. *Lancet* **341:** 1159.

Davies K, Jackson J, Williamson R, Harper P, Ball S, Sarfarazzi M, Meredith L, Fey G. (1983) Linkage analysis of myotonic dystrophy and sequences on chromosome 19 using a cloned complement 3 gene probe. *J. Med. Genet.* **29:** 766–769.

Davies J, Yamagata H, Shelbourne P, Buxton J, Ogihara T, Nokelainen P, Nakagawa M, Williamson R, Johnson K, Miki T. (1992) Comparison of the myotonic dystrophy associated CTG repeat in European and Japanese populations. *J. Med. Genet.* **29:** 766–769.

Eiberg H, Mohr J, Nielson LS, Simonsen N. (1983) Genetics and linkage relationships of the C3 polymorphism: discovery of C3–Se linkage and assignment of LES-C3-DM-Se-PEPD-Lu synteny to chromosome 19. *Clin. Genet.* **24:** 159–170.

Flynn GA, Hirst MC, Knight SJL, Macpherson JN, Barber JCK, Flannery AV, Davies KE, Buckle VJ. (1993) Identification of the FRAXE fragile site in two families ascertained for X linked mental retardation. *J. Med. Genet.* **30:** 97–100.

Fu YH, Pizzuti A, Fenwick RG Jr, King J, Rsjnsrsysn S, Dunne PW, Dubel J, Nasser GA, Ashizawa T, de Jong P, Wieringa B, Korneluk R, Perryman MB, Epstein HF, Caskey CT. (1992) An unstable repeat in a gene related to myotonic muscular dystrophy. *Science* **255:** 1256–1258.

Fu Y-H, Friedman DL, Richards S, Pearlman JA, Gibbs RA, Pizzuti A, Ashizawa T, Perryman MB, Scarlato G, Fenwick RG Jr, Caskey CT. (1993) Decreased expression of myotonin–protein kinase messenger RNA and protein in adult form of myotonic dystrophy. *Science* **260:** 235–238.

Goodship J, Gibson DE, Burn J, Honeyman J, Cubey RB, Schofield I. (1992) Genetic risks for children of women with myotonic dystrophy. *Am. J. Hum. Genet.* **50:** 11340–11341.

Harley HG, Walsh KV, Rundle S, Brook JD, Sarfarazi M, Koch MC, Floyd JL, Harper PS, Shaw DJ. (1991a) Localisation of the myotonic dystrophy locus to 19q13.2–19q13.3 and its relationship to twelve polymorphic loci on 19q. *Hum. Genet.* **87:** 73–80.

Harley HG, Brook DJ, Floyd J, Rundle SA, Crow S, Walsh KV, Thibault M-C, Harper PS, Shaw DJ. (1991b) Detection of linkage disequilibrium between the myotonic dystrophy locus and a new polymorphic DNA marker. *Am. J. Hum. Genet.* **49:** 68–75.

Harley HG, Brook JD, Rundle SA, Crow S, Reardon W, Buckler AJ, Harper PS, Housman DE, Shaw DJ. (1992) Expansion of an unstable DNA region and phenotype variation in myotonic dystrophy. *Nature* **355:** 545–547.

Harley HG, Rundle SA, MacMillan JC, Myring J, Brook JD, Crow S, Reardon W, Fenton I, Shaw DJ, Harper PS. (1993) Size of the unstable CTG repeat sequence in relation to phenotype and parental transmission in myotonic dystrophy. *Am. J. Hum. Genet.* **52:** 1164–1174.

Harper PS. (1975) Congenital myotonic dystrophy in Britain. 2. Genetic basis. *Arch. Dis. Child* **50:** 514–552.

Harper PS. (1989) *Myotonic Dystrophy*, 2nd Edn. Saunders, London.

Harper PS, Dyken PR. (1972) Early onset dystrophica myotonica evidence supporting a maternal environmental factor. *Lancet* **II:** 53–55.

Harper PS, Rivas M, Bias W, Hutchinson J, Dyken P, McKusick V. (1972) Genetic linkage confirmed between the locus for myotonic dystrophy and the ABH secretion and Lutheran blood group loci. *Am. J. Hum. Genet.* **24:** 310–316.

Hofmann-Radvanyi H, Lavedan C, Rabes JP, Savoy D, Duros C, Johnson K, Junien C. (1993) Myotonic dystrophy: absence of CTG enlarged transcript in congenital forms and low expression of the normal allele. *Hum. Mol. Genet.* **2:** 1263–1266.

Howeler CJ, Busch HFM. (1990) Congenital myotonic dystrophy and genetic aspects of myotonic dystrophy. *J. Neurol. Sci.* **98** (Suppl.): 1.13.1.

Howeler CJ, Busch HFM, Geraedts JPM, Niermeijer MF, Staal A. (1989). Anticipation in myotonic dystrophy: fact or fiction? *Brain* **112:** 779–797.

Hull K, Roses A. (1976) Stoichiometry of sodium and potassium transport in erythrocytes from patients with myotonic muscular dystrophy. *J. Physiol. Lond.* **254:** 169–181.

Hunter AGW, Jacob P, O'Hoy K, MacDonald I, Mettler G, Tsilfidis C, Korneluk RG. (1993) Decrease in the size of the myotonic dystrophy CTG repeat during transmission from parent to child: implications for genetic counselling and genetic anticipation. *Am. J. Med. Genet.* **45:** 401–407.

Huntington's Disease Collaborative Research Group (1993) A novel gene containing a trinucleotide repeat that is expanded and unstable on Huntington's disease chromosomes. *Cell* **72:** 971–983.

Imbert G, Kretz C, Johnson K, Mandel J-L. (1993) Origin of the expansion mutation in myotonic dystrophy. *Nature Genetics* **4:** 72–76.

Jansen G, Mahadevan M, Amemiya C, Workskskamp N,Segers B, Hendriks W, O'Hoy K, Baird S, Sabourin L, Lennon G, Jap PL, Iles D, Coerwinkel M, Hofker M, Carrano AV, de Jong PJ, Korneluk RG, Wieringa B. (1992) Characterisation of the myotonic dystrophy region predicts multiple protein isoform-encoding mRNAs. *Nature Genetics* **1:** 261–266.

Jansen G, Bartolomei M, Kalscheuer V, Merkx G, Wormskamp N, Mariman E, Smeets D, Ropers H-H, Wieringa B. (1993) No imprinting involved in myotonic dystrophy: expression of both allelic DM-kinase mRNAs in mouse and human tissues. *Hum. Mol. Genet.* **2:** 1221–1228.

Johnson KJ, Nimmo E, Jones P, Weiss M, Savonataus M-L, Anvret M, Bartlett R, Roses A, Shaw DJ, Harper P, Williamson R. (1988) Segregation of linked probes to myotonic dystrophy in a family demonstrating that 152 and APOC2 are on the same side of DM on 19q. *Hum. Genet.* **80:** 379–381.

Johnson KJ, Shelbourne P, Davies J, Buxton J, Nimmo E, Anvret M, Bonduelle M, Williamson R, Savontaus M-L. (1989) Recombination events that locate myotonic dystrophy distal to APOC2 on 19q. *Genomics* **5:** 746–751.

Knight SJL, Flannery AV, Hirst MC, Campbell L, Chrisodoulou Z, Phelps SR, Pointon J, Middleton-Price HR, Barnicoat A, Pembrey M, Holland J, Oostra BA, Bobrow M, Davies KE. (1993) Trinucleotide repeat amplification of a CpG island in FRAXE mental retardation. *Cell* **74:** 127–134.

Koch MC, Grimm T, Harley HG, Harper PS. (1991). Genetic risks for children of women with myotonic dystrophy. *Am. J. Hum. Genet.* **48:** 1084–1091.

Korneluk RG, MacKenzie AE, Nakamura Y, Dube I, Jacob P, Hunter AGW. (1989) A reordering of human chromosome 19 long-arm markers and identification of markers flanking the myotonic dystrophy locus. *Genomics* **4:** 146–151.

La Spada AR, Wilson EM, Lubahn DB, Harding AE, Fischbeck KH. (1991) Androgen receptor gene mutations in X-linked spinal and bulbar muscular atrophy. *Nature* **352:** 77–79.

Lavedan C, Hofmann-Radvanyi H, Rabes JP, Roume J, Junien C. (1993a) Different sex-dependent constraints in CTG length variation as explanation for congenital myotonic dystrophy. *Lancet* **341:** 237–238.

Lavedan C, Hofmann-Radvanyi H, Shelbourne P, Rabes J-P, Duros C, Savoy D, Dehupas I, Luce S, Johnson K, Junien C. (1993b) Myotonic dystrophy: size- and sex-dependent dynamics of CTG meiotic instability,and somatic mosaicism. *Am. J. Hum. Genet.* **52:** 875–883.

MacKenzie AE, MacLeod HL, Hunter AGW, Korneluk RG. (1989) Linkage analysis of the apolipoprotein C2 gene and myotonic dystrophy on human chromosome 19 reveals linkage disequilibrium in a French-Canadian population. *Am. J. Hum. Genet.* **44:** 140–147.

Mahadevan M, Tsilfidis C, Sabourin L, Shutler G, Ameiya C, Jansen G,Neville C, Narang M, Barcelo J, O'Hoy K, Leblond S, Earle-Macdonald J, de Jong PJ, Wieringa B, Korneluk RG. (1992) Myotonic dystrophy mutation: an unstable CTG repeat in the 3′ untranslated region of the gene. *Science* **255:** 1253–1256.

Mahadevan MS, Foitzik MA, Surh LC, Korneluk RG. (1993a) Characterisation and polymerase chain reaction (PCR) detection of an *Alu* deletion polymorphism in total linkage disequilibrium with myotonic dystrophy. *Genomics* **15:** 446–448.

Mahadevan MS, Amemiya C, Jansen G, Sabourin L, Baird S, Neville CE, Wormskamp N, Segers B, Batzer M, Lamerdin J, de Jong P, Wieringa B, Korneluk RG. (1993b) Structure and genomic sequence of the myotonic dystrophy (DM kinase) gene. *Hum. Mol. Genet.* **2:** 299–304.

Mohr J. (1954) *A Study of Linkage in Man.* Munskgaard, Copenhagen.

Moxley R, Corbett A, Minaker K, Rowe J. (1984) Whole body insulin resistance in myotonic dystrophy. *Ann. Neurol.* **15:** 157–162.

Mulley JC, Staples A, Donnelly A, Gedeon AK, Hecht BK, Nicholson GA, Haan EA, Sutherland GR. (1993) Explanation for exclusive maternal origin for congenital form of myotonic dystrophy. *Lancet* **341:** 236–237.

Mullner EW, Kuhn LC. (1988) A stem-loop in the 3′ untranslated region mediates iron-dependent regulation of transferrin receptor mRNA stability in the cytoplasm. *Cell* **53:** 815–825.

Mullner EW, Neuport B, Kuhn LC. (1989) A specific mRNA binding factor regulates the iron-dependent stability of cytoplasmic transferrin receptor mRNA. *Cell* **58:** 373–382.

Myring J, Meredith AL, Harley HG, Kohn G, Norbury G, Harper PS, Shaw DJ. (1992) Specific molecular prenatal diagnosis for the CTG mutation in myotonic dystrophy. *J. Med. Genet.* **29:** 785–788.

Nakagawa M, Yamada H, Higuchi I, Kaminishi Y, Miki T, Johnson K, Osame M. (1994) A family of congenital myotonic dystrophy associated with paternal inheritance of CTG repeat expansions. *J. Med. Genet.*, in press.

Nokelainen P, Alanen-Kurki L, Winqvist R, Falck B, Somer H, Leisti J, Johnson K, Savontaus M-L, Peltonen L. (1990) Linkage disequilibrium detected between dystrophia myotonica and APOC2 locus in the Finnish population. *Hum. Genet.* **85:** 541–545.

O'Hoy KL, Tsilfidis C, Mahedevan MS, Neville CE, Barcelo J, Hunter AGW, Korneluk RG. (1993) Reduction in size of the myotonic dystrophy trinucleotide repeat mutation during transmission. *Science* **259:** 809–812.

Orr HT, Chung M-Y, Banfi S, Kwiatkowski TJ Jr, Servadio A, Beaudet AL, McCall AE, Duvick LA, Ranum LPW, Zoghbi HY. (1993) Expansion of an unstable trinucleotide CAG repeat in spinocerebellar ataxia. *Nature Genetics* **4:** 221–226.

Redman JB, Fenwick RG Jr, Fu Y-H, Pizzuti A, Caskey T. (1993) Relationship between parental trinucleotide GCT repeat length and severity of myotonic dystrophy in offspring. *J. Am. Med. Assoc.* **269:** 1960–1965.

Renwick J, Bundey S, Ferguson-Smith M, Izatt M. (1971) Confirmation of the linkage analysis of the loci for myotonic dystrophy and ABH secretion. *J. Med. Genet.* **8:** 407–416.

Roses, AD, Appel SH. (1973) Protein kinase activity in erythrocyte ghosts of patients with myotonic muscular dystrophy. *Proc. Natl Acad. Sci. USA* **70:** 1855–1859.

Rudel R, Lehman-Horn F. (1985) Membrane changes in cells from myotonia patients. *Physiol. Rev.* **65:** 310–346.

Sabourin LA, Mahadevan MS, Narang M, Lee DSC, Surh LC, Korneluk RG. (1993) Effect of the myotonic dystrophy (DM) mutation on mRNA levels of the DM gene. *Nature Genetics* **4:** 233–238.

Shaw DJ, Meredith AL, Sarfarazi M, Harley HG, Huson SM, Brook JD, Bufton L, Litt M, Mohandas M, Harper PS. (1986) Regional localisations and linkage of seven RFLPs and myotonic dystrophy on chromosome 19. *Hum. Genet.* **74:** 262–266.

Shaw DJ, Chaudhary S, Rundle SA, Crow S, Brook JD, Harper PS, Harley HG. (1993) A study of DNA methylation in myotonic dystrophy. *J. Med. Genet.* **30:** 189–192.

Shelbourne P, Winqvist R, Kunert E, Davies J, Leisti J, Thiele H, Bachmann H, Buxton J, Williamson B, Johnson K. (1992) Unstable DNA may be responsible for the incomplete penetrance of the myotonic dystrophy phenotype. *Hum. Mol. Genet.* **1:** 467–473.

Shelbourne P, Davies J, Buxton J, Anvret M, Blennow E, Bonduelle M, Schmedding E, Glass I, Lindenbaum R, Lane R, Williamson R, Johnson K. (1993) Direct diagnosis of myotonic dystrophy with a disease-specific DNA marker. *N. Engl. J. Med.* **328:** 471–475.

Sutherland GR, Baker E. (1992) Characterisation of a new rare fragile site easily confused with the fragile X. *Hum. Mol. Genet.* **1:** 111–113.

Thornton C, Johnson K, Moxley R. (1994) Myotonic dystrophy patients have larger CTG expansions in skeletal muscle than in leucocytes. *Ann. Neurol.*, in press.

Thyagarajan D, Byrne E, Noer S, Lertrit P, Utthanophol P, Kapsa R, Marzuki S. (1993) Significance of mitochondrial DNA deletions in myotonic dystrophy. *Acta Neurol. Scand.* **83:** 32–36.

Verkerk AJMH, Peiretti M, Sutcliffe JS, Fu Y-H, Kuhl DPA, Pizzuti A, Reiner O, Richards S, Victoria MF, Zhang F, Eussen BE, van Ommen GJB, Blonden LAJ, Riggins GJ, Chastain JL, Kunst CB, Galjaard H, Caskey CT, Nelson DL, Oostra BA, Warren ST. (1991) Identification of a gene (FMR-1) containing a CGG repeat coincident with a breakpoint cluster region exhibiting length variation in fragile X syndrome. *Cell* **65:** 905–914.

Weber JL, Wong C. (1993) Mutation of human short tandem repeats. *Hum. Mol. Genet.* **2:** 1123–1128.

Whitehead A, Bundey S, Ferguson-Smith M, Izatt M. (1982) Assignment of a structural gene for the third component of human complement to chromosome 19. *Proc. Natl Acad. Sci. USA* **79:** 5021–5025.

Yamagata H, Miki T, Ogihara T, Nakagawa M, Higuchi I, Osame M, Shelbourne P, Davies J, Johnson K. (1992) Expansion of unstable DNA region in Japanese myotonic dystrophy patients. *Lancet* **339:** 692–693.

9

Length variation in fragile X

Mark C. Hirst

9.1 Introduction

The fragile X gene, as a major cause of X-linked mental retardation and as a mutation that directly alters the structure of the metaphase chromosome, has intrigued geneticists for many years. As its name suggests, it is associated with the presence of a fragile site on the X chromosome, termed FRAXA, first described by Lubs in 1969. It was not until 1977 however, that its association with mental impairment was recognized (Sutherland, 1977). The mutation results in a syndrome characterized phenotypically by moderate mental retardation, a long face with large everted ears and macroorchidism (Fryns, 1989). A prevalence which could be as high as 1 per 1000 in males and 0.6 per 1000 in females (Webb, 1989) means that up to 50% of all X-linked mental retardation (XLMR) may be attributable to the fragile X syndrome.

Genetic linkage analysis has positioned the disease gene to the fragile X site and highlighted some unusual features of its inheritance. The mutation can be transmitted through normal males (normal transmitting males; NTMs) and large kindred studies revealed that penetrance increased as the mutation progressed through several generations (Sherman *et al.*, 1984, 1985). These observations, which are inconsistent with classical X linkage and are collectively known as the 'Sherman paradox', are an example of genetic anticipation, which is now known to be a phenomenon associated with hereditary unstable DNA. An explanation for these features was found upon the isolation of the FMR1 (for fragile X-mental-retardation) gene and the discovery of an expanded CGG triplet repeat.

Whilst we now know a great deal about the expansion profile of the triplet array, we are only just beginning to understand the possible functions of the FMR1 gene and are a long way from understanding its role in mental impairment. The timing and mechanism of the CGG expansion are currently the subject of much research. Population studies now suggest that the fragile X mutation, or at least a predisposition toward it, may be very old. Many useful comparisons can be drawn with the triplet array expansions in myotonic dystrophy, Huntington's disease and spinocerebellar ataxia type I, and the recent identification of a second triplet repeat associated with the fragile site FRAXE has supported their role in chromosome fragility.

9.2 DNA studies

9.2.1 FMR1 mapping and isolation

The FMR1 gene was isolated by positional cloning from a candidate region which had been defined by intensive molecular investigations, including genetic linkage data, long range physical mapping, chromosome fragmentation and fluorescence *in situ* hybridization (FISH; reviewed in Hirst *et al.*, 1992). In brief, a single CpG island identified in a gene-deficient region was found to be unusually hypermethylated on the fragile X chromosome, confirming its previously suspected role in the fragile X syndrome. Yeast artificial chromosome (YAC) contigs were initiated from several flanking markers, the CpG island was isolated and the gene transcribed from it, FMR1, identified (Verkerk *et al.*, 1991). The CpG island is coincident with the 5′ end of the FMR1 gene and is assumed to be the gene promoter, although such activity has yet to be investigated.

A CGG triplet within FMR1: dynamics of expansion. The first exon of FMR1 lies within the CpG island and, on the normal X chromosome, carries between six and 52 copies of a CGG repeat. On normal chromosomes these behave like other simple tandem repeats (STRs) commonly used in genetic linkage analysis; they are stably inherited and demonstrate normal Mendelian inheritance. In contrast, on the fragile X chromosome, the number of repeats is much higher and new alleles are generated with an extremely high mutation rate (Fu *et al.*, 1991). Over several generations, this copy number increases in an almost unidirectional fashion, giving ever longer arrays and eventually leading to the loss of gene expression (Pieretti *et al.*, 1991). The degree of expansion provides both a molecular explanation for the puzzling genetics of the fragile X syndrome and an assay for its accurate diagnosis.

On the fragile X chromosome, the FMR1 CGG array exists in two states dependent upon copy number (see Table 9.1). Individuals with an array longer than normal, but below 200 copies, are said to carry a 'premutation' chromosome. This sized array is found in NTMs and their daughters. It is somatically stable within the individual but is unstable upon genetic passage, with 95% of mutations resulting in array expansion (Fu *et al.*, 1991). Affected individuals carry an array usually greater than 200 copies. In this size range, the FMR1 promoter becomes hypermethylated, mimicking the inactivated X chromosome and chromosome fragility occurs. This larger mutant allele frequently exhibits length variation within an individual, resulting in multiple or heterogeneous alleles.

Conversion from the premutation to full mutation occurs exclusively in the maternal line. Daughters of NTMs, who are unaffected and cytogenetically negative for fragile X expression, inherit a premutation allele from their fathers,

Table 9.1. Table of CGG expansion sizes, phenotypic effects and detection assays

Expansion size (CGG number)	DNA methylation	Detection assay	FMR1 mRNA	Chromosome fragility	Phenotype
6–52	No*	PstI PCR	Yes	No	Normal
50–200†	No‡	PstI PCR EcoRI or HindIII§	Yes	No	Normal
>200	Yes	EcoRI HindIII BglII PCR¶	No	Yes‖	Affected **

* The FMR1 promoter and CGG array is subject to normal X inactivation methylation.
† The exact boundary between pre- and full mutation is not precisely defined, and may be better assayed by the presence of methylation on arrays larger than 150 copies.
‡ Approximately 16% of carrier females in this range express low levels of fragility.
§ This assay will detect over 90% of premutation sized arrays.
¶ PCR detection of the full mutation is in some cases unreliable (see text).
‖ All males are positive, but only 77% of females are positive.
** Only 50% of females are affected.

but passage to the next generation results in expansion and a risk of conversion to the full mutation. The probability of this conversion occurring is dependent upon the allele length in the mother, such that any premutation allele over 90 copies has a 100% conversion rate, whereas an allele of 70 copies will convert only 30% of the time (Fu *et al.*, 1991). Clearly, the existence of the non-phenotypic premutation, the ever-increasing expansions and associated conversion risks provide a molecular explanation for the Sherman paradox.

Detection of the mutation by Southern blot and PCR-based tests. CGG copy number can be estimated by studying the increase in size of restriction fragment above the normal baseline size (often symbolized as Δ; Hirst *et al.*, 1991a; Kremer *et al.*, 1991; Nakahori *et al.*, 1991; Oberle *et al.*, 1991; Verkerk *et al.*, 1991; Yu *et al.*, 1991; Knight *et al.*, 1992). An example of this is shown in Figure 9.1. This shows a HindIII digestion of the peripheral blood lymphocytes from several normal and affected individuals. The normal fragment is 5.2 kb, and premutation arrays can be detected as discrete fragments of decreased mobility such as is seen in track 5. Premutation carrier females carry two discrete fragments, one from their normal X chromosome (tracks 3 and 11). The full mutation is present in two individuals as a heterogeneous smear of fragments, sometimes difficult to resolve on Southern blots (tracks 1 and 8). To aid resolution of the mutations, various restriction enzymes are used to provide a window of resolution based upon fragment size and array length (see Table 9.1). These can be used to size an array accurately within any class of expansion and can be used as predictive tests. Methylation may be assayed by performing a second restriction digest with any of the enzymes which cut within the CpG island, most commonly BssHII and EagI.

Polymerase chain reaction (PCR) tests have been developed to complement genomic restriction analysis (Pergolizzi *et al.*, 1992). Whilst they offer a rapid test, problems in efficient amplification of large full mutation arrays occur, particularly in the presence of smaller alleles (as in mosaics), requiring verification of some samples by the Southern blot methods. Additionally, the sizing of amplified products is hampered by heteroduplexes and secondary structure formation giving abnormal migration patterns. For normal and small array lengths, however, PCR is the optimum test and products can be sized accurately on standard denaturing sequencing gels, which removes the secondary structure problems. Figure 9.2 shows an example of such an amplification and illustrates polymorphism of the CGG array and its use for tracking normal alleles within families. In this family, two normal alleles confirm that the sister of the carrier female is not a carrier, and the normal segregation of the CGG_{19} is shown through her son to her granddaughters. Shadow bands, which are a feature of STR-PCR,

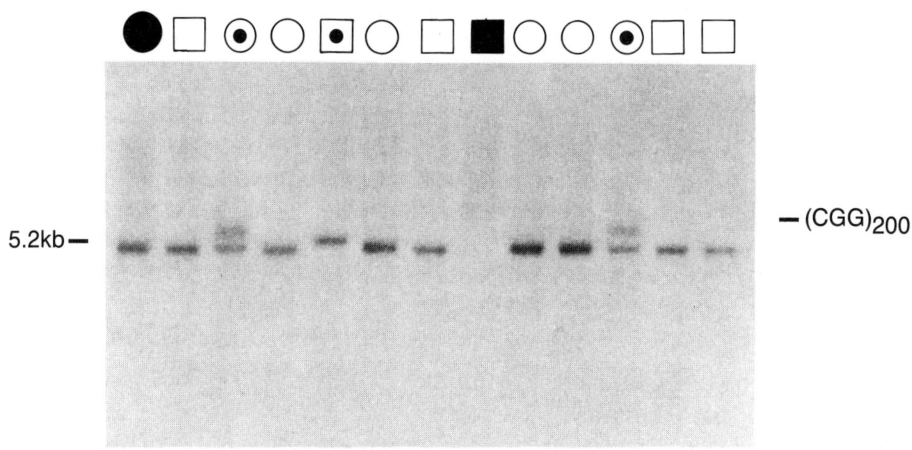

Figure 9.1. Southern blot analysis of FMR1 CGG expansions. Genomic lymphocyte DNA digested with HindIII and probed with DNA probe Ox1.9 gives a baseline fragment of 5.2 kb. Individuals with expansions at the fragile X CGG array have fragments of larger size which migrate higher up the gel. The position of the transition point to the full mutation $(CGG)_{200}$ is also shown. Carrier individuals have small increases in size, whereas affected individuals have a smear of hybridization over 200 copies in size corresponding to heterogeneity of array length between cells. ○ = normal female; □ = normal male; ● = affected female; ■ = affected male. Dots inside symbols indicate a premutation carrier individual.

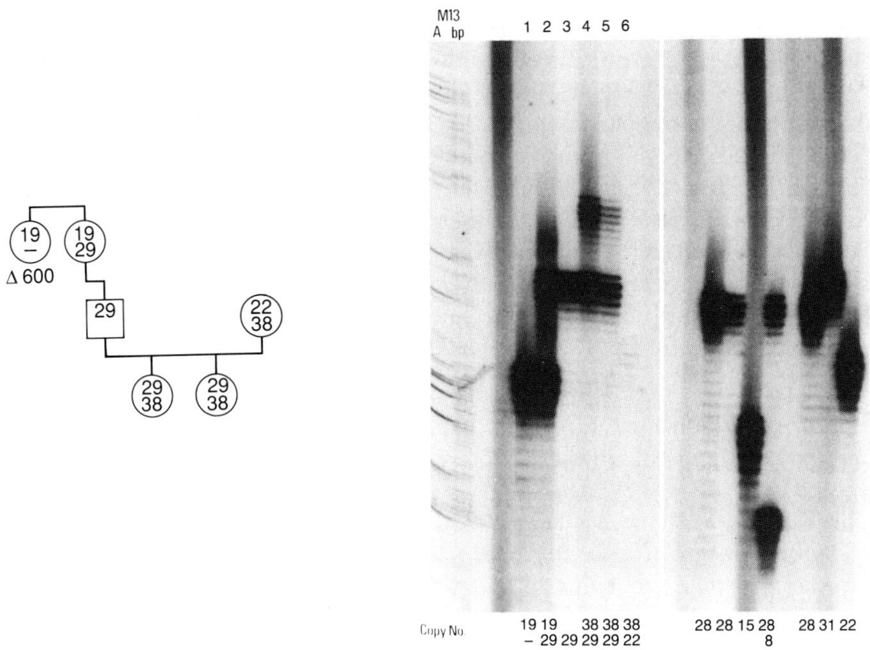

Figure 9.2. PCR analysis of a normal and a premutation allele at the CGG array. PCR across CGG arrays within the normal population range demonstrates the variation between individuals and its use to track the normal allele within families. A carrier female is shown (track 1) with the amplification of only her normal allele (19 copies). Δ symbolizes her fragile X premutation array (in base pairs) as judged by Southern blot analysis. The identification of two normal alleles in her sister (track 2) and the stable segregation of her $(CGG)_{29}$ allele is also shown (tracks 3, 4 and 5). The alleles contributed by the individual in track 6 are visible on a longer exposure (not shown). In the right hand panel, variation in CGG array length is shown in seven unrelated individuals. M13 is a size marker used to estimate PCR product size. PCR analysis was carried out as described by Fu *et al.* (1991).

show a three base separation, confirming that the variation in allele size is due to a variable copy number of the CGG triplet.

In diagnostic use, these assays provide an accurate test capable of identifying cytogenetically negative carriers. However, the correlation of genotype to phenotype is complicated by several factors. Where an individual carries several classes of CGG expansion, termed mosaicism, the degree of mental impairment is highly variable and is most likely a reflection of the body distribution of these two expansion classes. Similarly, mosaicism can occur with methylation, with full mutations undergoing variable methylation throughout the body (McConkie-Rosell *et al.*, 1993). These types of mosaicism are the most likely factors underlying the variable clinical phenotype of the fragile X syndrome. Secondly, X-inactivation within the female heterozygote carrying a full mutation can give rise to unaffected or affected individuals. Presumably, in this case, the proportion of cells in which the mutant chromosome is the 'active' chromosome and the distribution of these cells within the body (e.g. the brain) will greatly influence the penetrance of the phenotype.

One additional result of the large-scale screening of fragile X cohorts has been the identification of a small number of individuals who carry other fragile sites in Xqter, FRAXE and FRAXF (Nakahori *et al.*, 1991; Oberle *et al.*, 1992; Sutherland and Baker, 1992; Flynn *et al.*, 1993; Hirst *et al.*, 1993a). As these individuals were ascertained and subject to cytogenetic investigation because of mental impairment, their causative role in XLMR is still under investigation, although the case for FRAXE is quite strong (Knight *et al.*, 1993).

Hypermethylation of FMR1 in the full mutation. On the full mutation chromosome, hypermethylation is present across the FMR1 CpG island, including the CGG array (Bell *et al.*, 1991; Vincent *et al.*, 1991; Hansen *et al.*, 1992). This occurs when a threshold of CGG copy number is reached, and appears to mimic the process of X-inactivation, its presence correlating with loss of FMR1 expression. The examination of fetal tissue and chorionic villus (CV) DNA for methylation has shown that it is present from an early stage of development (Hirst *et al.*, 1991b; Rousseau *et al.*, 1991; Sutherland *et al.*, 1991; Worhle *et al.*, 1991a). Its presence in CV material is highly variable, causing problems for a predictive diagnosis in the case of a male fetus carrying a large premutation. In this situation it is advisable to assay a more representative DNA sample from fetal blood or an amniocentesis to ensure that no methylation is present. Where FMR1 transcription was assayed, expression was present in the unmethylated CV material but not in methylated fetal tissue (Sutcliffe *et al.*, 1992). This shows that the full mutation array alone does not interfere with transcription, and that methylation correlates with repression of FMR1 transcription.

9.2.2 The FMR1 gene, its structure and expression

The FMR1 gene covers 38 kb and consists of 17 exons (Eichler *et al.*, 1993). It has an unusually large first intron of 9.9 kb and the last exon, carrying the 17th coding exon and a long 3′ UTR, is 1.9 kb in length (see Figure 9.3). The first exon is relatively long (318 bp) when compared with other genes and is very GC rich, suggesting an involvement with post-transcriptional control. The CGG array lies within this first exon. The size of protein in NTMs, which would be predicted to be increased in size if the array was translated into a poly-arginine tract, is unaltered (Devys *et al.*, 1993; Verheij *et al.*, 1993). The conservation of an initiator methionine in the murine gene (Sutcliffe *et al.*, 1993) suggests that translation initiates 3′ to the CGG. Thus, most of the first exon remains untranslated and only contributes 17 amino acids of the mature protein. The gene is expressed predominantly as a 4.4 kb mRNA which translates into a 614 amino acid polypeptide of 69 kDa.

FMR1 mRNA is found at high levels in several adult tissues including the

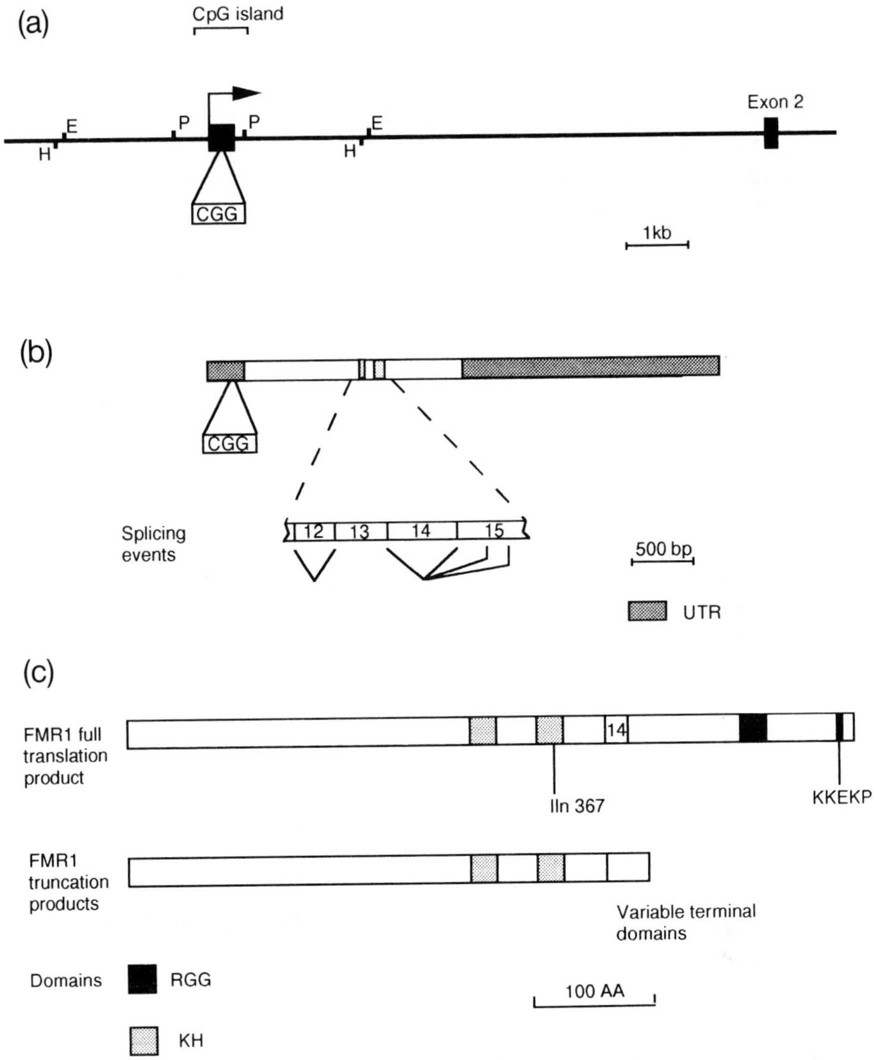

Figure 9.3. Structure and coding regions of the FMR1 gene. (a) Genomic structure of the 5′ region of the FMR1 gene showing the position of exons 1 and 2, the direction of transcription, the CGG repeat array, CpG island region and restriction enzyme sites used for diagnostic digests. H, HindIII; P, PstI; E, EcoRI. (b) Map of the FMR1 full length mRNA showing the proportion of the message translated into polypeptide. The positions of exons 12 and 14 are shown and the alternative splice events are shown enlarged below. Based upon data in Verkerk *et al.* (1993), Ashley *et al.* (1993) and Eichler *et al.* (1993). (c) Map of the FMR1 protein based upon the data of Ashley *et al.* (1993) and Siomi *et al.* (1993). RGG boxes are from amino acids 526 to 552 (note corrected from original manuscript); KH domains are from amino acids 286 to 321 and 347 to 382 (Siomi *et al.*, 1993). Amino acid iln-367 is the location of a point mutation found in a mentally retarded patient (De Boulle *et al.*, 1993) and lies within the highly conserved region of the second KH domain. The putative nuclear localization signal (KKEKP) is at amino acids 597–601. Alternative exon 12 deletes amino acids 375–396. Exon 14, which is also alternatively spliced in the mouse, represents the start of alternative phase isoforms of FMR1.

brain, lung, placenta, kidney and testes (Verkerk *et al.*, 1991; Hinds *et al.*, 1993). In human fetal brain, FMR1 is expressed in the proliferating and migrating cells of the nervous system, the retina and several non-nervous tissues at 8–9 weeks, and at 25 weeks in nearly all differentiated structures, highest in cholinergic neurons and the pyramidal neurons of the hippocampus (Abitbol *et al.*, 1993). A second transcript of 1.2 kb in heart tissue has been described, but the relevance of this is unknown. Expression of the gene in peripheral blood lymphocytes has allowed a study of transcription in fragile X individuals carrying the pre- and full mutation. It is transcribed in premutation individuals, but is not expressed in affected individuals (Pieretti *et al.*, 1991), a finding confirmed by the absence of FMR1 protein upon antibody investigations of full mutation males (Siomi *et al.*, 1993).

FMR1 has several sites of alternative splicing within the mRNA which have been variously studied by cDNA cloning and by PCR analysis (see Figure 9.3b) (Ashley *et al.*, 1993; Eichler *et al.*, 1993; Verkerk *et al.*, 1993). Exclusion of exon 12 removes a major hydrophobic segment of the protein. Other splicing events removing exon 14 use alternative splice acceptor sites within exon 15 in preference to the normal exon 15 acceptor site. Skipping of exon 14 would lead to the alteration of the FMR1 reading frame so, if these sites are used, polypeptides with variant carboxy-terminal ends will be produced (see Figure 9.3c). Several of these isoforms do not carry the putative nuclear localization signal or the RGG boxes (see below and Figure 9.3c) and may have an altered cellular location and function.

Little is known about the FMR1 protein, as comparisons with other proteins show only weak similarities. Multiple FMR1 protein isoforms have been detected (Devys *et al.*, 1993; Verheij *et al.*, 1993), but how these relate to the alternative splice events discussed above is not known. The FMR1 protein carries a putative nuclear localization signal, suggesting a role in the nucleus. It also contains two types of motifs found within RNA binding proteins; the RGG box and KH domain (Gibson *et al.*, 1993; Siomi *et al.*, 1993) (see Figure 9.3c). *In vitro* the protein demonstrates an RNA binding activity (Siomi *et al.*, 1993). RNA binding activity can influence many cellular processes including the regulation of mRNA splicing, mRNA stability, translation and compartmentalization. Studies of the protein distribution by immunohistochemistry show that it is predominantly cytoplasmic and is present in high levels in neurons and in dividing layers of epithelium (Devys *et al.*, 1993). This study also demonstrated that, in adult testes, FMR1 expression is found in spermatogonia suggesting a possible role in gametogenesis, a role which may be relevant to the timing of the array expansion events (see Section 9.4). None of the protein studies have been directed toward discriminating the possible FMR1 protein isoforms.

Comparative studies in the mouse. Mapping of the mouse FMR1 gene confirmed that it lay in the expected syntenic region of the mouse X chromosome (Faust *et al.*, 1992; Laval *et al.*, 1992). The gene shows a 95% homology at the cDNA level and 99% at the protein level with the human gene (Ashley *et al.*, 1993). Six copies of the CGG repeat array lie within the 5′ UTR, suggesting a conserved function for this element. Mapping of the sites of expression of the mouse FMR1 gene in the brain have indicated that it is expressed in such areas as the granular layers of the cerebellum and hippocampus, the cerebral cortex and the habenulae (Hinds *et al.*, 1992).

9.2.3 *Other mutations in FMR1 confirm its causative role in the fragile X phenotype*

Confirmatory evidence that FMR1 gene mutations cause the fragile X syndrome has come from studying males who are cytogenetically negative but who have the full clinical features of the fragile X syndrome. A *de novo* microdeletion removing proximal flanking DNA and the first four exons of the FMR1 gene, including the CpG island and the CGG array, was found in a mentally retarded male (Worhle *et al.*, 1991b). In the absence of the promoter region he probably does not express the FMR1 gene. In the second example, the deletion spans over 2500 kb of DNA and completely removes the gene (Gedeon *et al.*, 1992). A point mutation converting isoleucine-367 to an asparagine residue has been identified in an individual with some of the features of fragile X syndrome (De Boulle *et al.*, 1993). These features are more severe than found in the two deletion cases discussed above; a very low IQ (20) and extreme macroorchidism. This mutation lies within the highly conserved amino acids of the KH domain and is suggested to alter the putative RNA binding activity of the protein (Siomi *et al.*, 1993). Whether this single point mutation is enough to give rise to this severe phenotype, or whether he carries an additional alteration elsewhere in the gene has yet to be determined. Clearly the further identification of similar cases will greatly add to our understanding of FMR1 gene function.

9.3 Population dynamics of the fragile X mutation

Population studies suggest that the fragile X syndrome carrier rate could be as high as 1/800 (Brown, 1990). With low reproductive fitness in full mutation individuals, the mutation is constantly lost from the population, suggesting that either carriers may have a selective advantage (Vogel *et al.*, 1990) or that new mutations are common (Sherman *et al.*, 1984). Haplotype analysis with flanking $(AC)_n$ markers demonstrated linkage disequilibrium on the fragile X chromosome (Richards *et al.*, 1991; Hirst *et al.*, 1993b; Jacobs *et al.*, 1993; Oudet *et al.*, 1993a,b), which is highly suggestive that most current fragile X mutations descend from a small number of progenitor founder mutations. As no sporadic case of FMR1 expansions has been found, this suggests that the mutation is being maintained in the population in the form of a non-phenotypic mutation. How then do we reconcile the observations of an old mutation but with an apparently high new mutation rate? One attractive model which has been proposed, suggests that both these observations are correct and are linked in a multi-step process (Morton and MacPherson, 1992).

It is suggested that CGG arrays of larger number (perhaps 40–50) arise from normal alleles, possibly by sister chromatid exchange (see Section 9.5). These are non-phenotypic and would be stable for many generations. To account for the observed linkage disequilibrium, they must have arisen only on a particular haplotype, or were in themselves the rare founder events; in either case they exist long enough for an ancestral haplotype to be established. It is suggested that to account for the apparent high new mutation rate, these arrays would then convert at a frequency of 1–2% per generation to premutations, and then on to full mutations within four generations. Thus, a pool of predisposed carriers would constantly renew the pool of premutations lost from the population. No 'transition' from the top of the normal range into the premutation range has yet been observed, although early data seem to confirm that individuals with the high risk haplotype are more likely to have a higher CGG copy number (Richards *et al.*, 1992; Jacobs *et al.*, 1993). Some time in the future, correlating 'at risk' haplotypes and CGG array length may allow a predictive risk assessment of such individuals.

9.4 CGG expansion: when is it occurring?

CGG copy number mutations either maintain the premutation stage or convert to the full mutation. When these events actually occur, however, is not really known. Within the premutation range, the array length is clonal between individual cells of the body, but that array length is usually different from that present in the previous generation. In contrast, in the full mutation range the array differs both from the parental genotype and between individual cells in the body. Whilst initial studies interpreted these changes as being either meiotic or somatic events, there is growing evidence to support the idea that both these events occur post-fertilization.

9.4.1 Timing of expansion: germline or somatic?

Several lines of evidence suggest that conversion is a post-fertilization event (Nelson and Warren, 1993). Firstly, as mosaic individuals are a mixture of the pre- and the full mutation, this implies that the mutation is inherited as a premutation and converts to the full mutation in early embryogenesis. Mosaics arise when conversion does not occur early enough to be present in all progenitor cell lineages. Secondly, children of the few affected males who have successfully reproduced inherit a premutation sized allele (Willems *et al.*, 1992), and mothers carrying the full mutation in somatic tissue have also given rise to mosaic males (Rousseau *et al.*, 1991). This all suggests that the germline does not carry the full mutation. To address this question, sperm from full mutation males were examined and only premutation arrays were found (Reyneirs *et al.*, 1993). Oocytes of fragile X females have yet to be examined. Is this a process of selection, perhaps related to an essential function of FMR1 in gametogenesis, or are germ cells and their progenitors protected in some way from the conversion event? This argument is substantiated by the discovery that FMR1 is highly expressed in adult spermatogonia (see Section 9.2). The nature of the premutation allele in the sperm of an NTM may provide crucial information. This would reveal whether, in the sperm from such a male, the CGG array is of identical length or whether new mutations are occurring during the development of the male germ cells as is the case in Huntington's disease (Duyao *et al.*, 1993).

When is this conversion occurring in development? Samples from the first trimester show that the arrays have already undergone conversion to the full mutation (Hirst *et al.*, 1991b; Rousseau *et al.*, 1991; Sutherland *et al.*, 1991), and in the later stage fetus complete expansion and methylation occurs in all tissues (Worhle *et al.*, 1991). Monozygotic fragile X twins have mutations that are almost identical, suggesting that the process of expansion is occurring early in development before the twinning separation event (Devys *et al.*, 1992). A model for such early mutational STR mutations can be found with the mouse tetranucleotide repeat Hm-2 (GGCA; Gibbs *et al.*, 1993). New mutations occur in early pre-implantation development, particularly in the first two cell divisions after fertilization, and in several cases these have entered the germline, suggesting that the window of mutation is prior to the differentiation of the somatic and germline lineages.

To address the question further, studies have shown that full mutation FMR1 CGG array lengths in clonal fetal-derived cell lines are stable in culture (Worhle *et al.*, 1993), suggesting that changes in the array, including the conversion to the full mutation, once established, are quite stable. When taken together, these data imply that expansion is allowed to occur only within a short window of time in early embryological development. Once through this window, no further expansions occur. It will be of great interest to find out whether this window is closed by a cellular process which imposes stability via methylation of the CGG array.

9.4.2 Conversion is exclusive to the maternal lineage

Family studies appear to show that FMR1 arrays entering into fertilization behave differently dependent upon parental origin. Is this the result of an epigenetic parental effect, perhaps related to X-inactivation or due to the presence of two X chromosomes? As a similar effect is seen for the hypermutable minisatellites and in the expansion of other triplet repeats to the severe form, a common mechanism may underlie its nature. Alternatively, as we have already discussed, it may be a consequence of the exclusion of sperm bearing full mutation FMR1 arrays.

An understanding of the fate of cells within the developing embryo, particularly the progenitor germ cells, will be necessary to understand fully the timing of the instability of the mutations in the FMR1 array. As human female germline analysis is hampered by difficulties in obtaining material, early post-fertilization studies will be limited to the mouse or to human IVF studies. Many of these questions could be answered if we replicate *in vivo* mutations by the introduction of mutation sized arrays into early embryological cell lines such as ES (embryonic stem) and EG (embryonic germline progenitor) cells.

9.5 Mechanisms of mutation

Dramatic expansions to many times the size of the original array length are unseen in other STRs such as $(CA)_n$. They are restricted to the repeat itself, with no accompanying flanking elements being co-amplified, thus ruling out most models of amplification, for example those invoked to explain the amplification of the DHFR gene (reviewed by Stark *et al.*, 1989). Undoubtedly, expansion occurs because of the repetitive DNA and the way in which it escapes the normal checking processes within the cell. The most likely mechanisms invoke errors in the normal DNA processes of replication, genetic exchange and repair due to this repetitive nature. Mutations at the human minisatellite locus D1S8 are thought to happen in a similar manner (Jeffreys *et al.*, 1990). A second stage in the mutation pathway at FMR1 is methylation. As its imposition is only found with triplets carrying a CpG dinucleotide, it may well reflect the unusual nature of a CpG-bearing triplet.

9.5.1 Triplet expansion: errors in DNA processing

Sister chromatid exchange (SCE) can lead to unequal exchanges between repeat arrays (Latt, 1981; see Figure 9.4). However, the nature of this straightforward exchange does not account for the high mutation rate in the arrays. As the products of such exchanges are of complementary size, this would not account for 95% of changes being increases, unless the smaller arrays were selected against. By a similar mechanism of chromatid interaction, it may be possible for expansion to occur by inter-chromatid gene conversion. Additionally, SCE may have a role in generating larger predisposing alleles (n=40–50) and flanking elements may predispose to such events. A more likely mechanism of expansion is slippage synthesis.

Slippage synthesis of DNA occurs when an extending DNA strand becomes unpaired from the template strand and, because of the repetitive nature of the DNA, reanneals out of phase with the template (Levinson and Gutman, 1987; see Figure 9.4). This type of error may occur during replication or DNA repair. If the strand has a propensity to form a single-stranded DNA hairpin, this may increase its occurrence. To account for the proportions of expansion, the array must be repaired by fill-in, rather than excision. *In vitro* slippage assays have shown that short arrays of simple repeats are capable of slippage extension to give large arrays, although the repeat with the lowest slippage rate was the CGG repeat element (Schlotterer and Tautz, 1992). How well these *in vitro* models reflect *in vivo* situations is as yet unknown, but some attempts to study this problem have been made in yeast systems. STRs, being propagated in strains of yeast defective in mismatch repair, are highly unstable, with new array lengths being generated

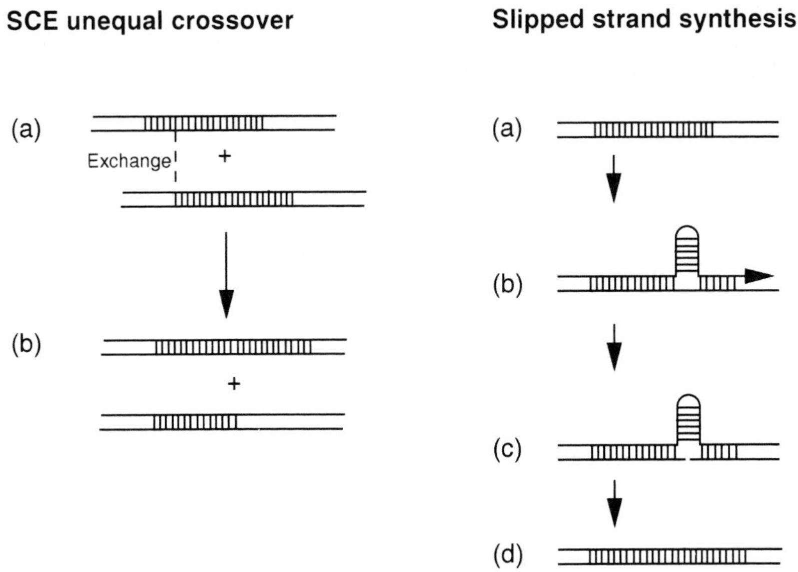

Figure 9.4. Schematic representation of SCE and slippage.
SCE unequal crossover. Two newly replicated chromatids are shown misaligned at a repeat array with 19 copies of repeat (a). If genetic exchange then occurs between the two misaligned pairs (b), new arrays of 13 and 25 copies are generated. Note that, as the exchange is reciprocal, one array expands as the other contracts. Additionally, as these are events between two newly replicated chromatids (i.e. intra-allelic), no flanking markers are affected.
Slipped strand synthesis. A double-stranded DNA strand is shown with a repeat array of 19 units (a). Upon extension of a strand, for example at replication, the newly extending strand becomes partly dissociated from the template and reanneals out of phase due to the repetitive nature of the template (b). In this example, this is exacerbated by the propensity of the strand to form a hairpin loop. Upon completion of the polymerization, repair of the loop is initiated with a nick on the opposite strand, followed by template-driven repair (d). The resulting array has increased its copy number to 25. Clearly, any event involving polymerization is susceptible to this type of error. Multiple events on the same strand may also occur.

by strand slippage during replication (Strand *et al.*, 1993).

If the mechanism driving repeat expansions at FMR1 is only present or tolerated in early embryogenesis it may be related to the processes of post-fertilization gene activation. This may be a specific factor present only at this stage of development or due to chromatin changes occurring in the newly formed diploid cell.

9.5.2 Methylation

Methylation may well be a downstream effect of altered chromatin structure or replication timing pattern. It appears to be imposed upon full mutation arrays as cells differentiate (Sutcliffe *et al.*, 1992) and may well mimic X-inactivation. Alternatively, methylation may be a cellular response to stabilize the unusual nature of the CGG repeat (Smith, 1991; Smith *et al.*, 1991) or it may represent a mechanism similar to MIP (methylation induced premeiotically; Rhounim *et al.*, 1992) or RIP (repeat induced point mutation; Selker, 1990), processes which are involved in the functional inactivation of repeated sequences. Indeed, such a mechanism may have played a role in the evolution of eukaryotic chromosomes (Kricker *et al.*, 1992). The enzymes responsible for *de novo* methylation are present at high levels in the pre-implantation mammalian embryo (for review see Adams, 1990).

As methylation is imposed upon array lengths which are extremely unstable, it may be that it is the stabilizing factor. The full mutation arrays which were demonstrated to be mitotically stable by Worhle *et al.* (1993) were indeed methylated. Experiments to dissect this further will most likely be carried out using mice carrying expanded CGG arrays. As murine X-inactivation is imposed at different times in different tissues (Tan *et al.*, 1993), the timing of methylation with regard to X-inactivation can be directly assessed.

9.5.3 Mechanism of chromosome fragility

Expansion of the CGG repeat and its subsequent methylation are involved intimately with chromosome fragility. This is most likely a specific property of chromatin behaviour induced by the methylation. Replication studies have shown that the full mutation FMR1 region replicates much later in S phase than usual (Hansen *et al.*, 1993), a characteristic of X-inactive chromatin. Laird *et al.* (1987) suggested that such behaviour on an otherwise normally replicating X chromosome would result in a failure to undergo correct chromatin condensation, leading to the cytological appearance of a fragile site. Further investigation of X-inactivation-related chromatin changes, such as histone H4 acetylation (Jepperson and Turner, 1993), and studying reactivation alteration (Sasaki *et al.*, 1992), will give useful insights into the state of the local chromatin structure around the full mutation FMR1.

9.5.4 The effect of the CGG expansion upon other genes

The activity of the IDS gene (iduronate sulphatase), which lies 1.6 Mb distal to FMR1, is decreased in activity by 25% on the full mutation chromosome (Clarke *et al.*, 1992). Whether this is a long-range effect of the CGG expansion, for example mediated through an alteration in local chromatin function, or a secondary effect of the fragile X phenotype is as yet unknown. Further investigation into the effect of CGG expansion on other nearby genes is necessary. Such effects, for example mediated through the gene at the FRAXE site, may contribute to the phenotypic variability of the fragile X phenotype.

9.6 Genome instability: other triplet repeat expansions

Expansions of triplet arrays have been found at a growing number of disease loci (see Table 9.2). Several features are common to these expansions (Mandel, 1993). All show a sex bias in the parental origin of large expansions leading to the most severe forms of the disease and most show a degree of genetic anticipation. They contrast in the degree of expansion which occurs and can be classified into two categories (see Table 9.2).

9.6.1 Large expansions: interference with gene expression

Triplets at the myotonic dystrophy (DM), fragile X and FRAXE loci frequently expand to several thousands of copies (Brook *et al.*, 1992; Fu *et al.*, 1991; Knight *et al.*, 1993). In DM and FMR1, the arrays lie within untranslated domains, and expansion alters the level of expression. With FRAXE (Knight *et al.*, 1993), no gene has yet been identified, but predictions based upon the nature of the expansion suggest it will behave like FMR1 and DM. In common with FMR1, methylation is imposed across the associated CpG island at a certain threshold, giving a similar discontinuous relationship between the expansion size and the phenotype.

9.6.2 Limited expansions: alteration of function

In Kennedy's disease, Huntington's disease and spinocerebellar ataxia type I (SCA-I), discrete expansions of a CAG encoding a polyglutamic acid stretch occur (La Spada and Fischbeck, 1991; Huntingtons Disease Collaborative Research Group, 1993; Orr *et al.*, 1993). These small expansions probably result in a protein with a altered function or one which interferes with another cellular process, leading to the death of specific neural cells. Other mutations which destroy the function of the protein may give rise to a completely different phenotype. For example, point mutations destroying the androgen receptor protein function lead to androgen insensitivity and testicular feminization (Jakubiczka *et al.*, 1992; Lobaccaro *et al.*, 1993).

Anonymous triplet arrays are being isolated to find novel expansions within coding regions as a method of searching for candidate disease genes (Li *et al.*, 1993). The suggestion, however, that all such events will be uniquely associated with neuropsychiatric disorders (Ross *et al.*, 1993) may simply reflect the number of genes which are expressed within these body systems and the bias in investigations into common neurological disorders in man. As polymorphism can only be tolerated within a coding region with triplet repeats, such expansions may allow the evolution of new functional variants of a protein to arise without loss of function. Thus, triplet arrays provide a previously unsuspected unit of genetic and functional variation between individuals.

Table 9.2. Triplet expansions at other disease loci. Triplet expansions found at these loci are divided into two categories dependent upon expansion size and the effect upon gene activity

Triplet	Position	Expression	Disease gene	Normal	Disease	Other effects
CGG/CCG	CpG island 5′ UTR	Loss of expression	FMR1 FRAXA site	6–52	50–150/200 >150/200	Hypermethylation Chromosome fragility. Late replication
GCC/GGC	CpG island 5′ UTR?	?	FRAXE site Gene?	6–25	130–850	Hypermethylation
CTG/CAG	3′ UTR	Suppression? Stability?	Myotonic dystrophy	5–37	50–1600	
CAG/CTG	Poly-Glu tract in androgen receptor	Altered function? Interference?	Spinal bulbar muscular atrophy	13–30	40–62	
CAG/CTG	Poly-Glu in Huntington's		Huntington's disease	12–34	37–86	
CAG/CTG	Poly-Glu? 10 kb mRNA		Spinocerebellar ataxia type 1 (SCA-1)	19–36	43–81	

9.7 Conclusions

The identification of the triplet array expansion at FMR1 has added greatly to our understanding of the genetics of the fragile X syndrome. Whilst we have yet to understand the function of the FMR1 protein and questions still exist about how and when these expansions occur, the detection of these events has added greatly to diagnostics. Clearly, future studies of unstable triplet arrays at other loci and at FMR1 will contribute greatly to our understanding of the human genome. Within these future realms lies access to an understanding of chromosome structure, X-inactivation, gene regulation and developmental switching as well as the further investigation of the most common cause of XLMR.

Acknowledgements

I would like to thank Professor Kay Davies and my colleagues within the IMM for many fruitful discussions. I would also like to acknowledge funding from the Wellcome Trust.

References

Abitbol M, Menini C, Delezoide AL, Rhyner T, Vekemans M, Mallet J. (1993) Nucleus basalis magnocellularis and hippocampus are the major sites of FMR-1 expression in the human fetal brain. *Nature Genetics* **4: 147–153.**

Adams RLP. (1990) DNA methylation. *Biochem. J.* **265:** 309–320.

Ashley CT, Sutcliffe JS, Kunst CB, Leiner HA, Eichler EE, Nelson DL, Warren ST. (1993) Human and murine FMR-1 alternative splicing and initiation downstream of the CGG-repeat. *Nature Genetics* **4:** 244–251.

Bell MV, Hirst MC, Nakahori Y, MacKinnon RN, Roche A, Flint TJ, Jacobs PA, Tommerup N, Tranjebaerg L, Froster-Iskenius U, Kerr B, Turner G, Lindenbaum R, Winter R, Pembrey M, Thibodeau S, Davies KE. (1991) Physical mapping across the fragile X: hypermethylation and clinical expression of the fragile X syndrome. *Cell* **64:** 861–866.

Brook JD, McCurrach ME, Harley HG, Buckler AJ, Church D, Aburatani H, Hunter K, Stanton VP, Thirion JP, Hudson T, Sohn R, Zemelman B, Snell RG, Rundle SA, Crow S, Davies J, Shelbourne P, Buxton J, Jones C, Juvonen V, Johnson K, Harper PS, Shaw DJ, Housman DE. (1992) Molecular basis of myotonic dystrophy: expansion of a trinucleotide (CTG) repeat at the 3′ end of a transcript encoding a protein kinase family member. *Cell* **68:** 799–808.

Brown WT. (1990) The fragile X: progress toward solving the puzzle. *Am. J. Hum. Genet.* **47:** 175–180.

Clarke A, Bradley D, Gillespie K, Rees D, Holland A, Thomas NST. (1992) Fragile X mental retardation and the iduronate sulphatase locus. *Am. J. Med. Genet.* **43:** 299–306.

De Boulle K, Verkerk AJMH, Reyniers E, Vits L, Hendrickx J, Van Roy B, Van Den Bos F, de Graaf E, Oostra BA, Willems PJ. (1993) A point mutation in the FMR-1 gene associated with fragile X mental retardation. *Nature Genetics* **3:** 31–35.

Devys D, Biancalana V, Rousseau F, Boue J, Mandel J. (1992) Analysis of full fragile X mutations in fetal tissues and monozygotic twins indicate that abnormal methylation and somatic heterogeneity are established early in development. *Am. J. Med. Genet.* **43:** 208–216.

Devys D, Lutz Y, Rouyer N, Bellocq JP, Mandel JL. (1993) The FMR-1 protein is cytoplasmic, most abundant in neurons and appears normal in carriers of a fragile X premutation. *Nature Genetics* **4:** 335–340.

Duyao M, Ambrose C, Myers R, Novelletto A, Persischetti F, Frontali M, Folstein S, Ross C, Franz M, Abbott M, Gray J, Conneally P, Young A, Penney J, Hollinsworth Z, Shoulson I, Lazzarini A, Falek A, Koroshetz W, Sax D, Bird E, Vonsattel J, Bonilla E, Alvir J, Bickham Conde J, Cha J-H, Dure L, Gomez F, Ramos M, Sanchez-Ramos J, Snodgrass S, de Young M, Wexler N, Moscowitz C, Penchaszadeh G, MacFarlane H, Anderson M, Jenkins B, Srinidhi J, Barnes G, Gusella J, MacDonald M. (1993) Trinucleotide repeat length and age of onset in Huntington's disease. *Nature Genetics* **4:** 387–392.

Eichler EE, Richards S, Gibbs RA, Nelson DL. (1993) Fine structure of the human FMR-1 gene. *Hum Mol. Genet.* **2:** 1147–1153.

Faust C, Verkerk A, Wilson P, Morris P, Hopwood J, Oostra B, Herman G. (1992) Genetic mapping on the mouse X chromosome of the cDNA clones for the fragile X and Hunter syndromes. *Genomics* **12:** 814–816.

Flynn GA, Hirst MC, Knight SJL, MacPherson JN, Barber JC, Flannery AV, Davies KE, Buckle VJ. (1993). Identification of the FRAXE fragile site in two families ascertained for X linked mental retardation. *J. Med. Genet.* **30:** 97–100.

Fryns JP. (1989) X-linked mental retardation and the fragile X syndrome: a clinical approach. In: *The Fragile X Syndrome* (ed. KE Davies). Oxford University Press, Oxford, pp. 1–39.

Fu Y-H, Kuhl DPA, Pizzuti A, Pieretti M, Sutcliffe JS, Richards S, Verkerk AJMH, Holden JJA, Fenwick RG, Warren ST, Oostra BA, Nelson DL, Caskey CT. (1991) Variation of the CGG repeat at the fragile X site results in genetic instability: resolution of the Sherman Paradox. *Cell* **67:** 1–20.

Gedeon AK, Baker E, Robinson H, Partington MW, Gross B, Manca A, Korn B, Poustka A, Yu S, Sutherland GR. (1992) Fragile X syndrome without CGG amplification has an FMR-1 deletion. *Nature Genetics* **1:** 341–344.

Gibbs M, Collick A, Kelly RG, Jeffreys AJ. (1993) A tetranucleotide repeat mouse minisatellite displaying substantial somatic instability during preimplantation development. *Genomics* **17:** 121–128.

Gibson T, Rice P, Thompson J, Heringa J. (1993) KH domains within the FMR1 sequence suggest that fragile X syndrome stems from a defect in RNA metabolism. *Trends Biochem. Sci.* **18:** 331–333.

Hansen RS, Gartler SM, Scott CR, Chen SH, Laird CD. (1992) Methylation analysis of CGG sites in the CpG island of the FMR1 gene. *Hum. Mol. Genet.* **1:** 571–578.

Hansen RS, Canfield TK, Lamb MM, Gartler SM, Laird CD. (1993) Association of fragile X syndrome with delayed replication of the FMR1 gene. *Cell* **73:** 1403–1409.

Hinds HL, Ashley CT, Sutcliffe JS, Nelson DL, Warren ST, Housman DE, Schalling M. (1993) Tissue specific expression of FMR-1 provides evidence for a functional role in fragile X syndrome. *Nature Genetics* **3:** 36–43.

Hirst MC, Nakahori Y, Knight SJL, Schwartz C, Thibodeau SN, Roche A, Flint TJ, Connor JM, Fryns J-P, Davies KE. (1991a) Genotype prediction in the fragile X syndrome. *J. Med. Genet.* **28:** 824–829.

Hirst M, Knight S, Cross G, O'Craft K, Raeburn S, Heeger S, Eunpu D, Jenkins E, Lindenbaum R, Davies K. (1991b) Prenatal diagnosis of the fragile X syndrome. *Lancet* **338:** 956–957.

Hirst MC, Knight SJL, Bell MV, Super M, Davies KE. (1992) The fragile X syndrome. *Clin. Sci.* **83:** 255–264.

Hirst MC, Barnicoat, A, Flynn G, Wang Q, Daker M, Buckle VJ, Davies KE, Bobrow M. (1993a) The identification of a third fragile site, FRAXF, in Xq27–q28 distal to both FRAXA and FRAXE. *Hum. Mol. Genet.* **2:** 197–200.

Hirst MC, Knight SJL, Christodoulou Z, Grewal PK, Fryns JP, Davies KE. (1993b) Origins of the fragile X syndrome. *J. Med. Genet.* **30:** 647–650.

Huntington's Disease Collaborative Research Group. (1993) A novel gene containing a trinucleotide repeat that is expanded and unstable on Huntington's disease chromosomes. *Cell* **72:** 971–983.

Jacobs PA, Bullman H, MacPherson J, Youings S, Rooney V, Watson A, Dennis NR. (1993) Population studies of the fragile X: a molecular approach. *J. Med. Genet.* **30:** 454–459.

Jakubiczka S, Werder EA, Wieacker P. (1992) Point mutation in the steroid binding domain of the androgen receptor gene in a family with complete androgen insensitivity syndrome. *Hum. Genet.* **90:** 311–312.

Jeffreys AJ, Neumann R, Wilson V. (1990) Repeat unit sequence variation in minisatellites: a novel source of DNA polymorphism for studying variation and mutation by single molecule analysis. *Cell* **60:** 473–485.

Jepperson P, Turner BM. (1993) The inactive X chromosome in female mammals is distinguished by a lack of histone H4 acetylation, a cytogenetic marker for gene expression. *Cell* **74:** 281–289.

Knight S, Hirst M, Roche A, Christodoulou Z, Huson S, Winter R, Fitchett M, McKinley M, Lindenbaum R, Nakahori Y, Davies K. (1992) Molecular studies of the fragile X syndrome. *Am. J. Med. Genet.* **43:** 213–217.

Knight SJL, Flannery AV, Hirst MC, Campbell L, Christodoulou Z, Phelps SR, Pointon J, Middleton-Price HR, Barnicoat A, Pembrey ME, Holland J, Oostra BA, Bobrow M, Davies KE. (1993) Trinucleotide repeat amplification and hypermethylation of a CpG island in FRAXE mental retardation. *Cell* **74:** 1–20.

Kremer EJ, Pritchard M, Lynch M, Yu S, Holman K, Baker E, Warren ST, Schlessinger D, Sutherland GR, Richards RI. (1991) Mapping of DNA instability at the fragile X to a trinucleotide repeat sequence $p(CCG)_n$. *Science* **252:** 1711–1718.

Kricker M, Drake J, Radman M. (1992) Duplication-targetted DNA methylation and mutagenesis in the evolution of eukaryotic chromosomes. *Proc. Natl Acad. Sci. USA* **89:** 1075–1079.

La Spada AR, Fischbeck KH. (1991) Variant androgen receptor gene in X-linked spinal and bulbar muscular atrophy. *Nature* **352:** 77–79.

Laird CD, Jaffe E, Karpen G, Lamb M, Nelson R. (1987) Fragile sites in human chromosomes as regions of late replicating DNA. *Trends Genet.* **3:** 274–281.

Latt SA. (1981) Sister chromatid exchange formation. *Annu. Rev. Genet.* **15:** 11–55.

Laval S, Blair H, Hirst M, Davies K, Boyd Y. (1992) Mapping of FMR1, the gene implicated in fragile X linked mental retardation, on the mouse X chromosome. *Genomics* **12:** 818–821.

Levinson G, Gutman GA. (1987) Slipped strand mispairing: a major mechanism for DNA strand evolution. *Mol. Biol. Evol.* **4:** 203–221.

Li SH, McInneis MG, Margolis RL, Antonarkis SE, Ross CA. (1993) Novel triplet repeat containing genes in human brain: cloning, expression and length polymorphism. *Genomics* **16:** 572–579.

Lobaccaro JM, Lumbruso S, Ktari R, Dumas R, Sultan C. (1993) An exonic point mutation creates a MaeIII site in the androgen receptor gene of a family with complete androgen insensitivity syndrome. *Hum. Mol. Genet.* **2:** 1041–1043.

Lubs HA. (1969) A marker X-chromosome. *Am. J. Hum. Genet.* **21:** 231–244.

Mandel JL. (1993) Questions of expansion. *Nature Genetics* **4:** 8–9.

McConkie-Rosell A, Lachiewicz A, Spirdigliozzi G, Tarleton J, Schoenwald S, Phelan M, Goonewardena P, Ding X, Brown W. (1993) Evidence that methylation of the FMR1 locus is responsible for variable phenotypic expression of the fragile X syndrome. *Am. J. Hum. Genet.* **53:** 800–809.

Morton NE, MacPherson J. (1992) Population genetics of the fragile X syndrome: multi-allelic model for the FMR1 locus. *Proc. Natl Acad. Sci. USA* **89:** 4215–4217.

Nakahori Y, Knight SJL, Holland J, Schwartz C, Roche A, Tarleton J, Wong S, Flint TJ, Froster-Iskenius U, Bentley D, Davies KE, Hirst MC. (1991) Molecular heterogeneity of the fragile X syndrome. *Nucleic Acids Res.* **19:** 4355–4359.

Nelson DL, Warren ST. (1993) Trinucleotide repeat instability: when and where? *Nature Genetics* **4:** 107–108.

Oberle I, Rousseau F, Heitz D, Kretz C, Devys D, Hanauer A, Boue J, Bertheas M, Mandel

JL. (1991) Instability of a 550 bp DNA segment and abnormal methylation in fragile X syndrome. *Science* **252:** 1097–1102.

Oberle I, Boue J, Croquette MF, Voelckel MA, Mattei MG, Mandel JL. (1992) Three families with high expression of a fragile site at Xq27.3, lack of anomalies at the FMR-1 CpG island, and no clear phenotypic association. *Am. J. Med. Genet.* **43:** 224–231.

Orr HT, Chung MY, Banfi S, Kwiatkowski TJ, Servadio A, Beaud AL, McCall AE, Duvick LA, Ranum LP, Zoghbi HY. (1993) Expansion of an unstable trinucleotide CAG repeat in spinocerebellar ataxia type 1. *Nature Genetics* **4:** 221–226.

Oudet C, Mornet E, Serre JL, Thomas F, Lentes-Zengerling S, Kretz C, Deluchat C, Tejada I, Boue J, Boue A, Mandel JL. (1993a) Linkage disequilibrium between the fragile mutation and two closely linked CA repeats suggests that fragile X chromosomes are derived from a small number of founder chromosomes. *Am. J. Hum. Genet.* **52:** 297–304.

Oudet C, von Huskell H, Nordstrom A, Peippo M, Mandel J. (1993b) Striking founder effect for the fragile X syndrome in Finland. *Eur. J. Hum. Genet.* **1:** 181–189.

Pergolizzi RG, Erster SH, Goonewardena P, Brown WT. (1992) Detection of the full fragile X mutation. *Lancet* **339:** 271–272.

Pieretti M, Zhang F, Fu Y-H, Warren ST, Oostra BA, Caskey CT, Nelson DL. (1991) Absence of expression of the FMR-1 gene in fragile X syndrome. *Cell* **66:** 817–822.

Reyniers E, Vits L, De Boulle K, Van Roy B, de Graffe E, Verkere AJMH, Jorens HZJ, Darby JK, Oostra BA, Willems PJ. (1993) The full mutation in the FMR-1 gene of male fragile X patients is absent in their sperm. *Nature Genetics* **4:** 143–146.

Rhounim L, Rossignol JL, Fougeron H. (1992) Epimutation of repeated genes in *Ascolobus immersus. EMBO J.* **11:** 4451–4457.

Richards RI, Holman K, Kozman H, Kremer E, Lynch M, Pritchard M, Yu S, Mulley J, Sutherland GR. (1991) Fragile X syndrome: genetic localization by linkage mapping of two microsatellite repeats *FRAXAC1* and *FRAXAC2* which immediately flank the fragile site. *J. Med. Genet.* **28:** 818–823.

Richards RI, Holman K, Friend K, Kremer E, Hillen D, Staples A, Brown WT, Goonewardena P, Tarleton J, Schwartz C, Sutherland GR. (1992) Evidence of founder chromosomes in fragile X syndrome. *Nature Genetics* **1:** 257–260.

Ross CA, McInnis MG, Margolis RL. (1993) Genes with triplet repeats: candidate mediators of neuropsychiatric disorders. *Trends Neurosci.* **16:** 254–260.

Rousseau F, Heitz D, Biancalana V, Blumenfeld S, Kretz C, Boue J, Tommerup N, Van der Hagen C, De-Lozier-Blanchet C, Croquette M-F, Gilgenkrantz S, Jalbert P, Voelckel M-A, Oberle I, Mandel J-L. (1991) Direct detection by DNA analysis of the fragile X syndrome of mental retardation. *N. Engl. J. Med.* **325:** 1673–1681.

Sasaki T, Hansen RS, Gartler SM. (1992) Hemimethylation and hypersensitivity are early events in transcriptional reactivation of human inactive X-linked genes in a hamster × human somatic cell hybrid. *Mol. Cell. Biol.* **12:** 3819–3826.

Schlotterer C, Tautz D. (1992) Slippage synthesis of simple sequence DNA. *Nucleic Acids Res.* **20:** 211–215.

Selker E. (1990) Premeiotic instability of repeated sequences in *Neurospora crassa. Annu. Rev. Genet.* **24:** 579–613.

Sherman SL, Morton NE, Jacobs PA, Turner G. (1984) The marker (X) syndrome: a cytogenetic and genetic analysis. *Ann. Hum. Genet.* **48:** 21–37.

Sherman SL, Jacobs PA, Morton NE, Froster-Iskenius U, Howard-Peebles PN, Nielsen KB, Partington MW, Sutherland GR, Turner G, Watson M. (1985) Further segregation analysis of the fragile X syndrome with special reference to transmitting males. *Hum. Genet.* **69:** 289–299.

Siomi H, Siomi MC, Nussbaum RL, Dreyfuss G. (1993) The protein product of the fragile X gene, FMR1, has characteristics of an RNA-binding protein. *Cell* **74:** 291–298.

Smith SS. (1991). DNA methylation in eukaryotic chromosome stability. *Mol. Carcinogen.* **4:** 91–92.

Smith SS, Kan JLC, Baker DJ, Kaplan BE, Dembek P. (1991) Recognition of unusual DNA structures by human DNA (cytosine-5) methyltransferase. *J. Mol. Biol.* **217:** 39–51.

Stark GR, Debatisse M, Giulotto E, Wahl GM. (1989) Recent progress in understanding mechanisms of mammalian DNA amplification. *Cell* **57:** 901–908.

Strand M, Prolla T, Liskay R, Petes T. (1993) Destabilisation of tracts of simple repetitive DNA in yeast mutations affecting DNA mismatch repair. *Nature* **365:** 274–276.

Sutcliffe JS, Nelson DL, Zhang F, Pieretti M, Caskey CT, Saxe D, Warren ST. (1993) DNA methylation represses FMR-1 transcription in fragile X syndrome. *Hum. Mol. Genet.* **1:** 397–400.

Sutherland GR. (1977) Fragile sites on human chromosomes: demonstration of their dependence on the type of tissue culture medium. *Science* **197:** 256–266.

Sutherland G, Baker E. (1992) Characterisation of a new rare fragile site easily confused with the fragile X. *Hum. Mol. Genet.* **1:** 111–113.

Sutherland GR, Gedeon A, Kornman L, Donnelly A, Byard RW, Mulley JC, Kremer E, Lynch M, Pritchard M, Yu S, Richards RI. (1991) Prenatal diagnosis of fragile X syndrome by direct detection of the unstable DNA sequence. *N. Engl. J. Med.* **325:** 1720–1722.

Tan SS, Williams EA, Tam PPL. (1993) X-chromosome inactivation occurs at different times in different tissues of the post-implantation mouse embryo. *Nature Genetics* **3:** 170–174.

Verheij C, Bakker CE, de Graaff E, Keulemans J, Willemsen R, Verkerk AJMH, Galjaard H, Reuser AJJ, Hoogeveen AT, Oostra BA. (1993) Characterisation and localisation of the FMR-1 gene product associated with fragile X syndrome. *Nature* **363:** 722–724.

Verkerk AJMH, Pieretti M, Sutcliffe JS, Fu Y-H, Kuhl DPA, Pizzuti A, Reiner O, Richards S, Victoria MF, Zhang F, Eussen BE, van Ommen G-JB, Blonden LJ, Riggins GJ, Chastain JL, Kunst CB, Galjaard H, Caskey CT, Nelson DL, Oostra BA, Warren ST. (1991) Identification of a gene (FMR-1) containing a CGG repeat coincident with a breakpoint cluster region exhibiting length variation in fragile X syndrome. *Cell* **65:** 905–914.

Verkerk AJMH, de Graaff E, De Boulle K, Eichler EE, Konecki S, Reyniers E, Manca A, Poustka A, Willems PJ, Nelson DL, Oostra BA. (1993) Alternative splicing of the fragile X gene FMR1. *Hum. Mol. Genet.* **2:** 399–404.

Vincent A, Heitz D, Petit C, Kretz C, Oberle I, Mandel J-L. (1991) Abnormal pattern detected in fragile X patients by pulsed field gel electrophoresis. *Nature* **329:** 624–626.

Vogel F, Crusio WE, Kovac C, Fryns JP, Freund M. (1990) Selective advantage of fra(X) heterozygotes. *Hum. Genet.* **86:** 25–32.

Webb T. (1989) The epidemiology of the fragile X syndrome. *The Fragile X Syndrome* (ed. KE Davies). Oxford University Press, Oxford, pp. 40–55.

Willems PJ, Van Roy B, De Boulle K, Vits L, Reyniers E, Beck O, Dumon JE, Verkerk A, Oostra B. (1992) Segregation of the fragile mutation from an affected male to his normal daughter. *Hum. Mol. Genet.* **1:** 511–515.

Worhle D, Hirst M, Davies K, Steinbach P. (1991a) Genotype variation in fragile X fetal tissues. *Hum. Genet.* **89:** 114–116.

Worhle D, Kotzot D, Hirst M, Manka A, Korn B, Schmidt A, Barbi G, Rott H-D, Poustka AM, Davies K, Steinbach P. (1991b) A microdeletion of less than 250 kb including the proximal part of the FMR-1 gene and the fragile site, in a male with the clinical phenotype of fragile X syndrome. *Am. J. Hum. Genet.* **51:** 299–306.

Worhle D, Hennig I, Vogel W, Steinbach P. (1993) Mitotic instability of fragile X mutations in differentiated cells indicates early post-conceptual trinucleotide repeat expansion. *Nature Genetics* **4:** 140–142.

Yu S, Pritchard M, Kremer E, Lynch M, Nancarrow J, Baker E, Holman K, Mulley JC, Warren ST, Schlessinger D, Sutherland GR, Richards RI. (1991) Fragile X genotype characterized by an unstable region of DNA. *Science* **252:** 1179–1181.

10

Somatic mosaicism, chimerism and X inactivation

Andrew O.M. Wilkie

10.1 Introduction

We are all somatic mosaics for countless mutations. That is the simple logic of the number of cells in the body (estimated at 10^{14}), the number of base pairs per diploid cell (6×10^9), the rate of cell division (10^7 sec^{-1}) and the base mutation rate per DNA replication (10^{-9}) (Prescott, 1988). Furthermore, programmed DNA rearrangements in lymphoid cells are responsible for immunological diversity and, in the female half of the species, X chromosome inactivation is required for dosage compensation. Thus it is not surprising that somatic mosaicism underlies a diverse range of phenotypes — ranging from commonplace examples such as moles, cancer and ageing, to more rarefied genetic phenomena like XX/XY hermaphroditism (a form of chimerism) and dominant lethal mutations that manifest only in mosaic form (Table 10.1). Although this diversity initially may appear confusing, a combined appreciation of the site and timing (Figure 10.1) and molecular mechanism of the mosaicism often enables a coherent picture of its phenotypic manifestation (Table 10.2) to be achieved. Each of these principal considerations will be discussed in turn; Hall (1988) has provided an excellent review of the subject, so this chapter emphasises more recent literature.

10.2 Site and timing of somatic mutation

The timing of mutation in relation to normal embryological processes is a crucial determinant of phenotype (Figure 10.1). The majority of cells in the blastocyst contribute to the trophoblast and other extra-embryonic tissues (Crane and Cheung, 1988; Bianchi *et al.*, 1993); chromosomal abnormalities arising during early cell divisions frequently result in confined placental mosaicism (Kalousek *et al.*, 1991; Section 10.4.6). It is estimated that three cells within the inner cell

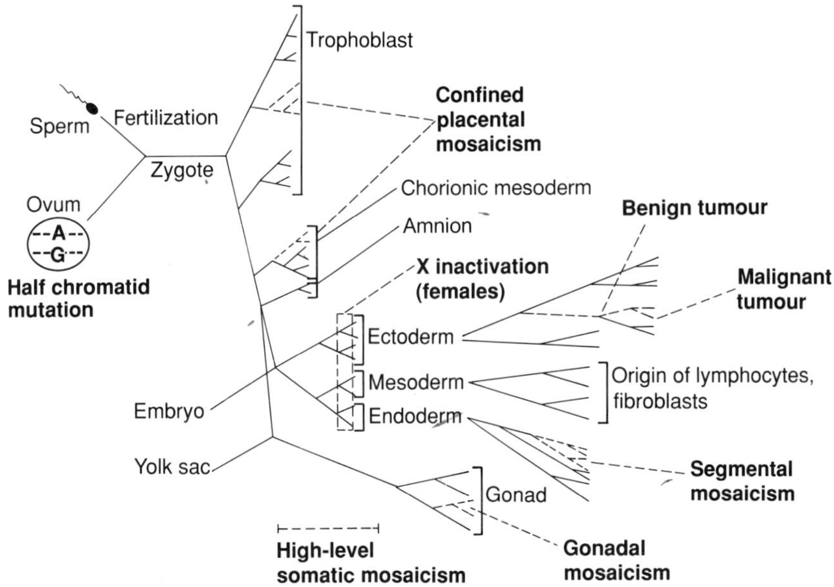

Figure 10.1. Diagrammatic representation of embryonic development, relating the manifestation of mosaicism to the site and timing of the mutational event. Normal processes are denoted in normal type, with solid lines; causes of mosaicism in bold type, with broken lines.

mass of the 64-celled blastocyst are committed to form the embryo (Markert and Petters, 1978), and differentiation into the definitive embryonic layers occurs during the third week. X inactivation in embryonic tissues takes place around this time (Gartler *et al.*, 1992; Tan *et al.*, 1993). These early divisions constitute a series of 'bottlenecks' and the manifestation of somatic mutations arising during this period will depend exquisitely on their exact site and timing. For instance, X inactivation in monozygotic twins may occur when there are relatively fewer inner cell mass cells; hence, X inactivation patterns in female monozygotic twins may be coarser than in singletons, with a greater tendency to a biased distribution (Nance, 1990; Richards *et al.*, 1990). Gibbs *et al.* (1993), in a study of mosaicism for length variation of a tetranucleotide repeat in the mouse, found evidence that most of the observed somatic variation arose during the first two cell divisions after fertilization. They speculate that this may reflect a critically limiting supply of one or more DNA replication factors within the early embryo.

Mutations occurring in germline progenitors give rise to gonadal mosaicism. This is an important phenomenon in clinical genetics, as it leads to uncertainty in the recurrence risk when a clinically unaffected parent has had a single offspring with a dominant or X-linked recessive condition. As pure gonadal mosaicism has no effect on phenotype, detailed consideration of this topic is, however, outside the scope of this chapter. Useful sources are the papers by Wijsman (1991),

Jeanpierre (1992), Passos-Bueno *et al.* (1992) and van Essen *et al.* (1992). When gonadal and somatic mosaicism are both present, this is termed 'gonosomal' mosaicism.

Simple mathematical considerations show that most detectable somatic mosaics must arise very early in development. A series of seven binary cell divisions produces 128 progeny: assuming that each cell contributes a roughly equal share to the adult organism, any mutation arising after the first seven embryonic cell divisions will be diluted out over 100-fold and be correspondingly hard to detect (the average cell divides 50–100 times). There are two important exceptions. First, mutations involving the skin may be detectable at much later stages: hence, the skin provides a valuable 'phenotypic window' into mutational processes. Second, mutations endowing the cell with a relative growth advantage will flourish: this is the mechanism of neoplasia.

At a practical level, it should be noted that sampling methods for mosaicism may bias significantly the likelihood of its detection. Cell culture after venepuncture or skin biopsy (the most frequently used tissue sources) will select for lymphocytes and fibroblasts respectively, both of which are of mesodermal origin: mosaicism confined to ectodermal or endodermal derivatives will be missed. Moss *et al.* (1993) have demonstrated the value of special culture conditions for keratinocytes (an ectodermal component) in the investigation of hypomelanosis of Ito (see Section 10.4.3).

10.3 Mechanisms of somatic mosaicism

10.3.1 Chimerism

Mosaicism covers the whole spectrum of mutational mechanisms (Table 10.1). Mosaics must be distinguished from true chimeras, which are derived from two or more distinct zygote lineages (Ford, 1969; Tippett, 1983). Chimerism in humans is rare but well described, and probably under-ascertained. It is most commonly detected by blood transfusion laboratories when an individual is found to have more than one blood group (Tippett, 1983). Most such cases arise by twin–twin transfusion of dizygotic twins (Figure 10.2a); the chimerism is confined to the haematopoietic cell line and there are no other phenotypic effects. (In cattle female/male twin pairs, the female gonads are partially masculinized, creating a sterile freemartin.) Tetragametic chimeras formed by the fusion of two independent zygotes (Figure 10.2a) are very rare, but a convincing case was described by Nyberg *et al.* (1992). The mother of the affected infant, which was stillborn with multiple malformations, carried a balanced (14;20) chromosome translocation; the infant was a mosaic 46,XY/46,XY,–14,+der(14)t(14;20), and must have originated from two separate oocytes. Other presentations of chimerism are with sexual ambiguity, including true hermaphroditism (46,XX/46,XY chimeras), or with pigmentary anomalies (for example, same-sex chimeras and diploid/triploid mosaics; Section 10.4.3). Most such chimeras are dispermic but appear to have a uniovular origin, and probably result from the additional

Table 10.1. General classification of somatic mosaicism. The mechanisms of origin of somatic mutations are discussed in the sections denoted

Type of mosaicism	Section	Examples
Chimerism	10.3.1	Blood group chimeras True hermaphrodites
Chromosomal	10.3.3	Mosaic trisomy 21 Pallister–Killian syndrome
Single gene	10.3.2, 10.3.4	Ubiquitous; includes dominant lethals (McCune–Albright) and neoplasia
Programmed DNA rearrangement	—	Immune system
X inactivation	10.3.5	All females
DNA methylation/ imprinting	10.3.5	Wilms' tumour Beckwith–Wiedemann syndrome
Unstable DNA repeats	10.3.7	Telomeres Fragile X syndrome
Uniparental disomy/ isoallelism	10.3.3, 10.3.6	Beckwith–Wiedemann syndrome

fertilization of one of the polar bodies or a prematurely divided ovum (Figure 10.2b; Tippett, 1983). Diploid/triploid mixoploidy occupies the borderline between true chimerism and mosaicism, and evidence that an important mechanism may be the incorporation of the maternal second polar body into one of the daughter cells has been suggested by several reports. Illustrative examples may be found in Bieber *et al.* (1981), Donnai *et al.* (1988), Müller *et al.* (1993) and Tuerlings *et al.* (1993).

10.3.2 Half chromatid mutation

A half chromatid mutation is a mutation (classically a single base change) introduced into one of a complementary pair of DNA strands. If no repair takes place before the subsequent round of DNA replication, the mutated strand will act as a template for DNA synthesis and a stable mutation will be introduced into one of the daughter cells (Gartler and Francke, 1975). A half chromatid mutation present in one of the gametes could thus give rise to somatic mosaicism by a pre-zygotic mechanism.

Although intriguing, it is very difficult to distinguish this mechanism from an early post-zygotic mutation, and therefore hard to demonstrate in practice. One possible example is the unusual patient with *de novo*, bilateral retinoblastomas described by Greger *et al.* (1990). This case was mosaic for two different deletions within the RB1 gene that shared one breakpoint in common but extended in opposite directions. There was no evidence for a co-existing normal cell line,

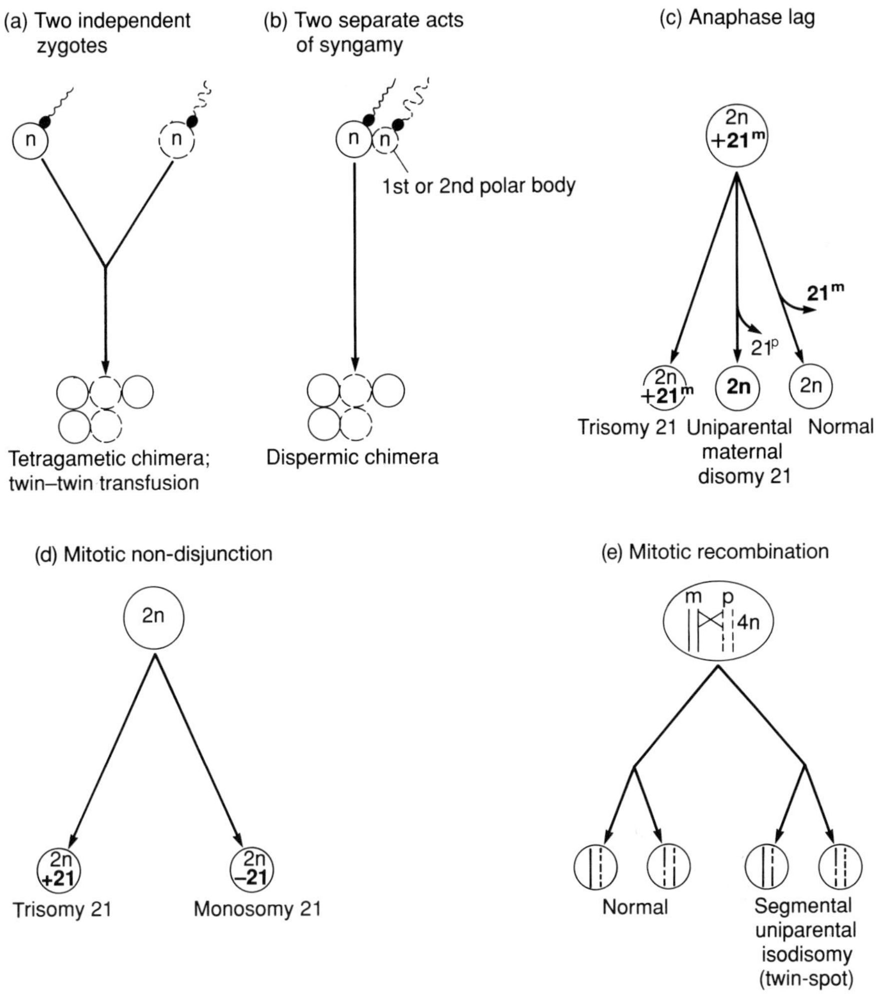

Figure 10.2. Some chromosomal mechanisms of mosaicism. (a) and (b), chimerism. (c) and (d), mosaic aneuploidy; chromosome 21 is used as an example, with chromosomes of maternal and paternal origin denoted 21^m and 21^p, respectively. (e) mitotic recombination; the individual chromatids of the paired chromosomes are shown. n represents the haploid genome content.

although blood was the only tissue source examined. However, it seems likely that the mutations arose very early in development, and the common breakpoint suggests that the two deletions result from the independent repair of each DNA strand following a single break.

10.3.3 Chromosomal aneuploidy

The major mechanisms of mosaic chromosomal aneuploidy are shown in Figure 10.2c,d. There are two broad possibilities. If the zygote is initially trisomic, loss of the extra chromosome from one cell by anaphase lag or mitotic non-disjunction will generate a euploid (i.e. with the normal chromosome complement) cell line (Figure 10.2c). The converse situation arises with an initially euploid zygote. Mitotic non-disjunction will yield a monosomic and trisomic daughter cell and, depending on their viability, mosaicism comprising two or three karyotypically different cell lines will arise (Figure 10.2d).

The trisomy→euploidy mechanism accounts for many cases of mosaic Down syndrome (trisomy 21), with consequent mitigation of the phenotype compared with full-blown aneuploidy (Bricarelli *et al.*, 1990; Pangalos *et al.*, 1993). A variation of this theme applies in fetuses with non-mosaic trisomies 13 and 18: Kalousek *et al.* (1989) have demonstrated a euploid line in the placenta of such cases, which may be important for fetal survival (Section 10.4.6). Although in these cases the euploid cell line (derived from the trisomy) appears to provide a 'rescuing' function, this cannot always be assumed. In one third of instances, the random loss of a chromosome present in three copies will give rise to uniparental disomy, which may itself be associated with phenotypic abnormality if the chromosome is imprinted (Figure 10.2c; Engel, 1993; and Chapter 1). Thus, loss of the single paternal chromosome 15 from fetuses with maternally originating trisomy 15 results in Prader–Willi syndrome (Cassidy *et al.*, 1992). Mosaicism for co-existing trisomy and uniparental disomy has been described for several chromosomes (Willatt *et al.*, 1992; Antonarakis *et al.*, 1993; Kalousek *et al.*, 1993).

The converse process, with an initially euploid zygote and subsequent mitotic non-disjunction also occurs (Figure 10.2d). Some cases of mosaic trisomy 21 arise in this fashion (Pangalos *et al.*, 1993). Patterns of sex chromosome mosaicism like 45,X/47,XXX and 45,X/46,XY/47,XYY are most simply explained by this mechanism, and are not unusual, probably because both the monosomic and trisomic cell lines are viable in the non-mosaic state. Localization of an aneuploid line to a malformed area of an otherwise normal fetus, also suggests a post-zygotic origin (Seely *et al.*, 1984; Schwartz *et al.*, 1992).

Whatever the mechanism of the initial mosaic event, the proportion of euploid cells tends to increase with time, presumably as a result of selective advantage. Gravholt *et al.* (1991) found an increase in the normal cell line of >10% in 16/32 unselected chromosomal mosaic cases followed up over 1–20 years. The degree of selection may vary between tissues: for instance, in Pallister–Killian syndrome (a distinctive condition with abnormal facial features, diaphragmatic hernia and severe mental retardation, caused by an additional mosaic isochromosome (12p)) the abnormal chromosome is lost more readily from lymphocytes than fibroblasts, as documented by Priest *et al.* (1992). Petersen *et al.* (1992) described two cases mosaic for structural abnormalities of chromosome 21, in which follow-up studies showed replacement by euploid cells in lymphocytes (but not fibroblasts). Analysis of DNA polymorphisms from blood samples demonstrated uniparental isodisomy for chromosome 21, indicating that in both cases a mitotic duplication of the normal 21 was responsible for the euploid line.

10.3.4 Single gene mutation

This is the most pervasive of all forms of mosaicism. Indeed the DNA sequence in every individual cell may be unique, as a result of mutations introduced during DNA replication and/or accumulated during the interphase life of the cell. The need for organisms to reproduce partly arises from this inevitable degeneration of the genome: both the production of gametes and the subsequent growth of the embryo select intensively for cells carrying a 'healthy' complement of DNA. Rarely, germinal mutations may be advantageous, thus contributing to the evolutionary process.

Many somatic mutations will be without phenotypic effect, falling in intergenic DNA, introns, or causing silent codon substitutions; others, causing critical damage to essential genes, may result in cell death. A variety of evidence may be used to estimate the rate at which phenotypically detectable somatic mutation occurs. Knudson (1971), in his classic studies of retinoblastoma, recognized that bilateral cases arise from random 'second hit' somatic mutations in individuals carrying a germline mutation, whereas sporadic unilateral tumours could be explained by two independent somatic mutations. Using these assumptions, the gene mutation rate in susceptible retinal cells could be estimated at $0.2–5 \times 10^{-6}$ per year. Identification of the retinoblastoma susceptibility gene RB1 has enabled direct analysis of the mutational mechanisms involved (Section 10.4.7). Similar methods have been used to analyse the occurrence of the multiple independent congenital malformations that characterize the dominant mouse mutant Disorganization. The number of separate malformations per animal conforms to Poissonian statistics, as expected if a random 'second hit' somatic mutation is responsible (Crosby *et al.*, 1993).

Another valuable system for estimating somatic mutation rates is the hypoxanthine phosphoribosyltransferase (HPRT) gene, because mutations can be selected for *in vitro* both positively and negatively (reviewed by Lambert *et al.*, 1992). Mutant frequencies of $0.4–28 \times 10^{-6}$ in T lymphocytes have been measured. These values are comparable to those calculated for RB1 above, and the conclusion is that the average adult may carry more than 10^6 cells mutated at one or other of just these two loci. These calculations emphasize both the ubiquity of somatic mosaic phenomena, and the remarkable ability of the body to tolerate an increasing mutational load over many years.

Occasionally, germline mutations may revert to produce a mosaic normal phenotype. This appears the most likely explanation for the rare patches of dystrophin-positive fibres in the muscles of patients with Duchenne muscular dystrophy (Klein *et al.*, 1992). Most cases probably result from 'second site' mutations correcting an original frameshift, rather than true back mutation.

10.3.5 X inactivation and other epigenetic modification

In the XX female, the majority of X chromosomal genes are subject to dosage compensation so that only one allele is transcribed. This is achieved at the late blastocyst stage by the inactivation of one or other X chromosome in each inner cell mass cell (Section 10.2). Subsequently, the pattern of X chromosome inactivity is stably transmitted to daughter cells (reviewed by Lyon, 1988; Gartler *et al.*, 1992). Females are therefore natural X inactivation mosaics.

Usually, X inactivation is random, generating two distinct populations of cells in carrier females. Dimorphism of the red cell population may be demonstrated in carriers of glucose-6-phosphate dehydrogenase deficiency, and stripey skin patterns can be visualized in several other X-linked conditions (Happle, 1985; Section 10.4.3). More rarely, X inactivation is highly skewed. Sometimes (for example, with structural X-chromosome abnormalities, or in the heterozygous states for several X-linked immunological and haematological disorders), the abnormal X is preferentially inactivated, either as a primary event, or by negative selection against cells in which the abnormal X is expressed; the effect is to mitigate, or eliminate, any abnormal phenotype associated with the mutation (Fearon *et al.*, 1987; reviewed by Gibbons *et al.*, 1992).

Skewed X inactivation can also occur in the opposite direction, so that the abnormal X is preferentially active; the disease will then manifest, even in the heterozygous female. Sometimes this paradoxical effect represents the 'lesser of two evils'. For instance, if there is an X/autosome translocation, inactivation of the abnormal X would spread to the autosome, leading to functional autosomal monosomy (Mattei *et al.*, 1982). A further example is provided by a girl carrying mutations on opposite X chromosomes for the X-linked diseases incontinentia pigmenti (which is lethal in the hemizygous male) and haemophilia A: because of selective inactivation of the X bearing the incontinentia pigmenti mutation, she manifested the haemophilia (Coleman *et al.*, 1993). Sometimes, however, there is no obvious explanation for the deleterious skewing, which may either reflect genetic variability in the susceptibility of different X chromosomes to inactivation, or random factors (Nisen and Waber, 1989; Kling *et al.*, 1991). The curious association of monozygous twinning with skewed X inactivation was mentioned in Section 10.2.

X inactivation represents a particular form of epigenetic modification. Another, called imprinting, is manifest as a differing functional contribution of the maternal and paternal alleles at a single locus (Cattanach and Beechey, 1990; Hall, 1990). The molecular mechanisms of X inactivation and imprinting are not delineated fully, but it is likely that DNA methylation is important in maintaining both states. As with X inactivation, imprinting may occur naturally as a mosaic phenomenon (McGowan *et al.*, 1989). Moreover, mosaic 'relaxation' of imprinting may manifest disease states. For example, the insulin-like growth factor II (IGF2) gene is normally imprinted in most tissues so that only the paternal allele is expressed; in contrast, some cases of Wilms' tumour and Beckwith–Wiedemann syndrome show abnormal, biallelic IGF2 expression (Ogawa *et al.*, 1993; Rainier *et al.*, 1993; Weksberg *et al.*, 1993).

10.3.6 Somatic recombination

Although well documented in *Drosophila* and yeast, somatic recombination has proved difficult to demonstrate in mammals, and its biological importance was long in doubt (Panthier and Condamine, 1991). However, work on retinoblastoma (Cavenee *et al.*, 1983) and rhabdomyosarcoma (Scrable *et al.*, 1987) clearly demonstrated the role of mitotic recombination in the inactivation of recessive anti-oncogenes, resulting in tumorigenesis (see Section 10.4.7 for further discussion).

Somatic recombination could be responsible for other types of mosaic phenomena. As illustrated in Figure 10.2e, this process gives rise to segmental uniparental isodisomy (or 'isoallelism'), and mosaicism for uniparental paternal isoallelism of the 11p15.5 region has been implicated in Beckwith–Wiedemann syndrome, hemihypertrophy and Wilms' tumour (Chao *et al.*, 1993; Henry *et al.*, 1993). The demonstration that two loci mapping within this region (IGF2 and H19) are imprinted (Ogawa *et al.*, 1993; Rainier *et al.*, 1993) strengthens the evidence that these observations are causally linked.

Further evidence of mosaic somatic recombination is provided by the 'twin-spot' phenomenon. If heterozygous mutations for two syntenic recessive skin pigment genes are present in *trans*, a mitotic recombination occurring between the centromere and the more proximal locus may produce two daughter cells, each one homozygous for one of the two mutations (Figure 10.2e). Following clonal growth, the result will be two apposed skin patches, each different in colour from the normal background. This has been demonstrated in mice (Panthier and Condamine, 1991), and in humans a probable case was described in a patient with Bloom's syndrome, a recessively inherited DNA repair defect associated with an enhanced rate of sister chromatid exchange (Festa *et al.*, 1979). Happle *et al.* (1990) have suggested that vascular twin naevi may arise by the 'twin-spot' mechanism.

10.3.7 Unstable repeated DNA sequences

The human genome contains a variety of repeated simple sequence motifs, some of which tend to be transmitted stably at meiosis and mitosis, whilst others are unstable and vary in repeat number from cell to cell, as well as between generations. In this latter category of naturally mosaic sequences are the hexamer $(TTAGGG)_n$, present in approximately 300–3000 copies at the telomeres of all chromosomes, and the trinucleotide repeats $(CCG)_n$ and $(CAG)_n$, which are stable at low copy number but may show meiotic or mitotic instability as the copy number is raised above 30–50. Telomeric DNA is essential to maintain the stability of the chromosome, and progressive loss of telomeric DNA may be associated with ageing (Wright and Shay, 1992; see Section 10.4.8). Mosaic mitotic instability of trinucleotide repeats within specific genes has been observed in several genetic diseases, and is most striking in the fragile X syndrome and myotonic dystrophy. Further information on these diseases can be found in Chapters 8 and 9.

10.4 The phenotypic effects of mosaicism

10.4.1 No phenotypic effect

The varied phenotypic consequences of somatic mosaicism are summarized in Table 10.2. Mosaic mutations may often have no phenotypic effect, and may only be ascertained by chance, or when co-existing gonadal mosaicism results in affected offspring. The absence of phenotype may be the consequence of dilution and/or negative selection of the cells bearing the mutation, or masking by X inactivation. Examples include chimeras ascertained during blood grouping (Tippett, 1983), many prenatally diagnosed cases of 45,X/46,XY mosaicism (Chang *et al.*, 1990), low level mosaic carriers of trisomy 21 (Pangalos *et al.*, 1992), female carriers of X-linked recessive diseases, in some of which the X chromosome bearing the mutation is preferentially inactivated in some or all cells (Section 10.3.5), maternal somatic mosaicism for X-linked disorders with random X inactivation (e.g. haemophilia A: Bröcker-Vriends *et al.*, 1990) and mosaic autosomal dominant mutations (osteogenesis imperfecta: Bonaventure *et al.*, 1992; Ehlers–Danlos syndrome type III: Kontusaari *et al.*, 1992).

Table 10.2. Phenotypic manifestations of somatic mosaicism. See Section 10.4 for further details

Phenotype	Comments
None	Ascertained coincidentally or after birth of affected child; female carriers of X-linked recessive disorders
Diluted	Mild version of full-blown phenotype; transmission to offspring shows 'anticipation'
Streaky changes (following Blaschko's lines)	Frequent, non-specific manifestation of chimerism and mosaicism of many types
Localized segments/ patches; body asymmetry	Diffuse or well defined; patches may be demarcated by dermatomes, or cross midline; includes twin-spotting
Unique phenotype (mutations lethal in pure form)	Genetically sporadic; several well characterized chromosomal and X-linked dominant conditions; autosomal single gene disorders difficult to identify, McCune–Albright is prototype
Confined placental mosaicism	Abnormal placental clone may cause intrauterine growth retardation; normal clone may rescue abnormal fetus
Neoplasia	Specific single or combination of multiple mutations, confers growth advantage on cell
Ageing	Accumulated burden of multiple somatic mutations of all types

10.4.2 Diluted phenotype

A dispersed admixture of normal and mutant cells in various tissues may result in a milder version of the full-blown phenotype. The mosaicism may come to light either directly, or when the mildly affected individual has a more severely affected offspring (thus providing an alternative molecular explanation for the genetic phenomenon of 'anticipation', more usually associated with instability of trinucleotide repeat sequences). Individuals with high level mosaic trisomy 21 show the clinical features of Down syndrome, but with a tendency to milder expression. Males with an unexpectedly mild phenotype for an X-linked recessive disease are sometimes found to be mosaics. Examples include Duchenne muscular dystrophy (Lebo *et al.*, 1990), ornithine transcarbamylase deficiency (Maddalena *et al.*, 1988; Legius *et al.*, 1990) and haemophilia B (Taylor *et al.*, 1991). A man with mild features of focal dermal hypoplasia (an X-linked lethal disease, not usually encountered in males) probably survived because he was a somatic mosaic; he transmitted the mutation to his daughter who had the classic phenotype (Gorski, 1991). For autosomal dominant disease, somatic mosaic mutations of the type 1 collagen gene COL1A1 causing osteogenesis imperfecta have been demonstrated in several mildly affected parents with severely affected offspring (reviewed by Constantinou-Deltas *et al.*, 1993). The description of a type II collagen (COL2A1) mutation associated with a mild 'Stickler syndrome' phenotype when present in mosaic form, and more severe 'Kniest dysplasia' when the mutation was transmitted to the patient's daughter (Winterpacht *et al.*, 1993), illustrates both the qualitative differences in phenotype that may be associated with mosaicism, and the semantic problems that can arise when naming and classifying syndromes in clinical genetics.

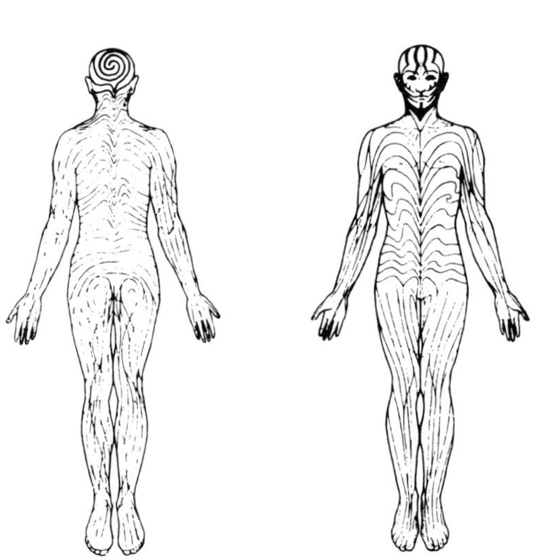

Figure 10.3. Blaschko's lines. See Section 10.4.3 for explanation, and Figure 10.4a,b for examples. Reproduced from Happle (1991) with permission from the University of Chicago Press.

10.4.3 Streaky skin phenotypes: Blaschko's lines

The tendency of striate naevi to follow a reproducible pattern over the body was first recognized by the Berlin dermatologist Alfred Blaschko in 1901 (see Jackson, 1976). Blaschko mapped out these patterns systematically to generate the diagram of lines that bears his name (Figure 10.3). They are clearly distinct from other 'systems' of lines over the body, such as the dermatomes, Langer's and Voight's lines. Particularly characteristic are the swirling S-shape over the abdomen and the V-shape over the centre of the back; along the limbs the lines tend to be longitudinal. Streaky pigment variation following these lines may reflect a wide variety of underlying types of mosaicism. Examples include true chimerism (Findlay and Moores, 1980), diploid/triploid mixoploidy and miscellaneous chromosomal aneuploidy (Donnai *et al.*, 1988; Thomas *et al.*, 1989), mosaicism for lethal dominant mutations, notably McCune–Albright syndrome (Happle, 1986; see Figure 10.4a and Section 10.4.6) and Lyonization patterns in X/autosome translocations (Sybert *et al.*, 1990) and X-linked dominant diseases like incontinentia pigmenti, focal dermal hypoplasia and hypohidrotic ectodermal dysplasia (Happle, 1985; Figure 10.4b).

Clinical geneticists always scrutinize the skin carefully for abnormal pigmentation, using an ultraviolet lamp in suspicious cases: the occurrence of hyper- or de-pigmentation following Blachko's lines is a strong indication of some type of mosaicism. When other clinical clues do not suggest a more specific diagnosis, the term 'hypomelanosis of Ito' is often used (although the more general label 'pigmentary dysplasia' is more appropriate, as it is sometimes the hyperpigmented clone that is abnormal). Careful investigation including karyotype of blood and multiple skin biopsies is indicated (reviewed by Thomas *et al.*, 1989; Flannery, 1990), and the specific culture of keratinocytes may improve diagnostic yield (Moss *et al.*, 1993). The reason why mosaic genetic abnormalities so often cause pigmentary dysplasia is not explained adequately; few of the karyotypic anomalies are associated with abnormal pigmentation when present in pure form, suggesting that the effect often arises from a non-specific 'incompatibility' between genetically distinct cell populations.

The embryological origin of Blaschko's lines probably reflects the migration of neural crest melanoblast or keratinocyte precursors of different pigmentary potentials. Analogous, but simpler, patterns may be observed in mice chimeric for coat colour (Mintz, 1967), heterozygous for X-linked pigmentation genes (Cattanach *et al.*, 1972), or containing clones individually marked with retroviruses (Huszar *et al.*, 1991). Mintz (1967) estimated that the patterns arose from just 17 clonal progenitors on each side, at about 8 days of development (equivalent to approximately 22 days in the human). A finer level of patterning may be imposed by clones of dermal cells that control melanocyte expression (Goudie *et al.*, 1985). The widespread manifestation of pigmentary abnormalities therefore implies an early embryological origin for the mosaicism; the breadth and pattern of pigmentation depending on the chance distribution of the two clones of cellular precursors, the direction of coherent growth during embryogenesis of the skin and the relative rates of migration of different cell layers (discussed by Happle, 1985; Thomas *et al.*, 1989; Flannery, 1990).

10.4.4 Patchy or segmental phenotypes

Single or multiple patches or localized dysplasias are another frequent manifestation of mosaicism. This distribution has two potential origins. First, mosaicism arising early in development may distribute unequally to different parts of the body, depending on the exact timing, tissue origin and migratory potential of the mutated cell. This may result in diffuse asymmetric abnormalities, or, if occurring slightly later in development, more well demarcated segmental or patchy skin changes. Second, an underlying germline mutation may render cells susceptible to 'second hit' somatic mutations, occurring independently in multiple cells.

A particularly striking example of the effect of unequal cellular distribution of mosaicism is the case described by Papenhausen *et al.* (1991). The patient (phenotypically male) presented with a leg length discrepancy of 4.4 cm, and initial blood lymphocyte analysis showed mosaicism of 45,X (36%) and 46,XY (64%) cells. Subsequent skin fibroblast cultures from biopsies of both legs showed 100% 46,XY in the normal leg, but 90% 45,X/10% 46,XY in the shorter leg, establishing a causal connection between the aneuploid cell line and the limb asymmetry. In general, chromosomal mosaicism should be suspected in any patient with asymmetrical changes like hemihypertrophy, hemiatrophy, scoliosis, asymmetric eye or brain abnormalities, and joint contractures, particularly if accompanied by patchy cutaneous signs. The distinction between 'normal' and 'abnormal' tissue is not always clear cut, so chromosome analysis from multiple biopsy sites may be indicated. Mosaicism for uniparental paternal isoallelism of chromosome 11p15.5 constitutes an additional specific cause of hemihypertrophy (Chao *et al.*, 1993).

Examples of more localized mosaic effects include a patient with true hermaphroditism (male and female gonads) due to a mosaic mutation of the testis determining gene SRY (Braun *et al.*, 1993), and localized congenital malformations due to chromosomal mosaicism (Seely *et al.*, 1984; Schwartz *et al.*, 1992), or 'second hit' mutations in the mouse disorganization mutant (Crosby *et al.*, 1993; Section 10.3.4). Rarely, genetic reversion (more likely due to gene conversion, somatic recombination or second site mutation, rather than simple back mutation) can give a small normal patch on an abnormal background: probable human examples are the case of achondroplasia described by Rimoin and McKusick (1969) and the scattered dystrophin-positive fibres documented in Duchenne muscular dystrophy (Klein *et al.*, 1992).

Although pigmentary dysplasia following Blachko's lines represents the classic cutaneous pattern of mosaicism, there are a number of other possibilities (Gorlin, 1993). The segmental skin changes in Sturge–Weber syndrome and segmental type 1 neurofibromatosis (Figure 10.4c) clearly follow the dermatomes (related to distribution of nerve innervation). 'Twin-spotting' was described in Section 10.3.6. Remarkable harlequin or checkerboard patterns have been described in

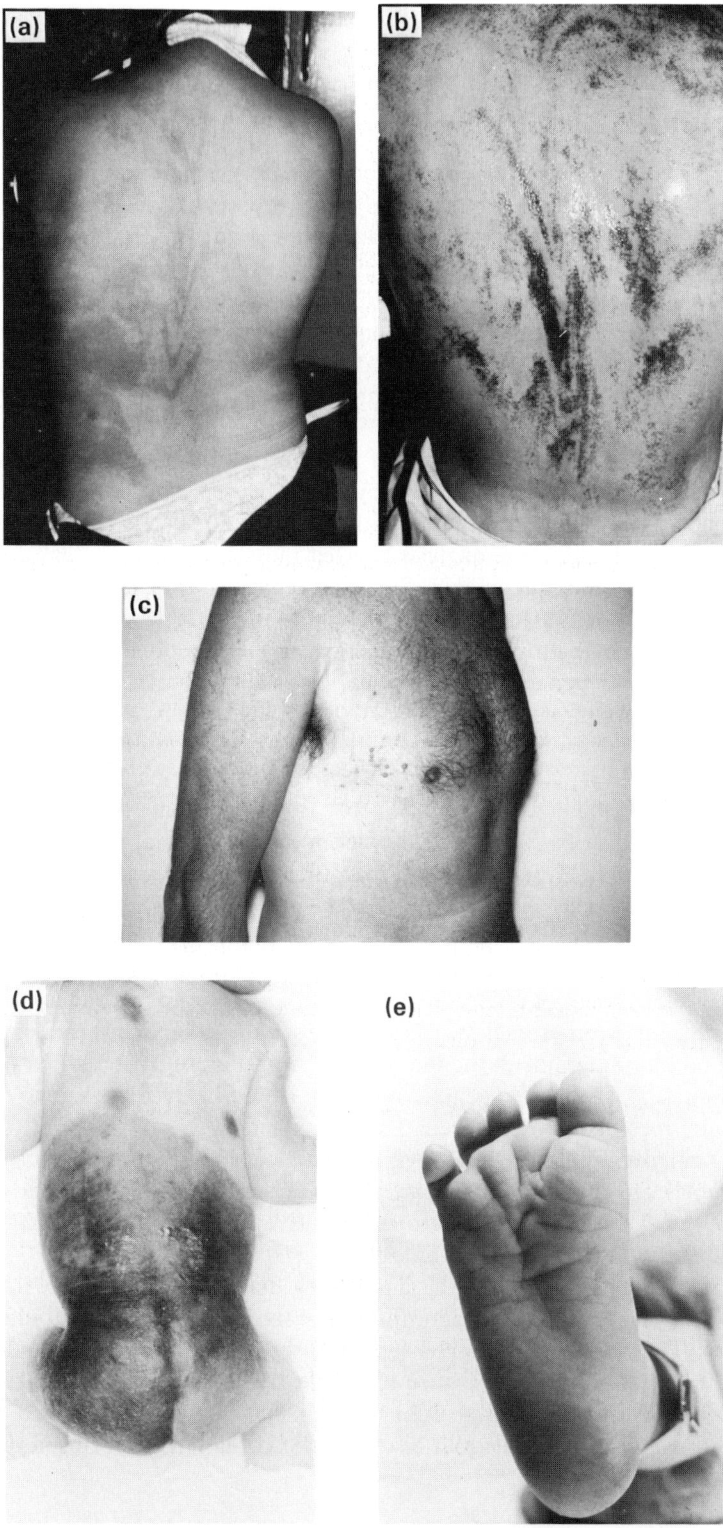
(a)
(b)
(c)
(d)
(e)

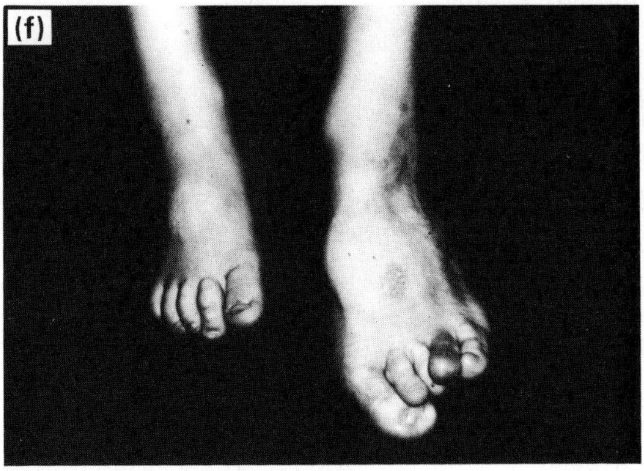

Figure 10.4. Phenotypic manifestations of mosaicism. (a) McCune–Albright syndrome, an autosomal dominant mosaic lethal disorder: predominantly left sided hyperpigmentation following Blaschko's lines. (b) Female carrier of X-linked hypohidrotic ectodermal dysplasia, showing mosaic pattern of reduced sweating following Blaschko's lines, due to random X inactivation. This is revealed by painting iodine on the skin, sprinkling with starch, and placing in a hot environment to induce sweating (Clarke *et al.*, 1987). (c) Segmental type 1 neurofibromatosis. Distribution of neurofibromas follows the thoracic dermatome. (d) Giant pigmented hairy naevus. Note the absence of midline demarcation. (e) Trisomy 8 mosaicism. Deep plantar and palmar creases are characteristic of this condition. (f) Proteus syndrome, a presumed dominant lethal. The left foot is grossly enlarged. Photographs courtesy of Michael Baraitser and Robin Winter (a, d, f), Angus Clarke (b), Susan Huson (c) and Helen Hughes (e).

multiple lentigines (a condition associated with numerous small darkly pigmented spots), with a series of alternating dark and light patches down one side of the body and mirror-image light and dark patches on the other side, demarcated at the midline (Goudie *et al.*, 1985). Another category of skin mosaicism comprises large patches showing no midline demarcation: giant pigmented hairy naevus (Figure 10.4d) is an example.

10.4.5 Unique phenotypes: dominant lethal mosaicism

It is easy to conceive that certain mutations that would be lethal in pure form, may survive in mosaic form if 'supported' by a population of normal cells. This is well documented for chromosomal mosaicism. In some cases, the clinical features may be sufficiently distinctive for the diagnosis to be suspected before the karyotype is obtained. Relatively common and well delineated examples include diploid/triploid mosaicism (syndactyly, wide space between 1st and 2nd toes), mosaic trisomy 8 (deep palmar and plantar creases; Figure 10.4e) and mosaic tetrasomy 12p (Pallister–Killian syndrome; see Section 10.3.3). In some cases, notably the last, only the normal cell line is usually present in blood, and cutaneous signs of the Ito type (Section 10.4.3) may be absent, so skin fibroblast karyotype is essential to confirm the diagnosis.

Dominant lethal single gene mosaicism is generally much harder to prove because, by definition, its occurrence will be sporadic and therefore not amenable to classic genetic analysis. An exception is when the mutation is present on the X chromosome. Although lethal in males, the mutation may be transmitted through heterozygous females, in whom its deleterious effects may be mitigated by selective inactivation of the X chromosome bearing the mutation. This mechanism has been documented most clearly in incontinentia pigmenti (Migeon *et al.*, 1989; Coleman *et al.*, 1993); other likely examples include focal dermal hypoplasia and dominant X-linked chondrodysplasia punctata (reviewed by Happle, 1985).

The best documented example of a somatic lethal autosomal disorder is McCune–Albright syndrome, a condition characterized by polyostotic fibrous dysplasia (multiple cystic lesions in the bones), skin pigmentation sometimes following Blaschko's lines (Figure 10.4a) and a variety of endocrine disorders including precocious puberty. Mosaicism for constitutively active point mutations in the gene for the α subunit of the stimulatory G protein of adenylyl cyclase (Gsα) has been documented in six patients (Weinstein *et al.*, 1991; Schwindinger *et al.*, 1992). The varied clinical features depend on the tissue distribution of the mosaicism, with activation of a variety of hormonal pathways. Other disorders that are strong candidates for mosaicism either for dominant lethal mutations or possibly uniparental disomy include the Proteus (Figure 10.4f), Klippel–Trenaunay–Weber and multiple linear sebaceous naevus syndromes (reviewed by Happle, 1986; Gorlin, 1993).

10.4.6 Placental mosaicism

The majority of cells of the early blastocyst contribute to extraembryonic tissues, notably the trophoblast and chorionic mesoderm (Section 10.2). Mutations may arise in these tissues and indirectly affect the embryonic phenotype, presumably by altering the nutritive function of the placenta, even though the embryo itself does not carry the mutation. (Another possibility, if the zygote is initially trisomic, is that the embryo may develop uniparental disomy, as discussed in Section 10.3.3.) Chromosomal mosaicism arising in this fashion is of considerable practical importance, because it is detected by the routine prenatal diagnostic procedure of chorionic villus sampling (CVS), usually performed at 10 weeks of pregnancy, and may lead to false positive or (much more rarely) negative diagnoses of fetal chromosomal abnormality. The frequency of this 'confined placental mosaicism' at CVS is about 2%, and the mosaicism can still be detected in term placentae in about half of these cases (Kalousek *et al.*, 1991). An increased rate of intrauterine growth retardation is claimed by some (Kalousek *et al.*, 1991), but not others (Kennerknecht *et al.*, 1993), when the fetal karyotype is normal whereas the placenta contains an aneuploid clone.

The converse situation, when the fetus is aneuploid but the placenta contains a normal cell line, was documented in all 14 cases of liveborn trisomies 13 and 18, but none of 12 cases of trisomy 21, studied by Kalousek *et al.* (1989). These authors suggested that the presence of a normal placental cell line is a prerequisite for the survival of trisomies 13 and 18 to full term. 'False-negative' CVS diagnosis (i.e. when the CVS is normal but the fetus has a chromosome abnormality) is clearly possible, but occurs remarkably infrequently (reviewed by Birkebaek *et al.*, 1992).

10.4.7 Neoplasia

Our growing understanding of the molecular basis of neoplasia is one of the most exciting themes in biology (Bishop, 1991). It is worth remembering that neoplasia represents a specific form of somatic mosaicism, in which the mutated cells have a growth advantage. Some seminal mechanistic concepts, developed from the study of particular tumours, are applicable to somatic mosaicism in general. Three of these will be summarized briefly here.

The 'two-hit' mutational hypothesis of Knudson (1971) in retinoblastoma was mentioned in Section 10.3.4. Subsequently, molecular studies on retinoblastoma and rhabdomyosarcoma (Cavenee *et al.*, 1983; Scrable *et al.*, 1987) elucidated the wide variety of mutational mechanisms responsible: these are summarized in Figure 10.5. Particularly important was the substantiation that the 'second hit' mutation could be a complete or partial chromosome loss (leading to monosomy), loss with duplication, or somatic recombination (leading to uniparental isodisomy or segmental isoallelism); the importance of the latter two processes in somatic mutation previously had been underestimated.

In contrast to retinoblastoma, four to six independent mutational events are required for much neoplastic transformation. Here the work on mechanisms of colorectal tumorigenesis (Vogelstein *et al.*, 1989; reviewed by Fearon and Vogelstein, 1990) has illustrated the multi-step nature and importance of timing of different mutations during this process. The conclusion is that identical, potentially harmful somatic mutations occur in many other cells, but do not manifest because the background cellular and/or mutational environment is wrong.

A novel mutational mechanism in colon cancer is also of interest (Aaltonen *et al.*, 1993; Thibodeau *et al.*, 1993). In some families segregating for hereditary non-polyposis colon cancer (HNPCC), mitotic instability of a wide variety of simple sequence repeats is observed in the tumour DNA, but apparently not in blood. This instability reflects an inherited deficiency of a DNA repair pathway homologous to bacterial MutS, which is manifest in tumours because of their clonal origin. Such HNPCC carriers are therefore highly mosaic, but most of the simple sequence DNA variation will be without phenotypic effect. A similar increase in rate of somatic mosaicism is likely to be associated with other DNA repair defects, such as Bloom's syndrome, Fanconi anaemia, ataxia telangectasia and xeroderma pigmentosum. These conditions are all associated with an increased rate of neoplasia, and some have features suggesting premature ageing (Warner and Price, 1989).

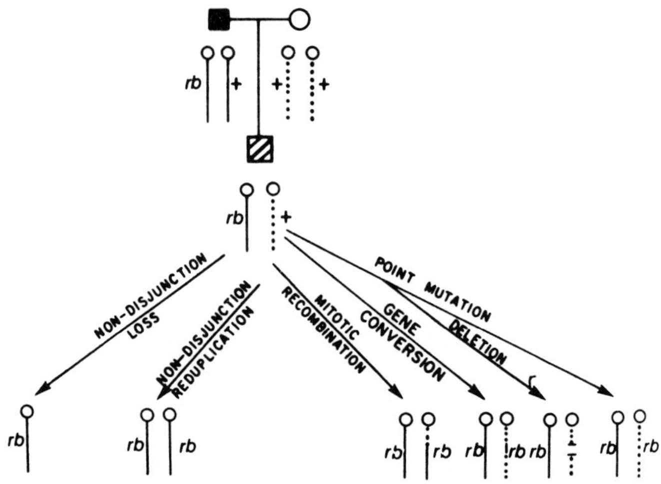

Figure 10.5. Mutational mechanisms resulting in retinoblastoma in an individual carrying a germline mutation on one chromosome. The normal and mutated alleles are denoted + and rb, respectively; retinoblastoma mutations behave in a recessive fashion. Reproduced from Cavenee *et al.* (1983) with permission from *Nature*. © 1983 Macmillan Magazines Limited.

10.4.8 Ageing

It is generally assumed that the ever-increasing burden of deleterious somatic mutations is the cause of ageing. Measurement of mutation rate at the HPRT locus suggests a progressive increase with age, and some of the DNA repair defects may be associated with features of premature ageing. However, since all classes of genes will accumulate mutations with time, it is more problematical to elucidate which are the important determinants of the ageing process. A detailed discussion of this topic is beyond the scope of this chapter, but recently, particular interest has focused on the possible roles of accumulated mutations in mitochondrial DNA (reviewed by Wallace, 1992, and see Chapter 11) and progressive telomere shortening. Simple sequence telomeric DNA (Section 10.3.7) is required to stabilize the ends of eukaryotic chromosomes and, due to the requirement of a 5′ RNA primer during DNA replication, some 50 bp of telomeric DNA is lost during each cell division. A corresponding shortening of telomeric sequence both with age and, *in vitro*, with successive rounds of cell culture has been observed (Wright and Shay, 1992). After many replications, some chromosomes may lose their telomeres altogether and become unstable, leading to senescence or neoplasia. Age-related relaxation of normal X inactivation has been observed (Lyon, 1988), but this seems unlikely to be quantitatively important in view of the human female's greater longevity!

10.5 Future research

This chapter has emphasized the ubiquity of somatic mosaic phenomena in human biology. Many fundamental questions remain. A study of chimerism using modern DNA fingerprinting methods is overdue and would yield valuable insights into this rare, but intriguing, phenomenon. Systematic study of cases of single gene and chromosomal mosaicism would yield a more detailed understanding of their mechanisms of origin, embryological derivation, and selective forces determining the final phenotype. Autosomal dominant lethal mutations are an instructive, but particularly difficult, group to characterize. Encouraged by successes in identifying the molecular basis of McCune–Albright sydrome and some cases of hemihypertrophy, efforts to elucidate the cause of other presumed dominant lethals may be rewarding. Finally, concerning the mechanisms of ageing, the mystery is not so much why this occurs, as why it occurs so slowly when somatic mutations appear to accumulate at such an alarming rate. A more quantitative and functional analysis may improve our understanding of this, even if (perhaps fortunately) it does not allow us to avoid the inevitability of the process.

References

Aaltonen LA, Peltomaki P, Leach FS, Sistonen P, Pylkkanen L, Mecklin J-P, Jarvinen H, Powell SM, Jen J, Hamilton SR, Petersen GM, Kinzler KW, Vogelstein B, de la Chapelle A. (1993) Clues to the pathogenesis of familial colorectal cancer. *Science* **260:** 812–816.

Antonarakis SE, Blouin J-L, Maher J, Avramopoulos D, Thomas G, Talbot CC. (1993) Maternal uniparental disomy for human chromosome 14, due to loss of a chromosome 14 from somatic cells with t(13;14) trisomy 14. *Am. J. Hum. Genet.* **52:** 1145–1152.

Bianchi DW, Wilkins-Haug LE, Enders AC, Hay ED. (1993) Origin of extraembryonic mesoderm in experimental animals: relevance to chorionic mosaicism in humans. *Am. J. Med. Genet.* **46:** 542–550.

Bieber FR, Nance WE, Morton CC, Brown JA, Redwine FO, Jordan RL, Mohanakumar T. (1981) Genetic studies of an acardiac monster: evidence of polar body twinning in man. *Science* **213:** 775–777.

Birkebaek NH, Andreasen EE, Ramsing M, Henriques UV, Bruun-Petersen G. (1992) Normal karyotype in cultured chorionic villus cells, but mosaicism in amniotic and fetal cells. *Prenat. Diagn.* **12:** 951–953.

Bishop JM. (1991) Molecular themes in oncogenesis. *Cell* **64:** 235–248.

Bonaventure J, Cohen-Solal L, Lasselin C, Maroteaux P. (1992) A dominant mutation in the COL1A1 gene that substitutes glycine for valine causes recurrent lethal osteogenesis imperfecta. *Hum. Genet.* **89:** 640–646.

Braun A, Kammerer S, Cleve H, Löhrs U, Schwarz H-P, Kuhnle U. (1993) True hermaphroditism in a 46,XY individual, caused by a postzygotic somatic point mutation in the male gonadal sex-determining locus (SRY): molecular genetics and histological findings in a sporadic case. *Am. J. Hum. Genet.* **52:** 578–585.

Bricarelli FD, Pierluigi M, Grasso M, Strigini P, Perroni L. (1990) Origin of extra chromosome 21 in 343 families: cytogenetic and molecular approaches. *Am. J. Med. Genet.* **7** (Suppl.): 129–132.

Bröcker-Vriends AHJT, Briët E, Dreesen JCFM, Bakker B, Reitsma P, Pannekoek H, van de Kamp JJP, Pearson PL. (1990) Somatic origin of inherited haemophilia A. *Hum. Genet.* **85:** 288–292.

Cassidy SB, Lai L-W, Erickson RP, Magnuson L, Thomas E, Gendron R, Herrmann J. (1992) Trisomy 15 with loss of the paternal 15 as a cause of Prader-Willi syndrome due to maternal disomy. *Am. J. Hum. Genet.* **51:** 701–708.

Cattanach BM, Beechey CV. (1990) Autosomal and X-chromosome imprinting. *Development* (Suppl.): 63–72.

Cattanach BM, Wolfe HG, Lyon MF. (1972) A comparative study of the coats of chimaeric mice and those of heterozygotes for X-linked genes. *Genet. Res.* **19:** 213–228.

Cavenee WK, Dryja TP, Phillips RA, Benedict WF, Godbout R, Gallie BL, Murphree AL, Strong LC, White RL. (1983) Expression of recessive alleles by chromosomal mechanisms in retinoblastoma. *Nature* **305**: 779–784.

Chang HJ, Clark RD, Bachman H. (1990) The phenotype of 45,X/46,XY mosaicism: an analysis of 92 prenatally diagnosed cases. *Am. J. Hum. Genet.* **46:** 156–167.

Chao L-Y, Huff V, Tomlinson G, Riccardi VM, Strong LC, Saunders GF. (1993) Genetic mosaicism in normal tissues of Wilms' tumour patients. *Nature Genetics* **3:** 127–131.

Clarke A, Phillips DIM, Brown R, Harper PS. (1987) Clinical aspects of X-linked hypohidrotic ectodermal dysplasia. *Arch. Dis. Child.* **62:** 989–996.

Coleman R, Genet SA, Harper JI, Wilkie AOM. (1993) Interaction of incontinentia pigmenti and factor VIII mutations in a female with biased X inactivation, resulting in haemophilia. *J. Med. Genet.* **30:** 497–500.

Constantinou-Deltas CD, Ladda RL, Prockop DJ. (1993) Somatic cell mosaicism: another source of phenotypic heterogeneity in nuclear families with osteogenesis imperfecta. *Am. J. Med. Genet.* **45:** 246–251.

Crane JP, Cheung SW. (1988) An embryogenic model to explain cytogenetic inconsistencies observed in chorionic villus versus fetal tissue. *Prenat. Diagn.* **8:** 119–129.

Crosby JL, Varnum DS, Nadeau JH. (1993) Two-hit model for sporadic congenital anomalies in mice with the disorganization mutation. *Am. J. Hum. Genet.* **52:** 866–874.

Donnai D, Read AP, McKeown C, Andrews T. (1988) Hypomelanosis of Ito: a manifestation of mosaicism or chimerism. *J. Med. Genet.* **25:** 809–818.

Engel E. (1993) Uniparental disomy revisited: the first twelve years. *Am. J. Med. Genet.* **46:** 670–674.

van Essen AJ, Abbs S, Baiget M, Bakker E, Boileau C, van Broeckhoven C, Bushby K, Clarke A, Claustres M, Covone AE, Ferrari M, Ferlini A, Galluzzi G, Grimm T, Grubben C, Jeanpierre M, Kääriäinen H, Liechti-Gallati S, Melis MA, van Ommen G-JB, Poncin JE, Scheffer H, Schwartz M, Speer A, Stuhrmann M, Verellen-Dumoulin C, Wilcox DE, ten Kate LP. (1992) Parental origin and germline mosaicism of deletions and duplications of the dystrophin gene: a European study. *Hum. Genet.* **88:** 249–257.

Fearon ER, Winkelstein JA, Civin CI, Pardoll DM, Vogelstein B. (1987) Carrier detection in X-linked agammaglobulinemia by analysis of X-chromosome inactivation. *N. Engl. J. Med.* **316:** 427–431.

Fearon ER, Vogelstein B. (1990) A genetic model for colorectal tumorigenesis. *Cell* **61:** 759–767.

Festa RS, Meadows AT, Boshes RA. (1979) Leukemia in a black child with Bloom's syndrome. Somatic recombination as a possible mechanism for neoplasia. *Cancer* **44:** 1507–1510.

Findlay GH, Moores PP. (1980) Pigment anomalies of the skin in the human chimaera: their relation to systematized naevi. *Br. J. Dermatol.* **103:** 489–498.

Flannery DB. (1990) Pigmentary dysplasias, hypomelanosis of Ito, and genetic mosaicism. *Am. J. Med. Genet.* **35:** 18–21.

Ford CE. (1969) Mosaics and chimaeras. *Br. Med. Bull.* **25:** 104–109.

Gartler SM, Francke U. (1975) Half chromatid mutations: transmission in humans? *Am. J. Hum. Genet.* **27:** 218–223.

Gartler SM, Dyer KA, Goldman MA. (1992) Mammalian X chromosome inactivation. In:

Molecular Genetic Medicine, Vol. 2 (ed. T. Friedmann). Academic Press, New York, pp. 121–160.

Gibbons RJ, Suthers GK, Wilkie AOM, Buckle VJ, Higgs DR. (1992) X-linked α-thalassemia/mental retardation (ATR-X) syndrome: localization to Xq12–q21.31 by X inactivation and linkage analysis. *Am. J. Hum. Genet.* **51:** 1136–1149.

Gibbs M, Collick A, Kelly RG, Jeffreys AJ. (1993) A tetranucleotide repeat mouse minisatellite displaying substantial somatic instability during early preimplantation development. *Genomics* **17:** 121–128.

Gorlin RJ. (1993) Second Robert J. Gorlin Conference on Human Dysmorphology, Minneapolis, USA. 7–9 November 1992. *Clin. Dysmorph.* **2:** 278–282.

Gorski JL. (1991) Father-to-daughter transmission of focal dermal hypoplasia associated with nonrandom X-inactivation: support for X-linked inheritance and paternal X chromosome mosaicism. *Am. J. Med. Genet.* **40:** 332–337.

Goudie RB, Jack AS, Goudie BM. (1985) Genetic and developmental aspects of pathological pigmentation patterns. In: *Dermatopathology* (eds. CL Berry). *Curr. Topics Pathol.* **74**: 103–139.

Gravholt CH, Friedrich U, Nielsen J. (1991) Chromosomal mosaicism: a follow-up study of 39 unselected children found at birth. *Hum. Genet.* **88:** 49–52.

Greger V, Passarge E, Horsthemke B. (1990) Somatic mosaicism in a patient with bilateral retinoblastoma. *Am. J. Hum. Genet.* **46:** 1187–1193.

Hall JG. (1988) Somatic mosaicism: observations related to clinical genetics. *Am. J. Hum. Genet.* **43:** 355–363.

Hall JG. (1990) Genomic imprinting: review and relevance to human diseases. *Am. J. Hum. Genet.* **46:** 857–873.

Happle R. (1985) Lyonization and the lines of Blaschko. *Hum. Genet.* **70:** 200–206.

Happle R. (1986) Cutaneous manifestation of lethal genes. *Hum Genet* **72:** 280.

Happle R. (1991) Association of pigmentary anomalies with chromosomal and genetic mosaicism and chimerism. *Am. J. Hum. Genet.* **48:** 1014–1016.

Happle R, Koopman R, Mier PD. (1990) Hypothesis: vascular twin naevi and somatic recombination in man. *Lancet* **335:** 376–378.

Henry I, Puech A, Riesewijk A, Ahnine L, Mannens M, Beldjord C, Bitoun P, Tournade MF, Landrieu P, Junien C. (1993) Somatic mosaicism for partial paternal isodisomy in Wiedemann–Beckwith syndrome:a post-fertilization event. *Eur. J. Hum. Genet.* **1:** 19–29.

Huszar D, Sharpe A, Hashmi S, Bouchard B, Houghton A, Jaenisch R. (1991) Generation of pigmented stripes in albino mice by retroviral marking of neural crest melanoblasts. *Development* **113:** 653–660.

Jackson R. (1976) The lines of Blaschko: a review and reconsideration. *Br. J. Dermatol.* **95:** 349–360.

Jeanpierre M. (1992) Germinal mosaicism and risk calculation in X-linked diseases. *Am. J. Hum. Genet.* **50:** 960–967.

Kalousek DK, Barrett IJ, McGillivray BC. (1989) Placental mosaicism and intrauterine survival of trisomies 13 and 18. *Am. J. Hum. Genet.* **44:** 338–343.

Kalousek DK, Howard-Peebles PN, Olson SB, Barrett IJ, Dorfmann A, Black SH, Schulman JD, Wilson RD. (1991) Confirmation of CVS mosaicism in term placentae and high frequency of intrauterine growth retardation associated with confined placental mosaicism. *Prenat. Diagn.* **11:** 743–750.

Kalousek DK, Langlois S, Barrett I, Yam I, Wilson DR, Howard-Peebles PN, Johnson MP, Giorgiutti E. (1993) Uniparental disomy for chromsome 16 in humans. *Am. J. Hum. Genet.* **52:** 8–16.

Kennerknecht I, Krämer S, Grab D, Terinde R, Vogel W. (1993) A prospective cytogenetic study of third-trimester placentae in small-for-date but otherwise normal newborns. *Prenat. Diagn.* **13:** 257–269.

Klein CJ, Coovert DD, Bulman DE, Ray PN, Mendell JR, Burghes AHM. (1992) Somatic reversion/suppression in Duchenne muscular dystrophy (DMD): evidence supporting a frame-restoring mechanism in rare dystrophin-positive fibers. *Am. J. Hum. Genet.* **50:** 950–959.

Kling S, Coffey AJ, Ljung R, Sjörin E, Nilsson IM, Holmberg L, Giannelli F. (1991) Moderate haemophilia B in a female carrier caused by preferential inactivation of the paternal X chromosome. *Eur. J. Haematol.* **47:** 257–261.

Knudson AG. (1971) Mutation and cancer: statistical study of retinoblastoma. *Proc. Natl Acad. Sci. USA* **68:** 820–823.

Kontusaari S, Tromp G, Kuivaniemi H, Stolle C, Pope FM, Prockop DJ. (1992) Substitution of aspartate for glycine 1018 in the type III procollagen (COL3A1) gene causes type IV Ehlers–Danlos syndrome: the mutated allele is present in most blood leukocytes of the asymptomatic and mosaic mother. *Am. J. Hum. Genet.* **51:** 497–507.

Lambert B, Andersson B, He S-M, Marcus S, Steen A-M. (1992) Molecular analysis of mutation in the human gene for hypoxanthine phosphoribosyltransferase. In: *Molecular Genetic Medicine*, Vol. 2 (ed. T. Friedmann). Academic Press, New York, pp. 161–188.

Lebo RV, Olney RK, Golbus MS. (1990) Somatic mosaicism at the Duchenne locus. *Am. J. Med. Genet.* **37:** 187–190.

Legius E, Baten E, Stul M, Marynen P, Cassiman J-J. (1990) Sporadic late onset ornithine transcarbamylase deficiency in a boy with somatic mosaicism for an intragenic deletion. *Clin. Genet.* **38:** 155–159.

Lyon MF. (1988) X-chromosome inactivation and the location and expression of X-linked genes. *Am. J. Hum. Genet.* **42:** 8–16.

Maddalena A, Sosnoski DM, Berry GT, Nussbaum RL. (1988) Mosaicism for an intragenic deletion in a boy with mild ornithine transcarbamylase deficiency. *N. Engl. J. Med.* **319:** 999–1003.

Markert CL, Petters RM. (1978) Manufactured hexaparental mice show that adults are derived from three embryonic cells. *Science* **202:** 56–58.

Mattei MG, Mattei JF, Ayme S, Giraud F. (1982) X-autosome translocations: cytogenetic characteristics and their consequences. *Hum. Genet.* **61:** 295–309.

McGowan R, Campbell R, Peterson A, Sapienza C. (1989) Cellular mosaicism in the methylation and expression of hemizygous loci in the mouse. *Genes Devel.* **3:** 1669–1676.

Migeon BR, Axelman J, de Beur SJ, Valle D, Mitchell GA, Rosenbaum KN. (1989) Selection against lethal alleles in females heterozygous for incontinentia pigmenti. *Am. J. Hum. Genet.* **44:** 100–106.

Mintz B. (1967) Gene control of mammalian pigmentary differentiation, I. Clonal origin of melanocytes. *Proc. Natl Acad. Sci. USA* **58:** 344–351.

Moss C, Larkins S, Stacey M, Blight A, Farndon PA, Davison EV. (1993) Epidermal mosaicism and Blaschko's lines. *J. Med. Genet.* **30:** 752–755.

Müller U, Weber JL, Berry P, Kupke KG. (1993) Second polar body incorporation into a blastomere results in 46,XX/69,XXX mixoploidy. *J. Med. Genet.* **30:** 597–600.

Nance WE. (1990) Do twin Lyons have larger spots? *Am. J. Hum. Genet.* **46:** 646–648.

Nisen PD, Waber PG. (1989) Nonrandom X chromosome DNA methylation patterns in hemophiliac females. *J. Clin. Invest.* **83:** 1400–1403.

Nyberg RH, Haapala AK, Simola KOJ. (1992) A case of human chimerism detected by unbalanced chromosomal translocation. *Clin. Genet.* **42:** 257–259.

Ogawa O, Eccles MR, Szeto J, McNoe LA, Yun K, Maw MA, Smith PJ, Reeve AE. (1993) Relaxation of insulin-like growth factor II gene imprinting implicated in Wilms' tumour. *Nature* **362:** 749–751.

Pangalos CG, Talbot CC, Lewis JG, Adelsberger PA, Petersen MB, Serre J-L, Rethoré M-O, de Blois M-C, Parent P, Schinzel AA, Binkert F, Boue J, Corbin E, Croquette MF, Gilgenkrantz S, de Grouchy J, Bertheas MF, Prieur M, Raoul O, Serville F, Siffroi JP, Thepot F, Lejeune J, Antonarakis SE. (1992) DNA polymorphism analysis in families with recurrence of free trisomy 21. *Am. J. Hum. Genet.* **51:** 1015–1027.

Pangalos C, Avramopoulos D, Blouin JL, Raoul O, de Blois MC, Prieur M, Schinzel A, Gika M, Abazis D, Antonarakis SE. (1993) Understanding the mechanism(s) of mosaic trisomy 21 using DNA polymorphism analysis. *Am. J. Hum. Genet.* **53** (Suppl.): abstract no. 253.

Panthier J-J, Condamine H. (1991) Mitotic recombination in mammals. *BioEssays* **13:** 351–356.

Papenhausen PR, Mueller OT, Bercu B, Salazar J, Tedesco TA. (1991) Cell line segregation in a 45,X/46,XY mosaic child with asymmetric leg growth. *Clin. Genet.* **40:** 237–241.

Passos-Bueno MR, Bakker E, Kneppers ALJ, Takata RI, Rapaport D, den Dunnen JT, Zatz M, van Ommen GJB. (1992) Different mosaicism frequencies for proximal and distal Duchenne muscular dystrophy (DMD) mutations indicate difference in etiology and recurrence risk. *Am. J. Hum. Genet.* **51:** 1150–1155.

Petersen MB, Bartsch O, Adelsberger PA, Mikkelsen M, Schwinger E, Antonarakis SE. (1992) Uniparental isodisomy due to duplication of chromosome 21 occurring in somatic cells monosomic for chromosome 21. *Genomics* **13:** 269–274.

Prescott DM. (1988) *Cells. Principles of Molecular Structure and Function.* Jones and Bartlett, Boston, p. 404.

Priest JH, Rust JM, Fernhoff PM. (1992) Tissue specificity and stability of mosaicism in Pallister–Killian +i(12p) syndrome: relevance for prenatal diagnosis. *Am. J. Med. Genet.* **42:** 820–824.

Rainier S, Johnson LA, Dobry CJ, Ping AJ, Grundy PE, Feinberg AP. (1993) Relaxation of imprinted genes in human cancer. *Nature* **362:** 747–749.

Richards CS, Watkins SC, Hoffman EP, Schneider NR, Milsark IW, Katz KS, Cook JD, Kunkel LM, Cortada JM. (1990) Skewed X inactivation in a female MZ twin results in Duchenne muscular dystrophy. *Am. J. Hum. Genet.* **46:** 672–681.

Rimoin DL, McKusick VA. (1969) Somatic mosaicism in an achondroplastic dwarf. *Birth Defects* V **4:** 17–19.

Schwartz S, Raffel LJ, Sun C-CJ, Waters E. (1992) An unusual mosaic karyotype detected through prenatal diagnosis with duplication of 1q and 19p and associated teratoma development. *Teratology* **46:** 399–404.

Schwindinger WF, Francomano CA, Levine MA. (1992) Identification of a mutation in the gene encoding the α subunit of the stimulatory G protein of adenyl cyclase in McCune–Albright syndrome. *Proc. Natl Acad. Sci. USA* **89:** 5152–5156.

Scrable HJ, Witte DP, Lampkin BC, Cavenee WK. (1987) Chromosomal localization of the human rhabdomyosarcoma locus by mitotic recombination mapping. *Nature* **329:** 645–647.

Seely JR, Seely BL, Bley R, Altmiller CJ. (1984) Localized chromosomal mosaicism as a cause of dysmorphic devopment. *Am. J. Hum. Genet.* **36:** 899–903.

Sybert VP, Pagon RA, Donlan M, Bradley CM. (1990) Pigmentary abnormalities and mosaicism for chromosomal aberration: association with clinical features similar to hypomelanosis of Ito. *J. Pediatr.* **116:** 581–586.

Tan S-S, Williams EA, Tam PPL. (1993) X-chromosome inactivation occurs at different times in different tissues of the post-implantation mouse embryo. *Nature Genetics* **3:** 170–174.

Taylor SAM, Deugau KV, Lillicrap DP. (1991) Somatic mosaicism and female-to-female transmission in a kindred with hemophilia B (factor IX deficiency). *Proc. Natl Acad. Sci. USA* **88:** 39–42.

Thibodeau SN, Bren G, Schaid D. (1993) Microsatellite instability in cancer of the proximal colon. *Science* **260:** 816–819.

Thomas IT, Frias JL, Cantu ES, Lafer CZ, Flannery DB, Graham JG. (1989) Association of pigmentary anomalies with chromosomal and genetic mosaicism and chimerism. *Am. J. Hum. Genet.* **45:** 193–205.

Tippett P. (1983) Blood group chimeras. *Vox Sang.* **44:** 333–359.

Tuerlings JHAM, Breed ASPM, Vosters R, Anders GJPA. (1993) Evidence of a second gamete fusion after the first cleavage of the zygote in a 47,XX,+18,/70,XXX,+18 mosaic. A remarkable diploid–triploid discrepancy after CVS. *Prenat. Diagn.* **13:** 301–306.

Vogelstein B, Fearon ER, Kern SE, Hamilton SR, Preisinger AC, Nakamura Y, White R. (1989) Allelotype of colorectal carcinomas. *Science* **244:** 207–211.

Wallace DC. (1992) Mitochondrial genetics: a paradigm for aging and degenerative diseases? *Science* **256:** 628–632.

Warner HR, Price AR. (1989) Involvement of DNA repair in cancer and aging. *J. Gerontol.* **44:** 45–54.

Weinstein LS, Shenker A, Gejman PV, Merino MJ, Friedman E, Spiegel AM. (1991) Activating mutations of the stimulatory G protein in the McCune–Albright syndrome. *N. Engl. J. Med.* **325:** 1688–1695.

Weksberg R, Shen DR, Fei YL, Song QL, Squire J. (1993) Disruption of insulin-like growth factor 2 imprinting in Beckwith–Wiedemann syndrome. *Nature Genetics* **5:** 143–150.

Wijsman EM. (1991) Recurrence risks of a new dominant mutation in children of unaffected parents. *Am. J. Hum. Genet.* **48:** 654–661.

Willatt LR, Davison BCC, Goudie D, Alexander J, Dyson HM, Jenks PE, Ferguson-Smith ME. (1992) A male with trisomy 9 mosaicism and maternal uniparental disomy for chromosome 9 in the euploid cell line. *J. Med. Genet.* **29:** 742–744.

Winterpacht A, Hilbert M, Schwarze U, Mundlos S, Spranger J, Zabel BU. (1993) Kniest and Stickler dysplasia phenotypes caused by collagen type II gene (COL2A1) defect. *Nature Genetics* **3:** 323–326.

Wright WE, Shay JW. (1992) Telomere positional effects and the regulation of cellular senescence. *Trends Genet.* **8:** 193–197.

11

Mitochondrial DNA-associated disease

Simon R. Hammans

11.1 Oxidative phosphorylation and mitochondrial DNA

The function of the mitochondrial oxidative phosphorylation system is to generate ATP, the energy 'currency' of the cell. Reducing equivalents, mainly derived from glycolysis and fatty acid oxidation, are passed down the four respiratory chain enzymes (complexes I–IV) embedded within the inner mitochondrial membrane. According to the chemiosmotic theory (Mitchell, 1976), this electron transfer is coupled at three sites to proton translocation across the inner mitochondrial membrane, thereby creating an electrochemical gradient, which is harnessed by ATP synthase (complex V) for ATP generation. It is thought that mitochondria were originally free-living organisms which were engulfed by anaerobic cells to become endosymbionts. During evolution the human nuclear genome appears to have incorporated most of the original mitochondrial genes. The reason for the survival of mitochondrial (mt) DNA is unclear, but it is likely that there is an evolutionary advantage in some polypeptides being synthesised *in situ*. Mitochondria are unique amongst cellular organelles in having their own genome, present in two to ten copies per mitochondrion. In total, mtDNA constitutes approximately 1% of cellular DNA. The genome is a closed circular double-stranded molecule of 16 569 bp, and encodes 13 of more than 70 polypeptide subunits of the respiratory chain, the remainder being encoded on nuclear DNA (Table 11.1). The nuclear-encoded respiratory chain subunits are translated on cytoplasmic polysomes, mostly as precursors with amino-terminal amphiphilic targeting presequences for import into the mitochondrion, prior to assembly into the respiratory chain enzyme complexes. Defects of either nuclear or mitochondrial DNA have the potential to impair respiratory chain function and cause disease. The association of mtDNA defects with human disease was established in 1988. The properties of mtDNA are important in determining the singular characteristics of mtDNA-determined disease.

The entire human mtDNA sequence has been determined (Anderson *et al.*, 1981). The mitochondrial genetic code differs only slightly from the nuclear code,

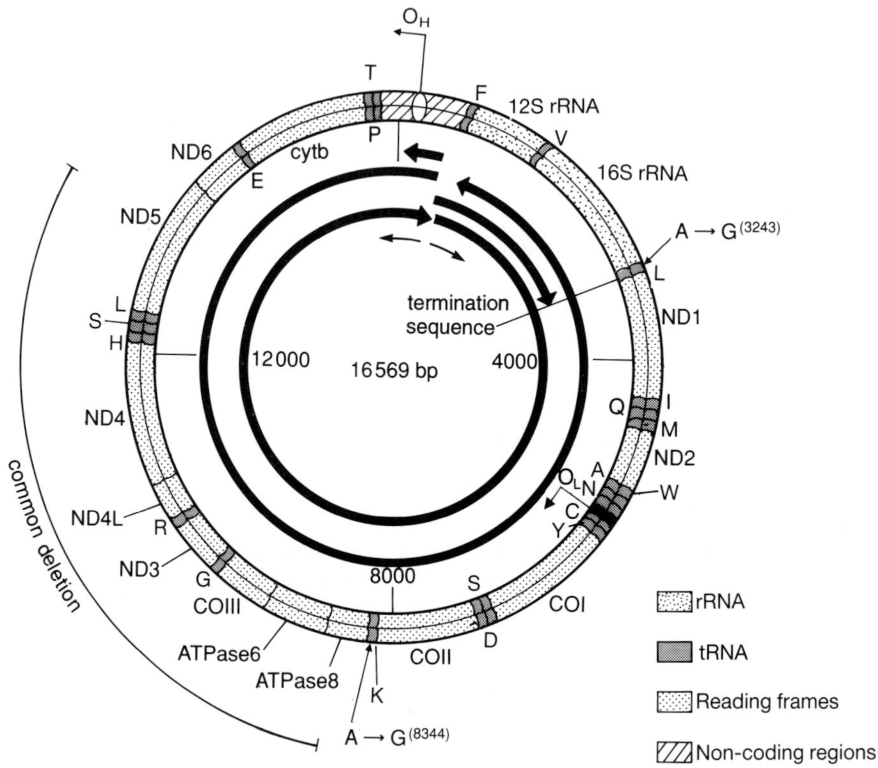

Figure 11.1. The mitochondrial genome. Transfer RNAs are shown at dark boxes with single letter codes. Arrows represent RNA transcripts. ND, NADH (complex I) subunits; cytb, cytochrome *b*; CO, cytochrome oxidase (complex IV); ATPase, ATP synthetase (complex V); O_H/O_L, origins of replication of the H-strand and L-strand. LSP/HSP show the sites of the light and heavy strand promoters. The sites of the three commonest pathogenic mutations are shown.

but the organization of mtDNA is strikingly different (Figure 11.1). The genes abut each other, or even overlap by one or two bases, and contain no introns. Protein-coding genes are punctuated by intervening transfer (t) RNA genes which are dispersed throughout the genome. The guanine-rich heavy (H)-strand encodes the 12S and 16S ribosomal (r) RNAs, 14 tRNAs, six subunits of complex I (ND), one of complex III, three of complex IV (COX) and two of complex V (ATPase). The cytosine-rich light (L)-strand encodes only one protein (ND6), but encodes the remaining eight of the full complement of 22 tRNAs. The only non-coding sequence occurs in the triple-stranded D (displacement)-loop region, which contains important sites of interaction between mtDNA, and nuclear-encoded proteins and riboproteins involved in transcription and replication. The D-loop is generated by the synthesis of a short strand of H-strand DNA, the 7S DNA. Both strands have a promoter within the D-loop region; each is associated

Table 11.1. Respiratory chain subunits

Complex	Enzyme	mtDNA-encoded subunits	Total subunits
I	NADH co-enzyme Q reductase (ND)	7	≈41
II	Succinate ubiquinone reductase	0	4
III	Ubiquinone–cytochrome *c* oxidoreductase	1	11
IV	Cytochrome *c* oxidase (COX)	3	13
V	ATP synthase (ATPase)	2	14

with an upstream binding site for a mitochondrial transcription factor (mtTF1). Polycistronic RNA is created as each strand is transcribed. Within the $tRNA^{Leu\,(UUR)}$ gene, a conserved tridecamer sequence binds a protein (mtTERM) which is capable of terminating transcription in either direction. Thus, although H-strand transcription may proceed around the genome to the $tRNA^{Thr}$ gene, it more frequently stops at the 3′ end of the 16S rRNA gene. This allows a smaller transcript containing 12S and 16S rRNAs to be synthesised 15–60 times more frequently than the larger H-strand transcript. Mature mRNAs are generated by cleavage of the transcripts at tRNAs. Apart from its role in transcription, the light strand promoter also creates RNA primers for H-strand replication. RNase MRP, containing a nuclear-encoded RNA, appears to cleave the RNA at one of three conserved sequence blocks within the D-loop (CSB–II), which is the most prevalent start site for 7S DNA synthesis. H-strand replication starts at the 7S DNA and continues around the L-strand. When the H-strand is displaced from the L-strand origin, L-strand replication is commenced.

Evolution of the mtDNA sequence is 6–17 times more rapid than that of nuclear DNA. In normal individuals, there is a high prevalence of sequence polymorphisms in coding regions, and higher still in non-coding regions of mtDNA. MtDNA is almost exclusively maternally inherited in mammals. In mice, the paternal contribution has been estimated to be at a ratio of 10^{-4} to the maternal contribution. The presence of two different mitochondrial genotypes within individuals (heteroplasmy) has been described in only one normal human family (through three generations) despite the high mutation rate of mtDNA (Howell *et al.*, 1992a). Heteroplasmy has also been observed in Holstein cows and some lower organisms. During mitotic division, mtDNA is transmitted to the progeny according to the distribution of the organelles. In the presence of heteroplasmy, the proportions of different populations of mtDNA can change in succeeding generations of cells. Furthermore, selection pressures on mtDNA may cause heteroplasmy to vary between cell lines.

During oogenesis the number of mitochondria per cell increases by approximately 100-fold, while the number of mtDNA molecules per mitochondrion falls to about one or two. MtDNA replication probably resumes at the blastocyst stage so that somatic cell numbers of mtDNA molecules are achieved. In Holstein cows heteroplasmy has been shown to vary dramatically between mother and offspring. These observations may be explained by a narrow 'bottleneck' at oogenesis, ensuring that selected mitochondria are of pure genetic type. However, any hypothesis must also explain the stable heteroplasmy sometimes observed in both silent and pathogenic mutations (Howell *et al.*, 1992a; Hammans *et al.*, 1993).

11.2 Mitochondrial myopathies

Historically, the first accounts of mitochondrial diseases followed the recognition of histological abnormalities of mitochondria in muscle (Olson *et al.*, 1972). In some patients, the modified Gomori trichrome stain allows a spectacular demonstration of accumulations of abnormal mitochondria in a proportion of muscle fibre segments, so called ragged red fibres (RRF; Figure 11.2). Diseases with this histological hallmark were termed mitochondrial myopathies. It became clear that the mitochondrial myopathies were a clinically, biochemically and geneti-

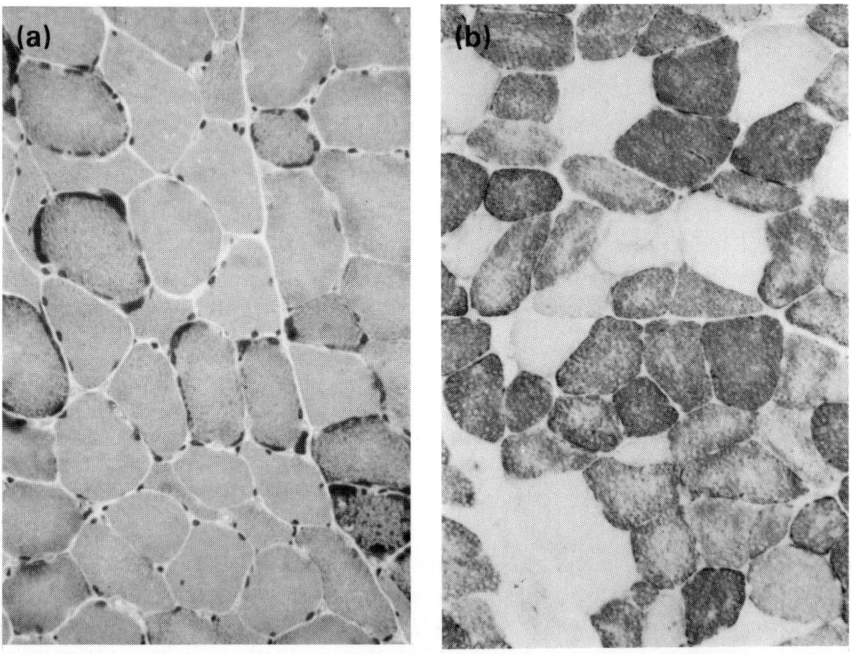

Figure 11.2. Transverse sections of muscle. (a) Ragged red fibres demonstrated by the modified Gomori trichome stain. Some fibres contain dark subsarcolemmal and intermyofibrillar deposits (abnormal accumulations of mitochondria. (b) Histochemical demonstration of cytochrome oxidase activity. Abnormal fibres show deficient or absent activity of this enzyme, which contains mtDNA-encoded subunits.

cally heterogeneous group of disorders. On genetic investigation of these disorders a striking correlation became apparent: genetic defects involving mitochondrial tRNAs, whether by deletion, point mutation or depletion, are typically associated with RRF (see Table 11.2). The precise stimulus for the proliferation of mitochondria to create ragged red change is unclear, but it appears that RRF may be provoked by defects of translation arising from tRNA dysfunction. Conversely, several disorders have been described which are associated with mtDNA defects not involving tRNAs. These disorders characteristically do not have RRF, and are not considered to be mitochondrial myopathies.

Table 11.2. Genetic classification of mtDNA-associated disease

	Inheritance	RRF?	Heteroplasmy?	Example
Primary mtDNA defects				
tRNA point mutations	Maternal	Yes	Yes	A→G$^{(3243)}$
Protein subunit reading frame point mutations	Maternal	No	Yes	G→A$^{(11778)}$
rRNA point mutations	Maternal	No	No	A→G$^{(1555)}$
mtDNA rearrangements				
Single large deletions	Sporadic	Yes	Yes	'Common' deletion (8470–13447)
Duplications	Sporadic	Yes	Yes	15056–6130
D-loop duplication	Maternal	?	Yes	302–567
Primary nuclear DNA defect with secondary effects on mtDNA				
Multiple deletions	Autosomal dominant (or recessive)	Yes	Pleioplasmy	
mtDNA depletion	?Autosomal recessive	Yes	n.a.	

Amongst mitochondrial myopathies, clustering of features may form distinctive clinical syndromes:

(i) Kearns–Sayre syndrome (KSS) is defined as progressive external ophthalmoplegia (PEO) and pigmentary retinopathy with onset earlier than the age of 20, with subsequent development of one of ataxia, heart block or raised CSF protein (Berenberg *et al.*, 1977).

(ii) Myoclonus epilepsy and ragged red fibres (MERRF) has the core features of myoclonic epilepsy and ataxia, with associated features of deafness, dementia and optic atrophy (Fukuhara *et al.*, 1980; Berkovic *et al.*, 1989). The syndrome is often familial, and analysis of several pedigrees suggests maternal inheritance.

(iii) Mitochondrial myopathy, encephalopathy, lactic acidosis and stroke-like episodes (MELAS), often associated with short stature (Pavlakis *et al.*, 1984). Transmission of MELAS occurs and is compatible with maternal inheritance.

Some cases never develop an encephalopathy, and continue with myopathic features alone, or have some, but not all, of the features of the syndromes described above. A substantial number of patients do not easily fit into any of the above syndromes (Petty *et al.*, 1986; Truong *et al.*, 1990). The value of clinical classification is controversial (Rowland *et al.*, 1991).

11.3 Classification of mtDNA-associated disease

Defects in mtDNA may be primary, or secondary to nuclear DNA defects. Primary mtDNA defects are maternally inherited and are qualitatively similar in mother and offspring, although proportions of mutant mtDNA may vary. In contrast, some disorders are inherited in Mendelian fashion, implying a causal *trans*-acting nuclear-encoded factor which induces secondary defects in mtDNA. These defects may be qualitative (multiple deletions) or quantitative (mtDNA depletion). Large-scale mtDNA rearrangements such as single mtDNA deletions and duplications generally are not inherited.

Some characteristics are common to most mtDNA-associated diseases. First, tissues have variable dependence on oxidative phosphorylation, and therefore vary in susceptibility to respiratory chain dysfunction. Organs most vulnerable include muscle, brain and heart. Second, respiratory chain function declines with age (Trounce *et al.*, 1989; Cooper *et al.*, 1992), predicting that while severe defects will be apparent in early life, more subtle respiratory chain defects may become clinically apparent only in later life. Third, the phenomenon of heteroplasmy and replicative segregation allows the possibility of variation of the proportion of mutant mtDNA and respiratory chain dysfunction between cells and tissues. Finally, defective mtDNA genomes may be subject to complex selection processes. For example, heteroplasmic mtDNA defects are generally less prevalent in leukocytes than muscle, perhaps reflecting the more rapid replicative activity of leukocytes. For the remainder of this chapter, genotype will be considered in relationship to phenotype according to a genetic classification (Table 11.2).

11.4 Primary mtDNA defects: tRNA point mutations

In 1990, Shoffner *et al.* first described a pathogenic, heteroplasmic point mutation within a tRNA gene. Since then an increasing number of point mutations in these genes have been described (Table 11.3). Diseases associated

Table 11.3. MtDNA point mutations*

Gene	Mutation (nucleotide)	Predominant clinical syndrome	Reference
Ribosomal RNA mutations			
12S rRNA	A→G$^{(1555)}$	Deafness	Prezant *et al.* (1993)
Transfer RNA mutations			
Leucine (UUR)	A→G$^{(3243)}$	MELAS and many others	See text
	A→G$^{(3260)}$	Myopathy and cardiomyopathy	Zeviani *et al.* (1991a), Sweeney *et al.* (1993)
	A→G$^{(3271)}$	MELAS	Goto *et al.* (1991)
Lysine	A→G$^{(8344)}$	MERRF	See text
	A→G$^{(8356)}$	MERRF/MELAS	Silvestri *et al.* (1992), Zeviani *et al.* (1993)
Protein subunit genes			
ND1	G→A$^{(3460)}$	LHON	Huoponen *et al.* (1991)
ND1	T→C$^{(4160)}$	LHON	Howell *et al.* (1991)
ND4	G→A$^{(11778)}$	LHON	Wallace *et al.* (1988)
ND6	T→C$^{(14484)}$	LHON	Howell *et al.* (1992b), Johns *et al.* (1992)
ATPase6	T→G$^{(8993)}$	NARP	Holt *et al.* (1990), Tatuch *et al.* (1992)

* Mutations included only if observed in more than one family.
For abbreviations see text.
Further point mutations may be implicated in LHON, but are not pathogenic alone.

with such mutations have many features in common, which are summarized below:

(i) tRNA mutations are heteroplasmic and are inherited through the maternal line. The proportion of mutant mtDNA may increase or decrease through generations.
(ii) Most, but not all, patients have RRF on muscle biopsy. In the author's experience, all clinically manifest tRNA defects are associated with histochemical abnormalities of succinic dehydrogenase and/or cytochrome oxidase activity in muscle (Figure 11.2).
(iii) Biochemical studies usually show diffuse respiratory chain defects.
(iv) Pathogenesis of disease associated with most tRNA mutations is thought to be a defect of translation caused by tRNA dysfunction (but see the A→G$^{(3243)}$ mutation below).
(v) For individual mutations the proportion of mutant mtDNA may correlate with biochemical or clinical severity, although this is not always the case. Carriers of the mutation may be asymptomatic.

(vi) The tissue distribution of the mutation is widespread, including leukocyte mtDNA. The level of heteroplasmy appears to be relatively uniform between different tissues, but is usually lower in leukocytes.

The $tRNA^{Leu(UUR)}$ $A \rightarrow G^{(3243)}$ and $tRNA^{Lys}$ $A \rightarrow G^{(8344)}$ mutations are the two most prevalent mitochondrial tRNA point mutations (Table 11.3), and have been the subject of intensive molecular and clinical investigation. The genotype–phenotype relationships of these two mutations will be discussed further in depth.

The $A \rightarrow G^{(8344)}$ mutation occurs at a conserved site in the TψC loop of $tRNA^{Lys}$, a part of the molecule thought to interact with the ribosome surface (Figure 11.3). Transfer of mitochondria carrying the mutation to mtDNA-less cell lines results in respiratory deficiency in these cells, implying that such mtDNA can induce the biochemical defect, independent of the nuclear genotype (Chomyn *et al.*, 1991). A study of mitochondrial translation in myotubes demonstrated a rough correlation between the protein size, lysine content and the magnitude of the decrease in the rate of translation (Boulet *et al.*, 1992). Thus, the mutation probably causes disease by $tRNA^{Lys}$ dysfunction, but the precise mechanism remains unclear. Interaction of the tRNA with the ribosome may be affected, charging of the tRNA may be impaired or $tRNA^{Lys}$ processing and stability may be altered.

The $A \rightarrow G^{(3243)}$ mutation affects a nucleotide within the dihydrouridine loop of the $tRNA^{Leu(UUR)}$ gene (Figure 11.3). This nucleotide is thought to form a hydrogen bond with a uridine nucleotide elsewhere in the tRNA, contributing to stabilization of the tertiary structure. There is a direct correlation between the presence of the defect and severe impairment of protein synthesis and of respiratory chain function (King *et al.*, 1992). It appears that this mutation alone is sufficient to cause these defects, although there is no correlation between the number of Leu(UUR) codons and severity of the depletion of the corresponding peptide (King *et al.*, 1992). Apart from altering $tRNA^{Leu(UUR)}$ structure, the mutation may also influence H-strand transcription. The mutation lies within the mtTERM binding site (see above). *In vitro*, the mutation impairs binding of the mtTERM protein and consequently influences termination of H-strand transcription at the 3′ end of the 16S rRNA gene (Hess *et al.*, 1991). While the ratio of rRNAs and mRNAs transcribed from the H-strand might be expected to be abnormal, this ratio is approximately constant in the presence of the mutation both *in vivo* (Hammans *et al.*, 1992b) and *in vitro* (King *et al.*, 1992). Suomalainen *et al.* (1993) confirmed that there was no deficiency of rRNA, but suggested that the defect of transcriptional control prevented effective compensation of the translational defect. Further, the defect may qualitatively affect processing of the 16S rRNA, the $tRNA^{Leu(UUR)}$ or ND1 (King *et al.*, 1992). Thus, the effects of the mutation may arise from impaired tRNA function, from defective transcriptional control or from a combination of these mechanisms.

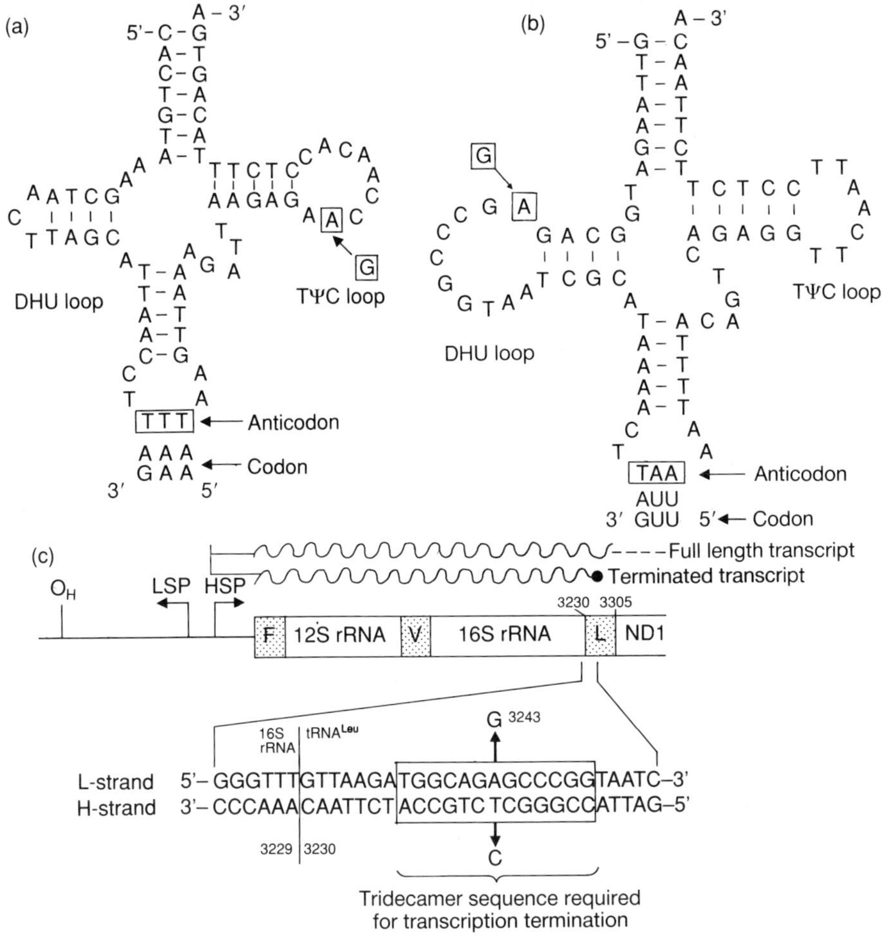

Figure 11.3. (a) tRNA^{Lys} secondary structure showing the A→G mutation at position 8344; (b) tRNA$^{Leu\ (UUR)}$ A→G$^{(3243)}$ mutation; (c) schematic of the transcription termination region, showing the A→G$^{(3243)}$ mutation within the conserved tridecamer sequence.

11.4.1 Relationship between genotype and phenotype: qualitative (Figure 11.4)

The tRNA^{Lys} A→G$^{(8344)}$ mutation has been shown by several authors to be characteristically associated with the MERRF phenotype, with the core features of myoclonus, ataxia and various types of seizure (Shoffner *et al.*, 1990, 1991; Hammans *et al.*, 1991, 1993; Zeviani *et al.*, 1991b; Silvestri *et al.*, 1993). However, it has become clear that many of these patients exhibit only some

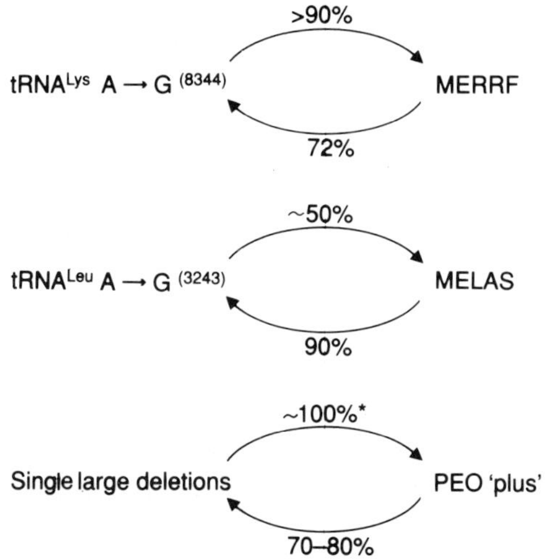

Figure 11.4. Genotype–phenotype correlation: estimated agreement between genotype and 'typical' phenotype. * Patients with Pearson's syndrome do not have PEO but may develop it if they survive.

features of the syndrome, or have additional features including almost any seen in the whole spectrum of mitochondrial myopathies, including PEO, stroke-like episodes or Leigh's syndrome. Some patients with the $A{\rightarrow}G^{(8344)}$ genotype have qualitatively quite different phenotypes, such as limb myopathy alone, multiple symmetrical lipomatosis or PEO (Berkovic *et al.*, 1991; Silvestri *et al.*, 1993; Suomalainen *et al.*, 1992a). RRF may be absent (Hammans *et al.*, 1991). Conversely, not all cases of MERRF are associated with the $A{\rightarrow}G^{(8344)}$ mutation. Both Zeviani *et al.* (1991b) and Hammans *et al.* (1991) found the mutation in five of seven MERRF families, and Silvestri *et al.* (1993) in 11 of 15, or 21 of 29 (72%) overall. Some of the remainder are accounted for by the $T{\rightarrow}C^{(8356)}$ (Silvestri *et al.*, 1992; Zeviani *et al.*, 1993) and $A{\rightarrow}G^{(3243)}$ (Hammans *et al.*, 1993) mutations; the others probably by hitherto undescribed mtDNA defects.

The $A{\rightarrow}G^{(3243)}$ mutation was first described in two series of patients with MELAS (Goto *et al.*, 1990; Kobayashi *et al.*, 1990) and it appears that the mutation is present in the vast majority of such patients (Hammans *et al.*, 1991; Ciafaloni *et al.*, 1992). Some of the MELAS patients without $A{\rightarrow}G^{(3243)}$ have been shown to have a $T{\rightarrow}C^{(3271)}$ transition (Goto *et al.*, 1991). The $A{\rightarrow}G^{(3243)}$ mutation appears to be the most prevalent of the tRNA point mutations, but of patients with this mutation only about half fit the criteria for the MELAS syndrome. Particularly prominent amongst the other phenotypes with this genotype is maternally inherited PEO, but many other diverse disorders have been described, such as myopathy alone (Hammans *et al.*, 1991), diabetes and deafness (Reardon *et al.*, 1992), and the MERRF syndrome (Hammans *et al.*, 1993).

11.4.2 Relationship between genotype and phenotype: quantitative

Shoffner *et al.* (1991) quantified the proportion of mutant mtDNA in muscle of ten patients with the A→G$^{(8344)}$ mutation. There was some correlation between the proportion of mutant mtDNA and clinical features, but only when age was taken into account. Silvestri *et al.* (1993) found no correlation between proportion of mutant mtDNA and clinical features. A clear correlation was found on analysis of leukocyte mtDNA from 22 subjects by Hammans *et al.* (1993). The proportion of mutant mtDNA was imperfectly related both to age of onset and a simple score of clinical severity (Figure 11.5), of uncertain clinical value. There are less data concerning the relationship between proportion of mutant mtDNA and phenotype in patients with the A→G$^{(3243)}$ mutation. Ciafaloni *et al.* (1992) concluded that patients had higher levels of heteroplasmy than relatives, and that symptomatic relatives had higher levels than asymptomatic relatives. The clinical value of any correlation appears doubtful.

These two transitional mutations pose tantalising questions concerning the relationship between genotype and phenotype. First, why should defects of tRNA genes correlate with characteristic phenotypes? Further, what other determinants perturb this correlation causing the disease spectrum from each mtDNA defect to overlap? For both the A→G$^{(8344)}$ mutation (Moraes *et al.*, 1992) and the A→G$^{(3243)}$ mutation (Ciafaloni *et al.*, 1991) the genetic defects appear almost uniformly distributed between affected and unaffected tissues. While this does not exclude variation in heteroplasmy between different cells in a tissue, it is unlikely that uneven segregation of mutant mtDNA can explain the differing

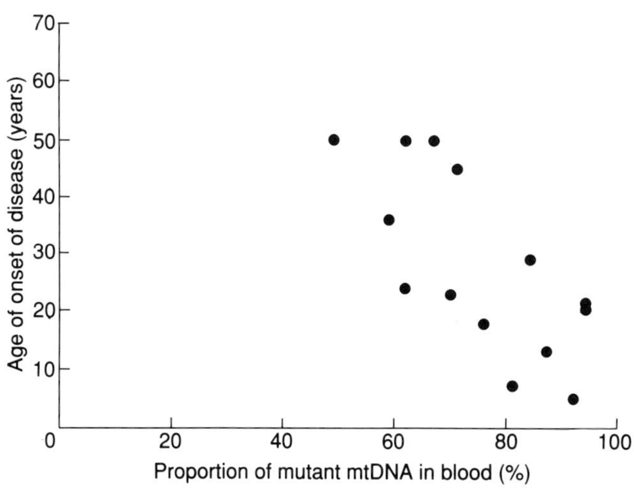

Figure 11.5. The tRNA^{Lys} A→G$^{(8344)}$ mutation: proportion of mutant mtDNA vs. age of onset. Reproduced from Hammans *et al.* (1993) with permission from Oxford University Press.

tissue specificities of the mutations. While both tRNA mutations induce translational defects of all mtDNA-encoded subunits of the respiratory chain, the effects of each mutation are likely to be unique, variably affecting each polypeptide both quantitatively and qualitatively. Evidence exists to support this. For example, in the context of the A→G$^{(8344)}$ mutation, both Chomyn *et al.* (1991) and Boulet *et al.* (1992) found a deficiency of translation of polypeptides that reflected their lysine content and both also noted the presence of a prominent abnormal translation product. Moreover, a tRNA^{Lys} T→C$^{(8356)}$ transition (Silvestri *et al.*, 1992; Zeviani *et al.*, 1993) is associated with the MERRF phenotype, and a tRNA$^{Leu(UUR)}$ T→C$^{(3271)}$ transition (Goto *et al.*, 1991) is associated with the MELAS phenotype. These observations allow speculation that mutations within specific tRNA genes are (imperfectly) associated with specific phenotypes. It is known that some of the respiratory chain subunits have tissue-specific isoforms. Thus, it is possible that individual tRNA defects produce a characteristic abnormal pattern in respiratory chain polypeptides which, in interaction with tissue-specific expression of nuclear-encoded proteins and other factors, produces dysfunction in certain cell types, e.g. in the cells of the brain cortex to produce myoclonus (A→G$^{(8344)}$ mutation). The variability of phenotype with each mutation is unexplained, but epigenetic factors are probably important. Some phenotypes (e.g deafness, PEO) are relatively constant in individual families, perhaps because of simultaneous inheritance of modifying genes on either genome.

11.5 Primary mtDNA defects: protein subunit reading frame point mutations

11.5.1 Leber's hereditary optic neuropathy

Leber's hereditary optic neuropathy (LHON) is an uncommon condition characteristically causing simultaneous or consecutive bilateral visual loss in young adult males. At presentation, the optic discs have a swollen appearance with tortuous retinal arterioles and peripapillary telangiectases, although no leakage occurs on fluorescein angiography. Optic atrophy follows, with persistent loss of visual acuity usually to 6/60 or less. LHON has long been recognized as showing maternal inheritance; in contrast to X-linked inheritance, descendants of males are never affected. In 1988, Wallace *et al.* described a G→A$^{(11778)}$ mutation in association with the disease, which may be either homoplasmic or heteroplasmic. Approximately two-thirds of LHON families carry this mutation, converting a highly conserved arginine to histidine within subunit 4 of complex I (ND4). Genetic heterogeneity has been confirmed by the description of further pathogenic point mutations (Table 11.3). Other mutations have been described in LHON pedigrees which are also found in the normal population. It has been suggested that such mutations, in combination, are sufficient to cause disease (Brown *et al.*, 1992). A biochemical study of a family with G→A$^{(11778)}$ homoplasmy showed a polarographically determined complex I defect in muscle (Larsson *et al.*, 1991).

Although clearly associated with LHON in many families, mtDNA mutations do not appear to be the sole determinant of the phenotype. In some families the disease affects other parts of the CNS, producing dystonia, ataxia and pyramidal deficit (Novotny *et al.*, 1986; Howell *et al.*, 1991). Some females with the G→A$^{(11778)}$ mutation have a multiple sclerosis-like illness (Harding *et al.*, 1992b). In many patients and relatives, the G→A$^{(11778)}$ mutation is heteroplasmic. Individuals with a lower proportion of mutant mtDNA have a lower risk of developing symptoms or transmitting the disease. This observation is of little clinical value; some males with homoplasmic mutant mtDNA remain unaffected. It can be concluded that the mtDNA defect is necessary but not sufficient for the disease to be expressed. One attractive hypothesis to explain the excess of affected males and peculiar inheritance of this disorder is a two locus model involving loci on mtDNA and the X chromosome. This is consistent with data from segregation analysis (Bu and Rotter, 1991). Vilkki *et al.* (1991) suggested linkage to a visual loss susceptibility allele close to DXS7, but this was subsequently excluded (Sweeney *et al.*, 1992). It remains possible that alternative loci on the X chromosome, or elsewhere on the genome, determine susceptibility. Alternative hypotheses include phenotype determination by hormonal, environmental or immunological factors. The importance of non-genetic influences is supported by the dramatic subacute onset which is unusual in a genetically determined disease. In favour of the immunological hypothesis is the appearance of the acutely symptomatic optic disc, suggestive of an inflammatory microangiopathy. Further, some complex I mutations cause an amino acid substitution in a hydrophilic loop of their respective complex I subunit, which could potentially provoke autoantibodies. Thus it seems likely that the disease is determined by bigenomic interaction as well as putative non-genetic determinants.

11.5.2 The ATPase subunit 6 mutation

Holt *et al.* (1990) described a T→G$^{(8933)}$ mutation in a family without RRF on muscle biopsy with the core features of neurogenic muscle weakness, ataxia and retinitis pigmentosa (NARP). A further family described by Tatuch *et al.* (1992) included members with similar clinical features, but also an infant dying with lesions typical of Leigh's syndrome. The Leigh's syndrome phenotype is common to several mtDNA defects when expressed at their most severe, including A→G$^{(8344)}$ (Hammans *et al.*, 1991; Berkovic *et al.*, 1991) and mtDNA deletions (Yamamoto *et al.*, 1991). The heteroplasmic T→G$^{(8933)}$ mutation converts a highly conserved hydrophobic leucine to a hydrophilic arginine, probably interfering with the H^+ channel formed by subunits 6 and 9 of ATPase, and thereby ATP synthesis. A study of oxidative phosphorylation in lymphoblast lines showed slowing of ATP production with all substrates, effectively reducing ATP production (Tatuch and Robinson, 1993). A reasonably close correlation of clinical severity with the proportion of mutant mtDNA has been observed, and this led to the performance of a prenatal test on a mother who had already had a severely affected child (Harding *et al.*, 1992a). The chorionic villus biopsy showed a high proportion of mutant mtDNA and the pregnancy was terminated. The fetal tissues showed a uniformly high level of mutant. This remains the only reported prenatal test in the context of a mtDNA-associated disease.

11.6 Primary mtDNA defects: mitochondrial rRNA point mutation

In China, up to one quarter of deaf mutes can relate their deafness to the use of aminoglycosides, which are in widespread use. Of those with other family members affected, inheritance is almost always matrilineal (Hu *et al.*, 1991). Because aminoglycosides target the bacterial ribosome, the evolutionary-related mitochondrial ribosome was a likely candidate for aminoglycoside ototoxicity. Sequencing of the mtDNA 12S rRNA gene showed a homoplasmic $A \rightarrow G^{(1555)}$ substitution in three unrelated Chinese pedigrees (Prezant *et al.*, 1993). This genotype was also found in a large Arab-Israeli family with non-syndromic deafness without aminoglycoside exposure. This nucleotide is evolutionarily conserved, and the mutation was not found in a large number of controls (Prezant *et al.*, 1993). The nucleotide substitution within the 12S rRNA molecule occurs at the aminoacyl site at which mRNAs are decoded, which is located at the ribosomal subunit interface. Aminoglycosides affect translational fidelity by binding to this decoding region and stabilizing mismatched aminoacyl tRNAs. The mutation has the effect of producing a new G–C bond, making the secondary structure similar to the bacterial small rRNA. Prezant *et al.* (1993) suggested that this increased the susceptibility of mitochondrial rRNA to the effects of aminoglycosides.

Members of the Arab-Israeli family had either severe or no deafness, without intermediate hearing impairment. Formal segregation analysis predicted that the phenotype occurs with the simultaneous occurrence of the mtDNA mutation and an autosomal recessive mutation. Prezant and colleagues suggested a ‘two hit’ model, with the mtDNA defect being necessary, but not sufficient, to cause the disease. Accordingly, expression of the phenotype additionally requires either the administration of aminoglycoside, or the presence of a hypothetical cochlea-specific ribosomal subunit.

11.7 Large-scale mtDNA rearrangements: single large deletions

Holt *et al.* (1988, 1989) first established the association of mtDNA defects with human disease when they described heteroplasmic large deletions of mtDNA of up to 7 kb. In muscle, the proportion of mutant mtDNA varied between 20 and 70%. In blood, deleted mtDNA is present in a lower proportion (usually undetectable by Southern blot). Almost all patients with single deletions, with notable exceptions (Bresolin *et al.*, 1991; Ballinger *et al.*, 1992), have the clinical feature of PEO, often with one or more other characteristics of the KSS. Heteroplasmic deletions of muscle mtDNA are found in 70–80% of patients with PEO, PEO 'plus' or KSS (Holt *et al.*, 1989; Moraes *et al.*, 1989). Pearson's marrow–pancreas syndrome presents in infancy as a refractory sideroblastic anaemia, with thrombocytopenia, neutropenia and exocrine pancreatic failure amongst other features. Although many patients die before the age of 3 years (Pearson *et al.*, 1979), some survive to develop KSS in later childhood (McShane *et al.*, 1991). In Pearson's syndrome large mtDNA deletions are present in muscle and blood, suggesting that this syndrome and KSS are age-specific manifestations of one genetic disease (Rotig *et al.*, 1988, 1990).

There are approximately 130 patients with mtDNA deletions described to date. About one-third of deletions are identical, being flanked by a 13 bp direct repeat existing at bp8470–8482 and bp13447–13459, and designated the 'common deletion'. The high incidence of this deletion may arise from the length of the flanking repeat. In a study of 28 patients with 17 different mtDNA deletions (Mita *et al.*, 1990), 12 had the common deletion and eight had deletions flanked by other direct repeats of 5–11 bp in length. The remaining eight were flanked by imperfect repeats or by non-homologous sequences, suggesting that pathogenetic mechanisms may be heterogeneous.

Apart from a family with several atypical features (Ballinger *et al.*, 1992), disease associated with single deletions is sporadic. There is only one other report of a deletion arising in more than one symptomatic member of a family (Ozawa *et al.*, 1988); the deletions, found in mother and daughter, differed in site and size. Using Southern blot or PCR, deleted molecules have not been detected consistently in muscle or blood of mothers or siblings of patients with deletions, suggesting that deletions arise as fresh mutational events. Brockington and colleagues (1993) found a low abundance heteroplasmic 165 bp tandem duplication within the D-loop region of mtDNA in 18/58 patients with deletions and 5/5 of their mothers. The insertion is probably relevant to deletion pathogenesis; it is located in a region implicated in both replication and transcription, is absent in controls and is heteroplasmic. Two main hypotheses may be used to explain the association of the D-loop duplication with deletion of mtDNA. RNA primers for replication of mtDNA commence at the L-strand promoter. Partial duplication of this and other elements may result in two displaced single-stranded H-strands, increasing the opportunity for slip replication and deletion formation. The scarcity of duplicated mtDNA compared to deleted mtDNA is against this hypothesis, but could be explained if the duplication was transient and spontaneously disappeared. Alternatively, both duplication and deletion could be one of a family of mtDNA rearrangements, present in some patients because of a propensity for mtDNA duplication or deletion engendered by mtDNA sequence, nuclear genes or by environmental influences.

Studies using *in situ* hybridization and other techniques have shown that, in muscle, deleted mtDNA genomes are concentrated within histochemically abnormal fibre segments, and are rare elsewhere (Hammans *et al.*, 1992b; Oldfors *et al.*, 1992; Moraes *et al.*, 1992). Since all reported deletions encompass tRNA genes, deficiency of these tRNAs would prevent translation. Hayashi and colleagues (1991) demonstrated that, when the proportion of deleted mtDNA exceeded 60%, the translation defect became increasingly severe. It could be predicted that deletions would cause a diffuse respiratory chain defect, which is indeed the case. However, a subgroup of patients with pure complex I defects have deletions that do not involve COX genes. In addition, such patients tend to have less severe histochemical COX deficiency (Holt *et al.*, 1989; Hammans *et al.*, 1992a). These findings support the idea that there is a limited correlation between mtDNA deletion site and biochemical phenotype, and imply the presence of at least some translation of mRNA from deleted mtDNA. Since deleted mtDNA lacks one or more tRNAs, sharing of tRNAs (complementation) between deleted and normal mtDNA must occur to some extent, indicating that such genomes must be in close proximity within the same mitochondrion. It has been suggested that complementation occurs in the junctional segment between ragged red and histochemically normal fibre segments (Hammans *et al.*, 1992b).

While there may be limited correlation of deletion site with biochemical phenotype, there is no correlation between the deletion site and the clinical phenotype. In general, the proportion of deleted mtDNA is also unrelated to clinical course, except for very early onset disease, which does tend to be associated with high proportions of deleted mtDNA (author's unpublished observations). PEO appears to be a constant manifestation of mtDNA deletions. Nevertheless, the phenotype varies hugely from the Pearson–KSS spectrum with early onset, severe disability and premature death, to late onset oculoskeletal myopathy, often with little disability. One explanation for this variation may be the distribution of the mtDNA defect between organs. Patients with KSS show widespread but uneven distribution of deleted mtDNA in tissues examined at autopsy (Moraes *et al.*, 1989; Shanske *et al.*, 1990; Zeviani *et al.*, 1990b). Patients with Pearson's syndrome tend to have higher proportions of deleted mtDNA, with deleted mtDNA easily detectable in blood. Tissue distribution of deleted mtDNA may be related to the timing of deletion formation. This probably occurs at or after oogenesis since deleted mtDNA has not been detected in mothers of patients. Modification of the proportion of deleted mtDNA by selection is likely in the case of mtDNA deletions since levels in leukocytes are usually less than 5%, indicating that deletions probably confer a selective disadvantage in this tissue.

11.8 Large-scale mtDNA rearrangements: mtDNA duplications

Duplications are ostensibly rare defects accompanying similar clinical phenotypes to mtDNA deletions, but a mtDNA duplication can be mistaken for a deletion on standard Southern blot analysis. Diabetes mellitus is more commonly associated with duplication than with deletions. Poulton *et al.* (1993) provided evidence that duplications may be a transient form in the origin of deletions. Ballinger and colleagues (1992) described a maternally inherited deletion in a family with diabetes and deafness. The deletion includes the origin of L-strand replication, making it difficult to understand the mode of replication of this genome. Poulton and colleagues suggested that the data presented by Ballinger *et al.* were consistent with the presence of a duplication, which also allows an explanation of replication of the molecule. Dunbar and colleagues (1993) described a maternally inherited duplication also associated with diabetes mellitus. The pathophysiology of these genetic defects is obscure. Unlike mtDNA deletions, there is no absence of specific tRNAs or protein-coding genes. Interference with transcription is one possible mechanism.

11.9 Primary nuclear genome defects: multiple deletions of mtDNA

The typical syndrome of multiple familial mtDNA deletions was first reported by Zeviani and colleagues (1989, 1990a) and since by others (Servidei *et al.*, 1991). The disorder is inherited in simple autosomal dominant fashion, with the characteristic clinical features of adult onset PEO, proximal myopathy and deafness. Cataracts, ataxia and peripheral neuropathy are also commonly observed. Muscle biopsy shows RRF and COX deficient fibres as well as neurogenic changes. Analysis of muscle mtDNA identifies the normal band of 16.5 kb, but also several smaller bands, which represent genomes with different deletions (pleioplasmy). The deletions were not detectable in blood. There may be an autosomal recessive form of the syndrome; two siblings with similar features were born to consanguineous healthy parents (Mizusawa *et al.*, 1988; Yuzaki *et al.*, 1989). Following the recognition of the typical clinical features associated with multiple deletions, a number of other phenotypes have been described. These include idiopathic dilated cardiomyopathy (Suomalainen *et al.*, 1992b), recurrent exertional myoglobinuria (Ohno *et al.*, 1991) and a syndrome of ataxia and ketoacidotic comas, with subsequent external ophthalmopegia, deafness and spasticity (Cormier *et al.*, 1991).

Sequencing of pleioplasmic deleted molecules generally identifies flanking direct repeats similar to those found in most single deletions but, in contrast to single deletions, flanking repeats often show imperfections. Candidate genes in the autosomal dominant multiple deletion syndrome are the nuclear-encoded genes that are involved in mtDNA replication, with the implicit hypothesis that unfaithful replication leads to accumulation of mtDNA deletions. Linkage analysis to localize the disease gene is underway (Zeviani, 1992). The late onset of the disorder is consistent with slow accumulation of mtDNA defects with increasing age. The clinical features of this syndrome resemble phenotypes observed in association with single deletions, suggesting that multiple deletions may cause pathophysiological processes similar to single deletions. The rare variant phenotypes have similar underlying mtDNA defects and the reason for the clinical variation is obscure.

11.10 Primary nuclear genome defects: mtDNA depletion

Moraes *et al.* (1991) studied two cousins with a fatal mitochondrial disease, affecting muscle in one and liver in the other. Quantitative analysis demonstrated a severe depletion of mtDNA in affected tissues. Similar findings of mtDNA depletion in affected tissues (muscle; muscle and kidney) were observed in two unrelated infants. Tritschler *et al.* (1992) studied five further children, with mitochondrial myopathy manifesting within or soon after the first year of life, all of whom had depletion of muscle mtDNA (2–34% of normal). Two of these patients were siblings born to healthy non-consanguineous parents. The authors speculated that the non-maternal mode of inheritance indicated a nuclear gene defect, probably with an autosomal recessive mode of inheritance. One hypothetical mechanism involves failure of the resumption of mtDNA replication after early embryogenesis. Resumption of replication at subtly varying stages of early embryo development would have a large effect on different groups of stem cells, explaining why differing tissue expression occurs in members of the same family.

While the responsible nuclear gene remains unknown, the pathophysiology of this group of disorders is easier to understand. There is proliferation of mitochondria in affected muscle, giving rise to RRF, even though the organelles are depleted of mtDNA. Such fibres are COX deficient, although expressing succinic dehydrogenase activity, which is nuclearly encoded. Biochemical analysis shows a diffuse respiratory chain defect. There is a close relationship between organ failure and the tissue-specific mtDNA depletion.

11.11 Conclusions

Defects of the mitochondrial genome exert their effects by disruption of the structure and function of the respiratory chain or ATP synthase. The common endpoint is impairment of oxidative phosphorylation and consequent deficiency of ATP, but the underlying pathophysiology and consequent biochemical and clinical phenotypes are remarkably diverse. There is an undoubted relationship between genotype and phenotype, but this is imperfect. Although many of the genetic defects of mtDNA-associated disease have been uncovered, these beg questions concerning pathophysiological mechanisms. Why should a single point mutation ($tRNA^{Leu(UUR)}$ $A \rightarrow G^{(3243)}$) typically cause disease with the central feature of stroke-like crises, while patients with the same mutation have quite different diseases such as the relatively benign syndrome of diabetes and deafness? Molecular biology has provided some clues to the relationship between genotype and phenotype, but this area promises to interest for years to come, not least because of the increased complexity of interactions between two genomes.

References

Anderson S, Bankier AT, Barrell BG, de Bruijn MH, Coulson AR, Drouin J, Eperon IC, Nierlich DP, Roe BA, Sanger F, Schreier PH, Smith AJ, Staden R, Young IG. (1981) Sequence and organization of the human mitochondrial genome. *Nature* **290:** 457–465.

Ballinger SW, Shoffner JM, Hadaya EV, Trounce I, Polak MA, Koontz DA, Wallace DC. (1992) Maternally transmitted diabetes and deafness associated with a 10.4 kb mitochondrial DNA deletion. *Nature Genetics* **1:** 11–15.

Berenberg RA, Pellock JM, DiMauro S, Schotland DL, Bonilla E, Eastwood A, Hays AP, Vicale CT, Behrens M, Chutorian A, Rowland LP. (1977) Lumping or splitting? 'Ophthalmoplegia-plus' or Kearns–Sayre syndrome? *Ann. Neurol.* **1:** 37–54.

Berkovic SF, Carpenter S, Evans A, Karpati G, Shoubridge EA, Andermann F, Meyer E, Tyler JL, Diksic M, Arnold D, Woolfe LS, Andermann E, Hakim AM. (1989) Myoclonus epilepsy and ragged-red fibres (MERRF). 1. A clinical, pathological, biochemical, magnetic resonance spectrographic and positron emission tomographic study. *Brain* **112:** 1231–1260.

Berkovic SF, Shoubridge EA, Andermann F, Andermann E, Carpenter S, Karpati G. (1991) Clinical spectrum of mitochondrial DNA mutation at base pair 8344. *Lancet* **338:** 457.

Boulet L, Karpati G, Shoubridge EA. (1992) Distribution and threshold expression of the $tRNA^{Lys}$ mutation in skeletal muscle of patients with myoclonic epilepsy and ragged red fibers (MERRF). *Am. J. Hum. Genet.* **51:** 1187–1200.

Bresolin N, Martinelli P, Barbiroli B, Zaniol P, Ausenda C, Montagna P, Gallanti A, Comi GP, Scarlato G, Lugaresi E. (1991) Muscle mitochondrial DNA deletion and ^{31}P–NMR spectroscopy alterations in a migraine patient. *J. Neurol. Sci.* **104:** 182–189.

Brockington M, Sweeney MG, Hammans SR, Morgan-Hughes JA, Harding AE. (1993) A tandem duplication in the D-loop of human mitochondrial DNA is associated with deletions in mitochondrial myopathies. *Nature Genetics* **4:** 67–71.

Brown MD, Voljavec AS, Lott MT, Torroni A, Yang CC, Wallace DC. (1992) Mitochondrial DNA complex I and III mutations associated with Leber's hereditary optic neuropathy. *Genetics* **130:** 163–173.

Bu X, Rotter JI. (1991) X-chromosome linked and mitochondrial gene control of Leber hereditary optic neuropathy: evidence from segregation analysis for dependence on X chromosome inactivation. *Proc. Natl Acad. Sci. USA* **88:** 8198–8202.

Chomyn A, Meola G, Bresolin N, Lai S, Scarlato G, Attardi G. (1991) *In vitro* genetic transfer of protein synthesis and respiration defects to mitochondrial DNA-less cells with myopathy-patient mitochondria. *Mol. Cell. Biol.* **11:** 2236–2244.

Ciafaloni E, Ricci E, Servidei S, Shanske S, Silvestri G, Manfredi G, Schon EA, DiMauro S. (1991) Widespread tissue distribution of a $tRNA^{Leu(UUR)}$ mutation in the mitochondrial DNA of a patient with MELAS syndrome. *Neurology* **41:** 1663–1665.

Ciafaloni E, Ricci E, Shanske S, Moraes CT, Silvestri G, Hirano M, Simonetti S, Angelini C, Donati A, Garcia C, Martinuzzi A, Mosewich R, Servidei S, Zammarchi E, Bonilla E, DeVivo DC, Rowland LP, Schon EA, DiMauro S. (1992) MELAS: clinical features, biochemistry, and molecular genetics. *Ann. Neurol.* **31:** 391–398.

Cooper JM, Mann VM, Schapira AHV. (1992) Analyses of mitochondrial respiratory chain function and mitochondrial DNA deletion in human skeletal muscle: effect of ageing. *J. Neurol. Sci* **113:** 91–98.

Cormier V, Rotig A, Tardieu M, Colonna M, Saudubray J-M, Munnich A. (1991) Autosomal dominant deletions of the mitochondrial genome in a case of progressive encephalopathy. *Am. J. Hum. Genet.* **48:** 643–648.

Dunbar DR, Moonie PA, Swingler RJ, Davidson D, Roberts R, Holt IJ. (1993) Maternally transmitted partial direct tandem duplication of mitochondrial DNA associated with diabetes mellitus. *Hum. Mol. Genet.* **2:** 1619–1624.

Fukuhara N, Tokiguchi S, Shirakawa K, Tsubaki T. (1980) Myoclonus epilepsy associated with ragged-red fibres (mitochondrial abnormalities): disease entity or a syndrome? Light- and electron-microscopic studies of two cases and review of literature. *J. Neurol. Sci.* **47:** 117–133.

Goto Y, Nonaka I, Horai S. (1990) A mutation in the tRNA$^{Leu(UUR)}$ gene associated with the MELAS subgroup of mitochondrial encephalomyopathies. *Nature* **348:** 651–653.

Goto Y, Nonaka I, Horai S. (1991) A new mtDNA mutation associated with mitochondrial myopathy, encephalopathy, lactic acidosis and stroke-like episodes (MELAS). *Biochim. Biophys. Acta* **1097:** 238–240.

Hammans SR, Sweeney MG, Brockington M, Morgan-Hughes JA, Harding AE. (1991) Mitochondrial encephalopathies: molecular genetic diagnosis from blood samples. *Lancet* **337:** 1311–1313.

Hammans SR, Sweeney MG, Holt IJ, Cooper JM, Toscano A, Clark JB, Morgan-Hughes JA, Harding AE. (1992a) Evidence for intramitochondrial complementation between deleted and normal mitochondrial DNA in some patients with mitochondrial myopathy. *J. Neurol. Sci.* **107:** 87–92.

Hammans SR, Sweeney MG, Wicks DAG, Morgan-Hughes JA, Harding AE. (1992b) A molecular genetic study of focal histochemical defects in mitochondrial encephalomyopathies. *Brain* **115:** 343–365.

Hammans SR, Sweeney MG, Brockington M, Lennox GG, Lawton NF, Morgan-Hughes JA, Kennedy CR, Harding AE. (1993) The mitochondrial DNA transfer RNA^{Lys} A–G$^{(8344)}$ mutation: clinical phenotype and relationship to proportion of mutant mitochondrial DNA. *Brain* **116:** 617–632.

Harding AE, Holt IJ, Sweeney MG, Brockington M, Davis MB. (1992a) Prenatal diagnosis of mitochondrial DNA^{8993} T–G disease. *Am. J. Hum. Genet.* **50:** 629–633.

Harding AE, Sweeney MG, Miller DH, Mumford CJ, Kellar-Wood H, Menard D, McDonald WI, Compston DAS. (1992b) Occurrence of a multiple sclerosis-like illness in women who have a Leber's hereditary optic neuropathy mitochondrial DNA mutation. *Brain* **115:** 979–989.

Hayashi J, Ohta S, Kikuchi A, Takemitsu M, Goto Y, Nonaka I. (1991) Introduction of disease-related mitochondrial DNA deletions into HeLa cells lacking mitochondrial DNA results in mitochondrial dysfunction. *Proc. Natl Acad. Sci. USA* **88:** 10614–10618.

Hess JF, Parisi MA, Bennett JL, Clayton DA. (1991) Impairment of mitochondrial transcription termination by a point mutation associated with the MELAS subgroup of mitochondrial encephalomyopathies. *Nature* **351:** 236–239.

Holt IJ, Harding AE, Morgan-Hughes JA. (1988) Deletions of muscle mitochondrial DNA in patients with mitochondrial myopathies. *Nature* **331:** 717–719.

Holt IJ, Harding AE, Cooper JM, Schapira AHV, Toscano A, Clark JB, Morgan-Hughes JA. (1989) Mitochondrial myopathies: clinical and biochemical features of 30 patients with major deletions of muscle mitochondrial DNA. *Ann. Neurol.* **26:** 699–708.

Holt IJ, Harding AE, Petty RKH, Morgan-Hughes JA. (1990) A new mitochondrial disease associated with mitochondrial DNA heteroplasmy. *Am. J. Hum. Genet.* **46:** 428–433.

Howell N, Kubacka I, Xu M, McCullough DA. (1991) Leber hereditary optic neuropathy: involvement of the mitochondrial NDI gene and evidence for an intragenic suppressor mutation. *Am. J. Hum. Genet.* **48:** 935–942.

Howell N, Halvorson S, Kubacka I, McCullough DA, Bindoff LA, Turnbull DM. (1992a) Mitochondrial gene segregation in mammals: is the bottleneck always narrow? *Hum. Genet.* **90:** 117–120.

Howell N, McCullough D, Bodis Wollner I. (1992b) Molecular genetic analysis of a sporadic case of Leber hereditary optic neuropathy. *Am. J. Hum. Genet.* **50:** 443–446.

Hu D-N, Qiu W-Q, Wu B-T, Fang L-Z, Zhou F, Gu Y-P, Zhang QH, Yan J-H, Ding Y-Q, Wong H. (1991) Genetic aspects of antibiotic induced deafness: mitochondrial inheritance. *J. Med. Genet.* **28:** 79–83.

Huoponen K, Vilkki J, Aula P, Nikoskelainen EK. (1991) A new mtDNA mutation associated with Leber hereditary optic neuroretinopathy. *Am. J. Hum. Genet.* **48:** 1147–1153.

Johns DR, Neufeld MJ, Park RD. (1992) An ND-6 mitochondrial DNA mutation associated with Leber hereditary optic neuropathy. *Biochem. Biophys. Res. Commun.* **187:** 1551–1557.

King MP, Koga Y, Davidson M, Schon EA. (1992) Defects in mitochondrial protein synthesis and respiratory chain activity segregate with the $tRNA^{Leu(UUR)}$ mutation associated with mitochondrial encephalopathy, lactic acidosis, and strokelike episodes. *Mol. Cell. Biol.* **12:** 480–490.

Kobayashi Y, Momoi MY, Tominaga K, Momoi T, Nihei K, Yanagisawa M, Kagawa Y, Ohta S. (1990) A point mutation in the mitochondrial $tRNA^{(Leu)(UUR)}$ gene in MELAS (mitochondrial myopathy, encephalopathy, lactic acidosis and stroke-like episodes). *Biochem. Biophys. Res. Commun.* **173:** 816–822.

Larsson NG, Andersen O, Holme E, Oldfors A, Wahlstrom J. (1991) Leber's hereditary optic neuropathy and complex I deficiency in muscle. *Ann. Neurol.* **30:** 701–708.

McShane MA, Hammans SR, Sweeney M, Holt IJ, Beattie TJ, Brett EM, Harding AE. (1991) Pearson syndrome and mitochondrial encephalomyopathy in a patient with a deletion of mtDNA. *Am. J. Hum. Genet.* **48:** 39–42.

Mita S, Rizzuto R, Moraes CT, Shanske S, Arnaudo E, Fabrizi GM, Koga Y, DiMauro S, Schon EA. (1990) Recombination via flanking direct repeats is a major cause of large-scale deletions of human mitochondrial DNA. *Nucleic Acids Res.* **18:** 561–567.

Mitchell P. (1976) Possible molecular mechanism of the proton motive function of cytochrome systems. *J. Theoret. Biol.* **62:** 327–367.

Mizusawa H, Watanabe M, Kanazawa I, Nakanishi T, Kobayashi M, Tanaka M, Suzuki H, Nishikimi M, Ozawa T. (1988) Familial mitochondrial myopathy associated with peripheral neuropathy: partial deficiencies of complex I and complex IV. *J. Neurol. Sci.* **86:** 171–184.

Moraes CT, DiMauro S, Zeviani M, Lombes A, Shanske S, Miranda AF, Nakase H, Bonilla E, Werneck LC, Servidei S, Nonaka I, Koga Y, Spiro AJ, Brownwell AKW, Schmidt B, Schotland DL, Zupanc M, DeVivo DC, Schon EA, Rowland LP. (1989) Mitochondrial DNA deletions in progressive external ophthalmoplegia and Kearns–Sayre syndrome. *N. Engl. J. Med.* **320:** 1293–1299.

Moraes CT, Shanske S, Tritschler HJ, Aprille JR, Andreetta F, Bonilla E, Schon EA, DiMauro S. (1991) mtDNA depletion with variable tissue expression: a novel genetic abnormality in mitochondrial diseases. *Am. J. Hum. Genet.* **48:** 492–501.

Moraes CT, Ricci E, Petruzzella V, Shanske S, DiMauro S, Schon EA, Bonilla E. (1992) Molecular analysis of the muscle pathology associated with mitochondrial deletions. *Nature Genetics* **1:** 359–367.

Novotny EJ Jr, Singh G, Wallace DC, Dorfman LJ, Louis A, Sogg RL, Steinman L. (1986) Leber's disease and dystonia: a mitochondrial disease. *Neurology* **36:** 1053–1060.

Ohno K, Tanaka M, Sahashi K, Ibi T, Sato W, Yamamoto T, Takahashi A, Ozawa T. (1991) Mitochondrial DNA deletions in inherited recurrent myoglobinuria. *Ann. Neurol.* **29:** 364–369.

Oldfors A, Larsson N-G, Holme E, Tulinius M, Kadenbach B, Droste M. (1992) Mitochondrial DNA deletions and cytochrome *c* oxidase deficiency in muscle fibres. *J. Neurol. Sci.* **110:** 169–177.

Olson W, Engel WK, Walsh GO, Einaugler R. (1972) Oculocraniosomatic neuromuscular disease with 'ragged-red' fibers; histochemical and ultrastructural changes in limb muscles of a group of patients with idiopathic progressive external ophthalmoplegia. *Arch. Neurol.* **26:** 193–211.

Ozawa T, Yoneda M, Tanaka M, Ohno K, Sato W, Suzuki H, Nishikimi M, Yamamoto M, Nonaka I, Horai S. (1988) Maternal inheritance of deleted mitochondrial DNA in a family with mitochondrial myopathy. *Biochem. Biophys. Res. Commun.* **154:** 1240–1247.

Pavlakis SG, Phillips PC, DiMauro S, De Vivo DC, Rowland LP. (1984) Mitochondrial myopathy, encephalopathy, lactic acidosis, and strokelike episodes: a distinctive clinical syndrome. *Ann. Neurol.* **16:** 481–488.

Pearson HA, Lobel JS, Kocoshis SA, Naiman JL, Windmiller J, Lammi AT, Hoffman R, Marsh JC. (1979) A new syndrome of refractory sideroblastic anaemia with vacuolisation of bone marrow precursors and exocrine pancreatic dysfunction. *J. Pediatr.* **95:** 976–984.

Petty RKH, Harding AE, Morgan-Hughes JA. (1986) The clinical features of mitochondrial myopathy. *Brain* **109:** 915–938.

Poulton J, Deadman ME, Bindoff L, Morten K, Land J, Brown G. (1993) Families of mtDNA re-arrangements can be detected in patients with mtDNA deletions: duplications may be a transient intermediate form. *Hum. Mol. Gen.* **2:** 23–30.

Prezant TR, Agapian JV, Bohlmann MC, Bu X, Öztas S, Qiu W-Q, Arnos KS, Cortopassi G, Jaber L, Rotter JI, Shohat M, Fischel-Ghodsian N. (1993) Mitochondrial ribosomal RNA mutation associated with both antibiotic induced and non-syndromic deafness. *Nature Genetics* **4:** 289–294.

Reardon W, Ross RJM, Sweeney MG, Luxon LM, Pembrey ME, Harding AE, Trembath RC. (1992) Diabetes mellitus associated with a pathogenic point mutation in mitochondrial DNA. *Lancet* **340:** 1376–1379.

Rotig A, Colonna M, Blanche S, Fischer A, Le Deist F, Frezal J, Saudubray JM, Munnich A. (1988) Deletion of blood mitochondrial DNA in pancytopenia. *Lancet* **2:** 567–568.

Rotig A, Cormier V, Blanche S, Bonnefont J-P, Ledeist F, Romero N, Schmitz J, Rustin P, Fischer A, Saudubray JM, Munnich A. (1990) Pearson's marrow–pancreas syndrome: a multisystem mitochondrial disorder in infancy. *J. Clin. Invest.* **86:** 1601–1608.

Rowland LP, Blake DM, Hirano M, DiMauro S, Schon EA, Hays AP, DeVivo DC. (1991) Clinical syndromes associated with ragged red fibres. *Rev. Neurol.* **147:** 467–473.

Servidei S, Zeviani M, Manfredi G, Ricci E, Silvestri G, Bertini E, Gellera C, Di Mauro S, Di Donato S, Tonali P. (1991) Dominantly inherited mitochondrial myopathy with multiple deletions of mitochondrial DNA: clinical, morphologic, and biochemical studies. *Neurology* **41:** 1053–1059.

Shanske S, Moraes CT, Lombes A, Miranda AF, Bonilla E, Lewis P, Whelan MA, Ellsworth CA, DiMauro S. (1990) Widespread tissue distribution of mitochondrial DNA deletions in Kearns–Sayre syndrome. *Neurology* **40:** 24–28.

Shoffner JM, Lott MT, Lezza AM, Seibel P, Ballinger SW, Wallace DC. (1990) Myoclonic epilepsy and ragged-red fiber disease (MERRF) is associated with a mitochondrial DNA tRNA(Lys) mutation. *Cell* **61:** 931–937.

Shoffner JM, Lott MT, Wallace DC. (1991) MERRF: a model disease for understanding the principles of mitochondrial genetics. *Rev. Neurol.* **147:** 431–435.

Silvestri G, Moraes CT, Shanske S, Oh SJ, DiMauro S. (1992) A new mtDNA mutation in the $tRNA^{Lys}$ gene is associated with myoclonic epilepsy and ragged red fibres (MERRF). *Am. J. Hum. Genet.* **51:** 1213–1217.

Silvestri G, Ciafaloni E, Santorelli FM, Shanske S, Servidei S, Graf WD, Sumi M, DiMauro S. (1993) Clinical features associated with the A–G transition at nucleotide 8344 of mtDNA ('MERRF mutation'). *Neurology* **43:** 1200–1206.

Suomalainen A, Ciafaloni E, Koga Y, Peltonen L, DiMauro S, Schon EA. (1992a) Use of single strand conformation polymorphism analysis to detect point mutations in human mitochondrial DNA. *J. Neurol. Sci* **111:** 222–226.

Suomalainen A, Paetau A, Leinonen H, Majander A, Peltonen L, Somer H. (1992b) Inherited idiopathic dilated cardiomyopathy with multiple deletions of mitochondrial DNA. *Lancet* **340:** 1319–1320.

Suomalainen A, Majander A, Pihko H, Peltonen L, Syvänen A-C. (1993) Quantification of $tRNA_{3243}{}^{Leu}$ point mutation of mitochondrial DNA in MELAS patients and its effects on mitochondrial transcription. *Hum. Mol. Genet.* **2:** 525–534.

Sweeney MG, Davis MB, Lashwood A, Brockington M, Toscano A, Harding AE. (1992) Evidence against an X-linked locus close to DXS7 determining visual loss susceptibility in British and Italian families with Leber's hereditary optic neuropathy. *Am. J. Hum. Genet.* **51:** 741–748.

Sweeney MG, Brockington M, Weston MJ, Morgan-Hughes JA, Harding AE. (1993) Mitochondrial DNA transfer mutation $Leu^{(UUR)}$A-G 3260: a second family with myopathy and cardiomyopathy. *Quart. J. Med.* **86:** 435–438.

Tatuch Y, Robinson BH. (1993) The mitochondrial DNA mutation at 8993 associated with NARP slows the rate of ATP synthesis in isolated lymphoblast mitochondria. *Biochem. Biophys. Res. Commun.* **192:** 124–128.

Tatuch Y, Christodoulou J, Feigenbaum A, Clarke JTR, Wherret J, Smith C, Rudd N, Petrova-Benedict R, Robinson BH. (1992) Heteroplasmic mtDNA mutation (T–G) at 8993 can cause Leigh disease when the percentage of abnormal mtDNA is high. *Am. J. Hum. Genet.* **50:** 852–858.

Tritschler HJ, Andreetta F, Moraes CT, Bonilla E, Arnaudo E, Danon MJ, Glass S, Zelaya BM, Vamos E, Telerman Toppet N, Shanske S, Kadenbach B, DiMauro S, Schon EA. (1992) Mitochondrial myopathy of childhood associated with depletion of mitochondrial DNA. *Neurology* **42:** 209–217.

Trounce I, Byrne E, Marzuki S. (1989) Decline in skeletal muscle mitochondrial respiratory chain function: possible factor in ageing. *Lancet* **i:** 637–639.

Truong DD, Harding AE, Scaravilli F, Smith SJ, Morgan-Hughes JA, Marsden CD. (1990) Movement disorders in mitochondrial myopathies. A study of nine cases with two autopsy studies. *Movement Disord.* **5:** 109–117.

Vilkki J, Ott J, Savontaus ML, Aula P, Nikoskelainen EK. (1991) Optic atrophy in Leber hereditary optic neuroretinopathy is probably determined by an X-chromosomal gene closely linked to DXS7. *Am. J. Hum. Genet.* **48:** 486–491.

Wallace DC, Singh G, Lott MT, Hodge JA, Schurr TG, Lezza AM, Elsas LJ, Nikoskelainen EK. (1988) Mitochondrial DNA mutation associated with Leber's hereditary optic neuropathy. *Science* **242:** 1427–1430.

Yamamoto M, Clemens PR, Engel AG. (1991) Mitochondrial DNA deletions in mitochondrial cytopathies: observations in 19 patients. *Neurology* **41:** 1822–1828.

Yuzaki M, Ohkoshi N, Kanazawa I, Kagawa Y, Ohta S. (1989) Multiple deletions in mitochondrial DNA at direct repeats of non-D-loop regions in cases of familial mitochondrial myopathy. *Biochem. Biophys. Res. Commun.* **164:** 1352–1357.

Zeviani M. (1992) Nucleus driven mutations of human mitochondrial DNA. *J. Inherit. Metab. Dis.* **15:** 456–471.

Zeviani M, Servidei S, Gellera C, Bertini E, DiMauro S, DiDonato S. (1989) An autosomal dominant disorder with multiple deletions of mitochondrial DNA starting at the D-loop region. *Nature* **339:** 309–311.

Zeviani M, Bresolin N, Gellera C, Bordoni A, Pannacci M, Amati P, Moggio M, Servidei S, Scarlato G, DiDonato S. (1990a) Nucleus-driven multiple large-scale deletions of the human mitochondrial genome: a new autosomal dominant disease. *Am. J. Hum. Genet.* **47**: 904–914.

Zeviani M, Gellera C, Pannacci M, Uziel G, Prelle A, Servidei S, DiDonato S. (1990b) Tissue distribution and transmission of mitochondrial DNA deletions in mitochondrial myopathies. *Ann. Neurol.* **28:** 94–97.

Zeviani M, Gellera C, Antozzi C, Rimoldi M, Morandi L, Villani F, Tiranti V, DiDonato S. (1991a) Maternally inherited myopathy and cardiomyopathy: association with mutation in mitochondrial DNA $tRNA^{Leu(UUR)}$. *Lancet* **338:** 143–147.

Zeviani M, Servidei S, Bresolin N, Antozzi C, Piccolo G, Toscano A, DiDonato S. (1991b) Rapid detection of the $A{\rightarrow}G^{(8344)}$ mutation in Italian families with myoclonus, epilepsy and ragged red fibres (MERRF). *Am. J. Hum. Genet.* **48:** 203–211.

Zeviani M, Muntoni F, Savarese N, Serra G, Tiranti V, Carrara F, Marrioti C. (1993) A MERRF/MELAS overlap syndrome associated with a new point mutation of mitochondrial $tRNA^{Lys}$ gene. *Eur. J. Hum. Genet.* **1:** 80–87.

12

Diabetes: from phenotype to genotype and back to phenotype

G.A. Hitman, M. Fennessy and K. Metcalfe

12.1 Introduction

In order to study the genetics of a disease, one has to first start with a disease definition. Diabetes is defined according to the results of an oral glucose tolerance test, in which the blood glucose of a subject is measured before drinking a 75 g load of glucose, and then half-hourly for 2 h after the drink. Using World Health Authority (WHO) criteria, defined by the blood glucose test results, patients can be categorized as diabetic, non-diabetic or as having impaired glucose tolerance (IGT). The sub-classification of diabetes is based on the observed clinical phenotype. A broad classification of diabetes is presented in Table 12.1.

The three main types of diabetes are insulin-dependent (type 1) diabetes mellitus (IDDM), non-insulin-dependent (type 2) diabetes mellitus (NIDDM) and malnutrition-related diabetes mellitus (MRDM). MRDM can be further sub-divided into fibrocalculous pancreatic diabetes (FCPD) and protein-deficient diabetes mellitus (PDDM).

Study of apparently monogenic diseases at the level of the genotype has, in some disorders, suggested disease heterogeneity; many examples are illustrated in the accompanying chapters. In multifactorial diseases such as diabetes, athero-sclerosis, schizophrenia, depression, hypertension, etc., it may be that once we have identified the genotypes involved in the disease, an aetiological, rather than a clinical, classification could be used to define the disease more accurately.

The progress made in identifying the genotypes and environmental determinants involved in diabetes has been excellent, although much work is still to be done. In IDDM, genes located to the major histocompatibility complex (MHC) and a gene on the short arm of chromosome 11 (11p15) explain over 50% of the genetic basis of the disease. In NIDDM, several genes have been shown to be involved in its aetiology, although combined they probably only account for less than 10% of disease susceptibility. Nevertheless, the different genetic associations

Table 12.1. Simplified classification of diabetes mellitus (adapted from National Diabetes Data Group, 1979)

Type I	Type 1 or insulin-dependent diabetes (IDDM)		
II	Type 2 or non-insulin-dependent diabetes (NIDDM):		
	(i) obese		
	(ii) non-obese		
III	Malnutrition-related diabetes (MRDM):		
	(i) protein deficient diabetes (PDDM)		
	(ii) fibrocalculous pancreatic diabetes (FCPD)		
Other types:			
	secondary diabetes:	(i)	pancreatic disease,
		(ii)	hormonal,
		(iii)	drug-induced,
		(iv)	insulin receptor abnormalities,
		(v)	genetic syndromes, including maturity onset diabetes of the young (MODY).
	Gestational diabetes		
	Impaired glucose tolerance		
	Potential abnormality of glucose tolerance (previously called 'prediabetes')		
	Previous abnormality of glucose tolerance		

are often characterized by particular phenotypic features, illustrating the heterogeneous nature of NIDDM (Table 12.2). It is also becoming apparent that mutation of a disease-associated locus (e.g. glucokinase) does not necessarily lead to diabetes as defined by the WHO. Furthermore, some genetic associations hold equally for IDDM and NIDDM, suggesting that the clinical classification may not be an aetiological classification. This is not a systematic review of the genetics of diabetes but rather will restrict itself to using the work on human leukocyte antigens (HLA) genes and the glucokinase gene to illustrate the complex genotype–phenotype interaction.

12.2 IDDM and the MHC

The clinical classification of IDDM was supported by a strong association of HLA found with IDDM but not with NIDDM. IDDM is a disease, probably autoimmune in nature, in which the final process involves the complete destruction of the insulin-secreting cells (the β-cells) in the pancreatic gland. As a consequence, the patient presents with the signs and symptoms of insulin deficiency. These typically occur over a short time course (weeks) and consist of thirst, polyuria, weight loss and metabolic decompensation leading to ketoacidosis. Whilst the clinical presentation of the disease is very short, the pathological process probably predates disease presentation by up to 10–15 years. Part of this 'pre-IDDM' stage can be identified by the presence of circulating antibodies to islet cells, insulin and glutamic acid decarboxylase and, finally, abnormalities of

Table 12.2. Examples of genetic associations with NIDDM and differences in phenotype

Gene	Phenotypic features
Glucokinase	Normal BMI, 'mild' hyperglycaemia, impaired insulin secretion, normal peripheral insulin sensitivity
Insulin receptor	Obesity, insulin resistance, hyperglycaemia (often severe), associated somatic deformities (e.g. acanthosis nigricans, leprechaunism)
Mitochondrial genome	Associated deafness and other neurological defects, maternal inheritance
Insulin	Hyperproinsulinaemia

insulin secretion and glucose intolerance.

It is very clear that not everyone with a genetic predisposition to IDDM develops the disease. This is best illustrated by the study of monozygotic twins to see how frequently both twins develop diabetes (concordance) or only one (discordance). In twin studies from the UK and Finland, concordance rates for IDDM varied from 20 to 30% (Barnet *et al.*, 1981; Kaprio *et al.*, 1992). Therefore, we can conclude that at least 70% of monozygotic twins, despite having an identical genetic background to their co-twin, do not develop the disease. This would suggest two main possibilities. Firstly, an environmental factor(s) may trigger the disease in those who develop IDDM and the discordant twin escapes this trigger. Alternatively, it may be that several genes lead to IDDM, some of which are highly penetrant (leading to the concordant twins) and some are of lower penetrance (leading to the discordant twins). This latter hypothesis assumes a common environmental trigger which affects all twins and is not necessarily confined to those who develop disease. This raises the possibility that some patients may undergo partial β-cell destruction but never present with IDDM. Evidence will be presented later that this group may indeed present with NIDDM.

12.3 MHC and HLA molecules

The MHC is located on the short arm of chromosome 6 (6p21.1–6p21.3) and is about 3500 kb in length. It is subdivided into three regions called classes I, II and III (Figure 12.1). Classes I and II contain highly polymorphic glycoproteins, HLA, which are involved in antigen presentation at the cell surface. Class III genes are less polymorphic proteins involved in immune responsiveness, e.g. four components of the complement system (C2, Bf, C4A and C4B) and tumour necrosis factors α and β. Both class I and II products form α/β heterodimers, the formation of which is essential for presentation of antigens.

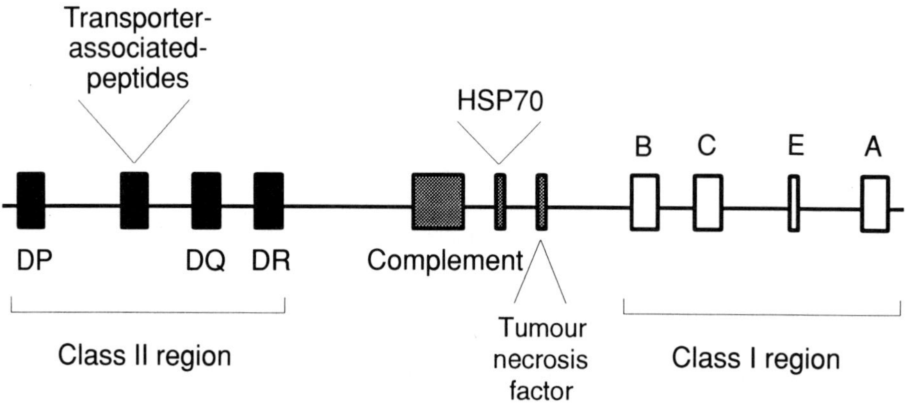

Figure 12.1. Simplified map of the human major histocompatibility complex (MHC) on the short arm of chromosome 6, extending over 3500 kb. HSP70, heat shock protein 70. Reproduced from Hitman and Metcalfe (1993) with permission from Elsevier Science Publishers BV.

Class I genes encode an α-chain and a β-chain; β_2-microglobulin, which is associated with the class I molecule, is not encoded by the MHC. Three regions of the α-chain are present extracellularly, α1, α2 and α3, and most of the polymorphism occurs in α1 and α2, which form the groove supporting the peptide for presentation. The walls of the groove are formed by two α-helices and the base by a β-pleated sheet. In contrast, α3 is conserved, probably because it is involved with binding to β_2-microglobulin. Class I products are expressed on all cells and present endogenous peptides to cytotoxic (CD8) T-lymphocytes. Expression can be up-regulated by interferon and tumour necrosis factor. Expression of class II products is limited to macrophages and β-cells which present processed phagocytosed material to helper (CD4) T-lymphocytes. Other cells at the site of an acute inflammatory reaction, when stimulated by interferon-γ, may also express HLA class II antigens. In recent years, several new class I genes have been described (HLA-E, -F, -G, -H), which are less polymorphic than the 'classical' HLA antigens (HLA-A, -B, -C) and may only be expressed at certain stages of cell development. Their role in antigen presentation, and therefore disease development, is not clear.

HLA class II genes are called DR, DQ and DP. Both the α- and β-chain of class II genes are encoded in pairs by the MHC. The three-dimensional structure of a class II antigen has been elucidated recently (Brown *et al.*, 1993). Class II product differs from class I product in that the ends of the groove are open, enabling it to bind larger peptides; 15–24 residues compared to the 9-residue peptides bound by class I. T-lymphocytes recognize peptides in conjunction with HLA α/β heterodimers. Class II heterodimers may be able to form dimers, leading to increased affinity for the CD4 co-receptor followed by cross-linking of

T-lymphocyte receptors and T-lymphocyte activation.

In the DR region, the number of DRβ genes varies between individuals, e.g. DR1 and DR8 regions possess only one functional gene – DRB1; other DR regions possess either DRB3, DRB4 or DRB5 in addition. DRA1 is the least polymorphic gene, DPA1 is slightly polymorphic but DQA1, DRB1, DQB1 and DPB1 are highly polymorphic. α- and β-chains from one chromosome can form α/β heterodimers with α- and β-chains from the same locus on the other chromosome, a process called *trans*-complementation (Kwok *et al.*, 1988). Strong associations exist between HLA alleles at different loci in the MHC (e.g. A1–Cw7–B8–DR3–DQA1*0501–DQB1*0201 or A2–Cw3–B62–DR4–DQA1*-0301–DQB1*0302), and these are referred to as haplotypes. Associations exist between HLA haplotypes and susceptibility to certain diseases regarded as being autoimmune, e.g. ankylosing spondylitis, coeliac disease and IDDM.

12.4 Autoimmunity and HLA association in IDDM

IDDM is thought primarily to be an autoimmune disease. Lymphocytic infiltration of the islets (insulitis) is common at disease onset, and both CD4 and CD8 T-lymphocytes are required for β-cell destruction. In some cases, other autoimmune endocrine disorders may be apparent. Most IDDM patients have antibodies to islet cell components and there is a clear association between certain HLA haplotypes and the risk of IDDM. HLA associations with IDDM were first described for HLA-B8 and B15 in Caucasoids (Cudworth and Woodrow, 1976) but, once the class II region had been defined in more detail, stronger associations were found with DR3 and DR4 (Sachs *et al.*, 1980), which are in linkage disequilibrium with B8 and B15, respectively. Later, even stronger associations were found with alleles of DQB1 (Festenstein *et al.*, 1986), for example two DQB1 alleles, DQB1*0301 and 0302, are associated with DR4, but only DR4–DQB1*0302 is associated with the disorder, implicating DQ as the susceptibility locus (Sheehy *et al.*, 1989). As yet, the role of HLA-DP in IDDM is not clear. In addition to DR3–DQB1*0201 and DR4–DQB1*0302, DR1–DQB1*0501 and DR2–DQB1*0502 were believed to confer susceptibility, and DR2–DQB1*0601, DR2–DQB1*0602 and DR4/DR5–DQB1*0301 were considered to be either protective or neutral (Todd *et al.*, 1987). Close examination of the DQB1 alleles revealed that the amino acid at position 57 was associated strongly with disease. DQB1 alleles with alanine, serine or valine at this position conferred susceptibility, whereas aspartate did not. A similar exercise carried out on DQA1 showed that arginine at position 52 conferred susceptibility whereas serine or histidine did not (Khalil *et al.*, 1990). Analysis of the three-dimensional structure of class II products illustrated that Asp57 is involved in forming a salt bridge at one end of the groove under the associated peptide (Brown *et al.*, 1993). Non-Asp57 alleles would be incapable of forming this salt bridge and this may affect peptide binding. Also, nearby residues 52 and 55 are involved in the dimerization of heterodimers and polymorphism at position 57 may affect the stability of these dimers.

Although the role of position 57 appears to be making a major impact in IDDM susceptibility, this is an oversimplification of the complex interactions involved. For instance, DQB1*0501 and DQB1*0604, although both non-Asp57, are not associated positively with IDDM but tend to be associated neutrally or negatively (Caillat-Zucman *et al.*, 1992; Cavan *et al.*, 1993). Non-Asp alleles therefore have a graded susceptibility to IDDM. Also, the hypothesis does not explain the interaction of DR3/DR4 which confers the greatest risk of IDDM. It seems unlikely that the interaction of DR3/DR4 is a result of *trans*-complementation alone, since one of the heterodimers can be formed by DR4/DR5 and the other one can be formed by DR4/DR7 or DR9 with either DR4 or DR7. The frequency of such combinations of haplotypes is not raised in IDDM. Neither does it explain how DR2 can eliminate the susceptibility of DR3 and DR4 (Thomson *et al.*, 1988) or why the susceptibility of DQB1*0302 varies depending on which DR4-related DRB1 gene is present on the haplotype (Figure 12.2) (Sheehy *et al.*, 1989; Caillat-Zucman *et al.*, 1992). The fact that 'Asp57' cannot explain disease susceptibility completely is well illustrated by the finding that in Japanese diabetic individuals the Asp57 allele DQB1*0401 is associated with disease (Todd *et al.*, 1990) and, furthermore, DQA1 appears to be the susceptibility locus because DQA1*0301 is most strongly associated with IDDM. This obviously raises again the possibility that HLA-DQ is in linkage disequilibrium with the real susceptibilty locus or, alternatively, may be evidence for 'locus heterogeneity', where different loci confer disease risk in different populations.

Some of the difficulties with the DQ hypothesis have been resolved by the 'peptide affinity model' (Nepom, 1990). Putative 'diabetogenic' peptides bind to different class II molecules with varying affinity. An individual with a suscepti-

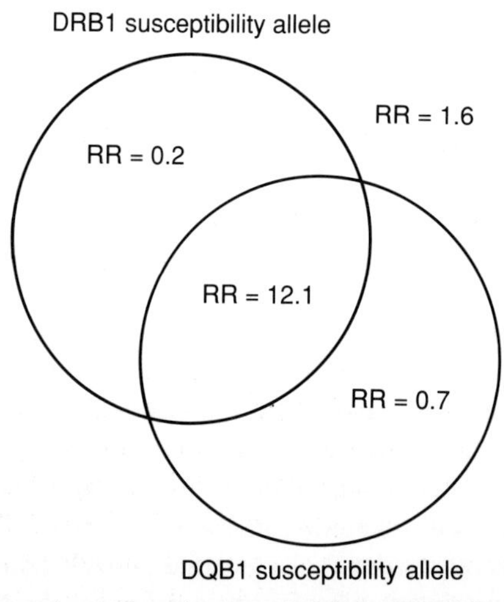

Figure 12.2. Illustration of how increased risk for IDDM, expressed as relative risk (RR), is seen only for haplotypes with IDDM-associated alleles at both DRB1 and DQB1. Adapted from Sheehy *et al.* (1989) by copyright permission of the American Society for Clinical Investigation.

bility class II gene, e.g. DQB1*0302, will be susceptible only if the protein product of that gene is the most efficient peptide binder amongst the different class II molecules expressed. An individual with a susceptible class II gene will not be susceptible when products of the other class II genes are present which bind the diabetogenic peptide with greater affinity. Hence, in a DR2/DR4 subject, one or more of the protein products from the DR2 haplotype, which is protective, would out-compete those from the DR4 haplotype for the diabetogenic peptide, reducing disease susceptibility. On DR5 and DR7 haplotypes which contain DQ alleles associated with diabetes on other haplotypes (DQA1*0501, DQB1*0201), the DRB1 gene may be providing the protection. The model also explains the differing association of DR4 subtypes on DQB1*0302 haplotypes. Protective subtypes may be providing protection by out-competing the HLA-DQ product for peptide. If the model operates during tolerization, normal tolerance to HLA-DQ autoantigens may not develop. Clearly, functional expression of susceptibility alleles relative to protective alleles is important and may explain why some individuals with the same class II genes go on to develop disease while others do not. There is evidence that such variation in expression may exist (Andersen *et al.*, 1991).

Up to now a direct role of HLA molecules in pathogenesis has been assumed. However, the recent discovery of non-HLA genes in the class II region, the peptide transporter genes (Spies *et al.*, 1990), provides an alternative explanation. Peptide transporters deliver peptide to the class I molecules prior to cell surface expression, hence the class II HLA genes may be acting as markers for defective peptide transporter genes.

12.5 Class I in IDDM

From the above evidence, haplotypes containing the same class II HLA genes might be expected to carry the same risk for diabetes, but this is in fact not the case. Amongst haplotypes typed at HLA-B, -DR and in the class III region, Raum *et al.* (1984) identified four that had stronger associations with IDDM than DR3 or DR4 alone, suggesting that MHC loci outside the class II region were influencing susceptibility. In Finland, which has the highest incidence of IDDM in the world (35/100 000 per year), there is a wide variation in IDDM predisposition in those subjects who possess the class II haplotype DR4–DQA1*0301–DQB1*0302, depending on the associated class I genes (Tienari *et al.*, 1992). For example, the haplotype A2–Cw4–B35–DR4–DQA1*0301–DQB1*0302 carries no greater risk for IDDM than the background population (35/100 000 per year), whereas the haplotype A2–Cw1–B56–DR4–DQA1*0301–DQB1*0302 has the highest absolute risk for IDDM (218/100 000 per year). The latter haplotype is specific for the Finnish population, not being found in eight other European populations. The Cw1–B56–DR4–DQA1*0301–DQB1*0302 haplotype is only associated with four HLA-A alleles (A2, A1, A3 and A11), and the risk of developing diabetes appears to vary depending on the HLA-A allele, with the highest absolute risk being with HLA-A2 (Fennessy *et al.*, 1992). This suggests that, on this haplotype at least, the HLA-A locus, or a gene in close proximity to it, contributes to genetic susceptibility to IDDM.

12.6 HLA and IDDM heterogeneity

HLA alleles may contribute to the phenotypic heterogeneity of IDDM. A strong immunological prediposition exists in DR3/DR4 subjects, as illustrated by DR3/DR4 monozygotic twins and sibling pairs who have the highest rate of disease concordance (Wolf *et al.*, 1983). Possessing DR3 and/or DR4 also leads to the development of disease early in life. In one study of IDDM with onset in childhood, 38% of individuals were DR3/DR4 heterozygotes and 97% possessed DR3 and/or DR4 as compared to 10% DR3/DR4 and 74% DR3 and/or DR4 in disease which presented in adult (>30 years) life (Caillat-Zucman *et al.*, 1992). Also, in non-DR3/non-DR4 diabetes, symptoms may be less severe, weight loss is less marked, the frequency of islet cell antibodies is lower and there tends to be less family history of IDDM but more history of NIDDM (Caillat-Zucman *et al.*, 1992). Subjects lacking the DR3/DR4 predisposition may require other genetic predisposition or greater exposure to environmental components or acquired factors, and the disease, often of milder onset, may be of non-immune origin. There is also a distortion in parental transmission of IDDM. Significantly more paternal than maternal DR4 haplotypes are inherited by IDDM probands (Clerget-Darpoux *et al.*, 1989). Although the explanation for transmission ratio distortion remains unclear, its effect might explain why the genes associated with IDDM have never been eliminated from the genetic pool by natural selection, as might be expected. The genes responsible for transmission ratio distortion could be in tight linkage disequilibrium with such deleterious genes, resulting in the protection of the latter from selection pressures, because the preferential transmission of the haplotype as a whole outweighs the disadvantage of carrying an IDDM susceptibilty gene. Concurrent with this, children of diabetic men have a higher prevalence of IDDM than children of diabetic women (Warram *et al.*, 1984). This observation may be due to 'genomic imprinting', where disease or inheritance depends on the difference in expression between the father's and mother's genes.

In IDDM in the Japanese population, both class I and class II contribute to the severity of the disease. HLA antigens A24, B54, DQA1*0301–DQB1*0302 and DQA1*0301–DQB1*0303 are only associated with acute onset, and not slow onset, IDDM. Acute onset was characterized by lower age of presentation, lower prevalence of islet cell antibodies, less preserved β-cell function and less family history of NIDDM (Kobayashi *et al.*, 1993). Further investigation of residual β-cell activity in IDDM revealed that HLA-A24, but no other HLA antigen, was associated with no residual β-cell activity (Nakanishi *et al.*, 1993). These studies show that HLA-A24 contributes to severity of disease.

12.7 HLA genes from genotype to phenotype

At the genotypic level, it is therefore clear that different combinations of genes present on MHC haplotypes are associated with differences in penetrance of IDDM, and that this is expressed by age of onset of the disease and, in some studies, the presence or absence of other autoimmune diseases. Is it possible that the original concept that the HLA associations are unique to IDDM may be incorrect? Several pieces of evidence point to this conclusion. A number of previous studies of NIDDM had shown weak associations between HLA and NIDDM (Serjeantson and Zimmet, 1990; Tuomilehto-Wolf *et al.*, 1993). Rich and colleagues have been studying the interesting group of IDDM patients who have relatives with NIDDM (mixed families). These investigators compared HLA distributions in: (i) sporadic cases of IDDM with parents of normal glucose tolerance, (ii) families with more than one sibling with IDDM and non-diabetic parents, (iii) mixed NIDDM/IDDM families and, (iv) a control panel. The IDDM probands from NIDDM parents had an increased prevalence of DR4 but not DR3 compared to the sporadic and multiplex siblings who have both DR3 and DR4 associations. A significantly older age of onset of the disease was found in the IDDM probands of NIDDM parents (18.3 $\pm$ 2 yr) compared to the sporadic cases (11.5 $\pm$ 0.8 yr) and multiplex siblings (11.6 $\pm$ 1.1 yr), suggesting a gene dosage effect (Rich *et al.*, 1991).

A number of other investigators have been examining the clinical interface between IDDM and NIDDM. In clinical practice, there is a small number of patients with established IDDM whose disease presentation was more like NIDDM, with an insidious onset of symptoms over many months or years in mid-life, but who eventually require insulin for maintenance of good glycaemic control. This group of patients has been found to have an increase in islet cell and glutamic acid decarboxylase antibodies and it has been suggested that they are a group of slowly progressing IDDM patients who have been misclassified as NIDDM. At the genetic level, Groop and colleagues (1988) have demonstrated that these patients have a DR4 but not a DR3 association, similar to the previously described studies of Rich and colleagues (1991). The concept of a reduced number of IDDM genes leading to a NIDDM-like illness is equally applicable to a subgroup of patients with MRDM, namely FCPD. FCPD is confined to patients in developing countries who have chronic pancreatitis of unknown aetiology. These patients are frequently treated with insulin but are rarely truly insulin dependent and prone to ketoacidosis, as compared to patients with IDDM. At the genetic level, several groups have described 'partial' or 'incomplete' HLA associations. Thus, in southern Indians with IDDM, both HLA-DQA1 and DQB1 associations exist, but with FCPD there is an association with DQB1 but not with HLA-DQA1 alleles (Kambo *et al.*, 1989). These studies therefore suggest that the genetic interface between IDDM and NIDDM, as defined by HLA genes, might be more extensive than previously considered. To investigate this hypothesis, we studied a group of elderly men in Finland. These subjects represented a cohort of men aged 70–89 who were first studied as part of a multinational prospective cardiovascular study. In 1984 and 1989, all surviving subjects were studied for diabetes using an oral glucose tolerance test and, based on these results, divided into diabetic, IGT and non-diabetic. The genetic markers studied were 57 diabetes-associated HLA haplotypes identified in a previous population-based study of IDDM in Finland (Tienari *et al*, 1992). The diabetes-associated haplotypes in the elderly men were present in 94% of the diabetic subjects, 79% with IGT and 13% of non-diabetic subjects. The mean 2 h glucose in those subjects possessing diabetes-associated haplotypes was 10.4 mmol l^{-1} ($\pm$ 2.7 mmol l^{-1}) compared to 6.4 mmol l^{-1} ($\pm$ 2.0 mmol l^{-1}) in those subjects without diabetes-associated haplotypes (Tuomilehto-Wolf *et al.*, 1993). The phenotype of those diabetics with the diabetes-associated HLA haplotypes is of interest. At the time of the last oral glucose tolerance test in 1989, 67 were treated on diet alone, 19 with oral agents, two with insulin and two with both oral agents and insulin. This phenotype therefore, does not resemble IDDM.

12.8 Summary of the phenotype of those diabetics possessing the IDDM-associated HLA genotypes

The HLA associations in diabetes have been found to a varying extent throughout the whole clinical spectrum of the disease, including insulin dependency (true IDDM), late-onset IDDM, the slowly progressing IDDM subjects on insulin, NIDDM subjects treated by oral agents and at least a sizeable proportion of elderly NIDDM subjects treated with diet. The MHC association with diabetes is unlikely to be due to the effect of a mutation in a single gene, but more likely to be the end result of variation in several genes on the HLA haplotype. The highly penetrant form of the disease is likely to be the result of subjects possessing those genes with the highest disease susceptibility. In other forms of diabetes in which the subjects are not insulin dependent, affected individuals are likely to possess fewer IDDM-associated genes on the MHC haplotype or less penetrant allelic variations of the disease-susceptibility genes. The diabetes phenotype is likely to be the product of several genes and, apart from the highly penetrant haplotypes (i.e. A2–Cw1–B56–DR4–DQA1*0301–DQB1*0302), will require non-MHC genes for expression of the disease. In IDDM there is an additional effect on disease predisposition from a gene on the short arm of chromosome 11 (11p15) (Owerbach and Nerup, 1982; Bell *et al.*, 1984; Hitman *et al.*, 1985) and several other genes are likely to be involved. Whereas 95% of IDDM subjects will possess at least one IDDM-associated HLA allele, in NIDDM HLA markers are present in less than a third of patients. Furthermore, HLA alone is unlikely to lead to the NIDDM phenotype; this requires the interaction of environmental factors and other genes (e.g. insulin receptor substrate-1, glucokinase, fatty acid-binding protein 2, insulin receptor, etc.) before the manifestation of disease.

12.9 Glucokinase

As mentioned earlier, NIDDM encompasses a number of different phenotypes identified by clinical parameters. One such phenotype, an early onset, mild and autosomal dominant subset of NIDDM, is 'maturity-onset diabetes of the young' (MODY). Recently it has been shown that about half of MODY patients carry mutations in the glucokinase (GCK) gene (Froguel *et al.*, 1992; Hattersley *et al.*, 1992).

The GCK gene has long been considered as a candidate gene for NIDDM. The enzyme is expressed in liver and pancreatic β-cells and is responsible for the phosphorylation of glucose to glucose-6-phosphate (G6P). In contrast to other hexokinases, GCK has a high K_m (8 mmol l^{-1}) and is not inhibited by G6P; thus, given that glucose transport into the cell is generally not rate limiting, rates of glucose phosphorylation vary in tandem with glucose levels within the physiological range (Figure 12.3). This has led to the suggestion that, in the β-cell, GCK acts as a 'glucose sensor' (Matschinsky, 1990) and that, in the liver, it is an important arbiter of glucose flux. Defects in GCK could therefore produce abormalities of insulin secretion as well as apparent insulin insensitivity at the hepatic level.

The GCK gene is thus a good candidate gene for family and population genetic studies. Studies have employed microsatellites close to GCK as they were defined. GCK1 (GCK 3′) lies 10 kb downstream of GCK (Matsutani *et al.*, 1992), GCK2 (GCK 5′) lies 6 kb upstream of GCK (Tanizawa *et al.*, 1992), and the recently discovered GCK3 lies 4.3 kb upstream of GCK (Stoffel and Bell, 1993).

In a series of 16 French families (Froguel *et al.*, 1992) and one large British family (Hattersley *et al.*, 1992) GCK1 was found to be linked to MODY. Sequencing of the GCK-coding regions revealed mutations segregating with disease in these families (Stoffel *et al.*, 1992; Vionnet *et al.*, 1992). A further study revealed that about half (18 of 32) of the MODY French families showed linkage to the GCK gene. Sixteen mutations were identified in the 18 families (Froguel *et al.*, 1993). The majority of mutations identified are missense that result in amino acid substitutions affecting enzyme kinetics (Gidh-Jain *et al.*, 1993).

The role of GCK in NIDDM is now being examined. In family studies there was no association of GCK with NIDDM in 21 French, 18 American Caucasian or 12 British Caucasian pedigrees (Cook *et al.*, 1992; Elbein *et al.*, 1993; Froguel *et al.*, 1993). In four populations, American Blacks, Mauritian Creoles, southern Indians and elderly Finnish men, GCK was positively associated with NIDDM (Chiu *et al.*, 1992a,b, McCarthy *et al.*, 1993, 1994). However, in British Caucasoids, Mauritian Indians and Welsh Caucasoids there was no such association (Chiu *et al.*, 1992b; Hattersley *et al.*, 1993; Tanizawa *et al.*, 1993). In the elderly Finnish men without NIDDM there was an association of GCK with impaired glucose tolerance (McCarthy *et al.*, 1993). In contrast, there was no association between GCK and plasma glucose in Welsh Caucasoids (Tanizawa *et al.*, 1993) or impaired glucose tolerance in the American Caucasoid pedigrees (Elbein *et al.*, 1993). Screening of the coding region in the American Blacks did

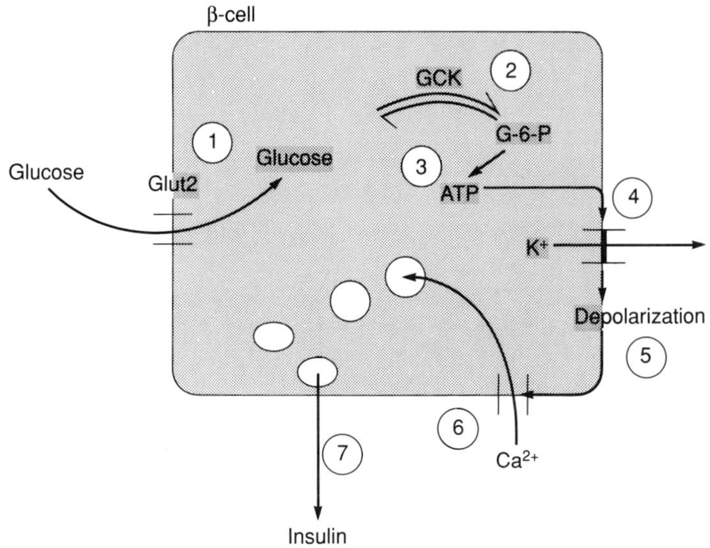

Figure 12.3. Simplified scheme of the biochemistry of the β-cell. Glucose enters the β-cell under the influence of the glucose transporter (GluT2) (1) and is phosphorylated to glucose-6-phosphate by glucokinase (2). By uncertain mechanisms, glucose metabolism leads to elevation of ATP (3), closure of ATP-sensitive potassium channels (4), depolarization of the plasma membrane (5), entry of calcium (6) and exocytosis of insulin (7).

not identify any mutations that were involved (Chiu *et al.*, 1993), however, this does not eliminate a role for the promoter region in susceptibility.

Unlike the typical subject with NIDDM, a patient with MODY has a normal body mass index. However, MODY itself is a heterogeneous disorder, with different phenotypes associated with different genotypic mutations. Diabetes in patients carrying GCK mutations is characterized by mild hyperglycaemia, impaired insulin secretion and normal insulin sensitivity. In MODY which is not caused by mutations of GCK, symptoms are more severe, with greater hyperglycaemia and higher prevalence of microvascular complications. Patients with GCK-linked MODY also have a distinctly different response to a glucose stimulus from those with late onset NIDDM or non-GCK MODY (Velho *et al.*, 1992). In one large MODY pedigree (Bell *et al.*, 1991) and possibly a French pedigree (Froguel *et al.*, 1992), linkage with the gene for adenosine deaminase has recently been demonstrated and, although such mutations may account for only a small proportion of cases, this illustrates the genetic heterogeneity of the disease. Recently, GCK mutations have also been implicated in gestational diabetes (glucose intolerance with onset or diagnosis during pregnancy; Stoffel *et al.*, 1993)

12.10 Conclusions

Diabetes is a heterogeneous disorder of multifactorial origin, probably resulting from exposure to some environmental agent(s) in those individuals with a genetically determined predisposition. Identification of environmental agents is complicated by the long temporal hiatus between exposure and disease onset. The recent rapid advances in the techniques of molecular biology have confirmed that the genetic predisposition to diabetes undoubtedly is due to variation in a number of different genes in different patients and introduces the concept of 'genetic load', i.e. the risk of diabetes is greatest in those individuals and families possessing multiple susceptibility genes. Furthermore, whilst it is clear that types of diabetes previously considered as well-defined may differ considerably genotypically, there is also overlap at the genetic level between the different clinical entities. Rigid classification of diabetes is neither clinically nor scientifically justified.

Though a seemingly ever more complicated task, definition of the genetic basis of diabetes does have real purpose. Intervention trials with nicotinamide are already under way in islet cell antibody (ICA)-positive siblings of children with insulin-dependent diabetes but cannot be considered in sporadic IDDM, which constitutes 90% of disease, as the predictive value of ICA for developing diabetes in children with no family history of disease is 6–7 times less than for those with a family history (Bonifacio *et al.*,1990; Bingley *et al.*, 1993). Identification of the genetic factors which determine progression from ICA positivity to the IDDM phenotype will hopefully facilitate a screening strategy which is applicable to the general population. Assessment of risk of diabetes on the basis of identification of susceptibility genes and a better understanding of the complex genotype–phenotype interaction provides an opportunity for prevention of diabetes in the future.

Acknowledgements

We would like to acknowledge the help of our collaborators Professor J. Tuomilehto and Professor E. Tuomilehto-Wolf of the National Public Health Institute, Helsinki, Finland, and The Childhood Diabetes in Finland (DiMe) Study Group, and funding from the Wellcome Trust and the British Diabetic Assoociation.

References

Andersen LC, Beaty JS, Nettles JW, Seyfried CE, Nepom GT, Nepom BS. (1991) Allelic polymorphism in transcriptional regulatory regions of HLA-DQB. *J. Exp. Med.* **173**: 181–192.

Barnet AH, Eff C, Leslie RDG, Pyke DA. (1981) Diabetes in identical twins. *Diabetologia* **20**: 87–93.

Bell GI, Horita S, Karam JH. (1984) A polymorphic locus near the human insulin gene is associated with insulin-dependent diabetes mellitus. *Diabetes* **31**: 176–183.

Bell GI, Xiang K-S, Newman MV, Wu SH, Wright LG, Fajans SS, Spielman RS, Cox NJ. (1991) Gene for non-insulin-dependent diabetes (maturity onset diabetes of the young subtype) is linked to DNA polymorphism on human chromosome 20q. *Proc. Natl Acad. Sci. USA* **88**: 1484–1488.

Bingley, PJ, Bonifacio E, Gale EAM. (1993) Can we really predict IDDM? *Diabetes* **42**: 213–220

Bonifacio E, Bingley PJ, Dean BM, Shattock M, Dungar D, Gale EAM, Bottazzo GF. (1990) Quantification of islet-cell antibodies and prediction of insulin-dependent diabetes. *Lancet* **335**: 147–149.

Brown JH, Jardetzky TS, Gorga JC, Stern LJ, Urban RG, Strominger JL, Wiley DC. (1993) Three-dimensional structure of the human class II histocompatibility antigen HLA-DR1. *Nature* **364**: 33–39.

Caillat-Zucman S, Garchon HJ, Timsit J, Assan R, Boitard C, Djilali-Saiah I, Bougneres P, Bach J-F. (1992) Age-dependent HLA genetic heterogeneity of Type 1 insulin-dependent diabetes mellitus. *J. Clin. Invest.* **90**: 2242–2250.

Cavan DA, Jacobs KH, Penny MA, Kelly MA, Mijovic C, Jenkins D, Fletcher JA, Barnett AH. (1993) Both DQA1 and DQB1 genes are implicated in HLA-associated protection from Type 1 (insulin-dependent) diabetes mellitus in a British Caucasian population. *Diabetologia* **36**: 252–257.

Chiu KC, Province MA, Permutt MA. (1992a) Glucokinase gene is genetic marker for NIDDM in American Blacks. *Diabetes* **41**: 843–849.

Chiu KC, Province MA, Dowse GK, Zimmet PZ, Wagner G, Serjeantson S, Permutt MA. (1992b) A genetic marker at the glucokinase gene locus for type 2 (non-insulin-dependent) diabetes mellitus in Mauritian Creoles. *Diabetologia* **35**: 632–638.

Chiu KC, Tanizawa Y, Permutt MA. (1993) Glucokinase gene variants in the common form of NIDDM. *Diabetes* **42**: 579–582.

Clerget-Darpoux F, Falk CT, MacCluer JW. (1989) Genetic analysis of complex traits: insulin-dependent diabetes and affective disorders. *Genet. Epidemiol.* **6**: 1–160.

Cook JTE, Hattersley AT , Christopher P, Bown E, Barrow B, Patel P, Shaw JAG, Cookson WOCM, Permutt MA,Turner RC. (1992) Linkage analysis of glucokinase gene with NIDDM in Caucasian pedigrees. *Diabetes* **41**: 1496–1500.

Cudworth AG, Woodrow JC.(1976) Genetic susceptibility in diabetes mellitus: analysis of the HLA association. *Br. Med. J.* **2**: 846–848.

Elbein SC, Hoffman M, Chiu K, Tanizawa Y, Permutt MA. (1993) Linkage analysis of the glucokinase locus in familial Type 2 (non-insulin-dependent) diabetic pedigrees. *Diabetologia* **36**: 141–145.

Fennessy M, Hitman GA, Tienari P, Biro A, Tuomilehto J, Tuomilehto-Wolf E. (1992) A gene in the HLA class I region contributes to susceptibility to type 1 diabetes. *Diabetologia* **35** (Suppl. 1): A135.

Festenstein H, Awad J, Hitman GA, Cutbush S, Groves AV, Cassell P, Ollier W, Sachs JA. (1986) New HLA DNA polymorphisms associated with autoimmune diseases. *Nature* **322**: 64–67.

Froguel P, Vaxillaire M, Sun F, Velho G, Zouali H, Butel MO, Lesage S, Vionnet N, Clément K, Fougerousse F, Tanizawa Y, Weissenbach J, Beckmann JS, Lathrop GM, Passa P, Permutt MA, Cohen D. (1992) Close linkage of glucokinase locus on chromosome 7p to early-onset non-insulin-dependent diabetes mellitus. *Nature* **356**: 162–165.

Froguel P, Zouali H, Vionnet N, Velho G, Vaxillaire M, Sun F, Lesage S, Stoffel M, Takeda J, Passa P, Permutt MA, Beckmann JS, Bell GI, Cohen D. (1993) Familial hyperglycaemia due to mutations in glucokinase. Definition of a subtype of diabetes mellitus. *N. Engl. J. Med.* **328**: 697–702.

Gidh-Jain M, Taleda J, Xu LZ, Lange AJ, Vionnet N, Stoffel M, Froguel P, Velho G, Sun F, Cohen D, Patel P, Lo YMD, Hattersley AT, Luthman H, Wedell A, Charles RS, Harrison RW, Weber IT, Bell GI, Pilkis SJ. (1993) Glucokinase mutations associated with non-insulin-dependent (type 2) diabetes mellitus have decreased enzymatic activity: implications for structure/function relationships. *Proc. Natl Acad. Sci. USA* **90**: 1932–1936.

Groop L, Miettenen A, Groop PH, Meri S, Koskimies S, Bottazzo GF. (1988) Organ-specific autoimmunity and HLA-DR antigens as markers for β-cell destruction in patients with β-cell destruction of the pancreas. *Diabetes* **37**: 99–103.

Hattersley AT, Turner RC, Permutt MA, Patel P, Tanizawa Y, Chiu KC, O'Rahilly S, Watkins PJ, Wainscoat JS. (1992) Linkage of type 2 diabetes to the glucokinase gene. *Lancet* **339**: 1307–1310.

Hattersley AT, Saker PJ, Cook JTE, Stratton IM, Patel P, Permutt MA, Turner RC, Wainscoat JS. (1993) Microsatellite polymorphisms at the glucokinase locus: a population association study in Caucasian Type 2 diabetic subjects. *Diabet. Med.* **10**: 694–698.

Hitman GA, Metcalfe KA. (1993) The genetics of diabetes–an update. In: *The Diabetes Annual*, Vol. 7 (eds SM Marshall, PD Home, KGMM Alberti, LP Krall). Elsevier Science Publishers, Amsterdam, pp. 1–17.

Hitman GA, Tarn AC, Winter RM, Drummond V, Williams LG, Jowett NI, Bottazzo GF, Galton DJ. (1985) Type 1 (insulin-dependent) diabetes and a highly variable locus near to the insulin gene on chromosome 11. *Diabetologia* **28**: 218–222.

Kambo PK, Hitman GA, Mohan V, Ramachandran A, Snehalatha C, Suresh S, Metcalfe K, Ryait BK, Viswanathan M. (1989) The genetic predisposition to fibrocalculous pancreatic diabetes. *Diabetologia* **32**: 45–51.

Kaprio J, Tuomilehto J, Koskenvuo M, Romanov K, Reunanen A, Eriksson J, Stengard J, Kesaniemi YA. (1992) Concordance for Type 1 (insulin-dependent) and Type 2 (non-insulin-dependent) diabetes mellitus in a population based cohort of twins in Finland. *Diabetologia* **35**: 1060–1067

Khalil I, d'Auriol L, Gobet M, Morin L, Lepage V, Deschamps I, Park MS, Degos L, Galibert F, Hors J. (1990) A combination of HLA-DQβ Asp 57-negative and HLA-DQα Arg52 confers susceptibility to insulin-dependent diabetes mellitus. *J. Clin. Invest.* **85**: 1315–1319.

Kobayashi T, Tamemoto K, Nakanishi K, Kato N, Okubo M. (1993) Immunogenetic and clinical characterization of slowly progressive IDDM. *Diabet. Care* **16**: 780–788.

Kwok WW, Schwarz D, Nepom B, Hock RA, Thurtle PS, Nepom GT. (1988) HLA-DQ molecules form α–β heterodimers of mixed allotype. *J. Immunol.* **141**: 3123–3127.

Matschinsky FM. (1990) Glucokinase as glucose sensor and metabolic signal generator in pancreatic β-cells and hepatocytes. *Diabetes* **39**: 647–652

Matsutani A, Janssen R, Donis-Keller H, Permutt MA. (1992) A polymorphic (CA) repeat element maps the human glucokinase gene (GCK) to chromosome 7p. *Genomics* **12**: 319–325.

McCarthy MI, Hitchins M, Hitman GA, Cassell P, Hawrami K, Morton N, Mohan V, Ramachandran A, Snehalatha C, Viswanathan M. (1993) Positive association in the absence of linkage suggests a minor role for the glucokinase gene in the pathogenesis of Type 2 (non-insulin–dependent) diabetes mellitus amongst South Indians. *Diabetologia* **36**: 633–641.

McCarthy MI, Hitman GA, Hitchins M, Riikonen A, Stengard J, Nissinen A, Tuomilehto-Wolf E, Tuomilehto J. (1994) Glucokinase gene polymorphisms: a genetic marker for glucose intolerance in a cohort of elderly Finnish men. *Diabet. Med.* **11**: 198–204.

Nakanishi K, Kobayashi T, Murase T, Nakatsuji T, Inoko H, Tsuji K, Kosaka K. (1993) Association of HLA-A24 with complete β-cell destruction in IDDM. *Diabetes* **42**: 1086–1093.

National Diabetes Data Group (1979) Classification and diagnosis of diabetes mellitus and other categories of glucose intolerance. *Diabetes* **28**: 1039–1047.

Nepom GT. (1990) A unified hypothesis for the complex genetics of HLA associations with IDDM. *Diabetes* **39**: 1153–1157.

Owerbach D, Nerup J. (1982) Restriction fragment length polymorphisms of the insulin gene in diabetes mellitus. *Diabetes* **32**: 275–277.

Raum D, Awdeh Z, Yunis EJ, Alper CA, Gabbay KH. (1984) Extended major histocompatibility complex haplotypes in type 1 diabetes mellitus. *J. Clin. Invest.* **74**: 449–454.

Rich SS, Panter SS, Goetz FC, Hedlund B, Barbosa J. (1991) Shared genetic susceptibility to Type 1 (insulin-dependent) and Type 2 (non-insulin-dependent) diabetes mellitus: contributions of HLA and haptoglobin. *Diabetologia* **34**: 350–355.

Sachs JA, Cudworth AG, Jaraquemada D, Gorsuch AN, Festenstein H. (1980) Type 1 diabetes and the HLA-D locus. *Diabetologia* **18**: 41–43.

Serjeantson S, Zimmet P. (1990) Genetics of non-insulin dependent diabetes. *Balliere's Clin. Endocrinol. Metab.* 5: 477–493.

Sheehy MJ, Scharf SJ, Rowe JR, Neme de Gimenez MH, Meske LM, Erlich HA, Nepom BS. (1989) A diabetes-susceptible HLA haplotype is best defined by a combination of HLA-DR and -DQ alleles. *J. Clin. Invest.* **83**: 830–835.

Spies T, Bresnahan M, Bahram S, Arnold D, Blanck G, Mellins E, Pious D, DeMars R. (1990) A gene in the human major histocompatibility complex class II region controlling the class I antigen presentation pathway. *Nature* **348**: 744–747.

Stoffel M, Bell GI. (1993) Characterization of a third simple tandem repeat polymorphism in the human glucokinase gene. *Diabetologia* **36**: 170–171.

Stoffel M. Patel P, Lo Y-MD, Hattersley AT, Lucassen AM, Page R, Bell JI, Bell GI, Turner RC, Wainscoat JS. (1992) Missense glucokinase mutation in maturity-onset diabetes of the young and mutation screening in late-onset diabetes. *Nature Genetics* **2**: 153–156.

Stoffel M, Bell KL, Blackburn CL, Powell KL, Seo TS, Takeda J, VIonnet N, Xiang K-S, Gidh-Jain M, Pilkis SJ, Ober C, Bell GI. (1993) Identification of glucokinase mutations in subjects with gestational diabetes mellitus. *Diabetes* **42**: 937–940.

Tanizawa Y, Matsutani A, Chiu KC, Permutt MA. (1992) Human glucokinase gene: isolation, structural characterization, and identification of a microsatellite repeat polymorphism. *Mol. Endocrinol.* **6**: 1070–1081.

Tanizawa Y, Chiu KC, Province MA, Morgan R, Owens DR, Rees A, Permutt MA. (1993) Two microsatellite repeat polymorphisms flanking opposite ends of the human glucokinase gene: use in haplotype analysis of Welsh Caucasians with Type 2 (non-insulin-dependent) diabetes mellitus. *Diabetologia* **36**: 409–413.

Thomson G, Robinson WP, Kuhner MK, Joe S, McDonald MJ, Gottschall JL, Barbosa J, Rich SS, Bertrams J, Baur MP, Partanen J, Tait BD, Schober E, Mayr WR, Ludvigsson J, Lindblom B, Farid NR, Thompson C, Deschamps I. (1988) Genetic heterogeneity, modes of inheritance, and risk estimates for a joint study of Caucasians with insulin-dependent diabetes mellitus. *Am. J. Hum. Genet.* **43**: 799–816.

Tienari PJ, Tuomilehto-Wolf E, Tuomilehto J, Peltonen L, The Childhood Diabetes in Finland (DiMe) Study Group. (1992) HLA haplotypes in type 1 (insulin-dependent) diabetes mellitus: molecular analysis of the HLA-DQ locus. *Diabetologia* **35**: 254–260.

Todd JA, Bell JI, McDevitt HO. (1987) HLA-DQβ gene contributes to susceptibility and resistance to insulin-dependent diabetes mellitus. *Nature* **329**: 599–604.

Todd JA, Fukui Y, Kita'awa T, Sasazuki T. (1990) The A3 allele of the HLA-DQA1 locus is associated with susceptibility to type 1 diabetes in Japanese. *Proc. Natl Acad. Sci. USA* **87**: 1094–1098.

Tuomilehto-Wolf E, Tuomilehto J, Hitman GA, Nissinen A, Stengard J, Pekkanen J, Kivenen P, Kaarsalo E, Karvonen MJ. (1993) Genetic susceptibility to non-insulin dependent diabetes mellitus and glucose intolerance are located in HLA region. *Br. Med. J.* **307**: 155–159.

Velho G, Froguel P, Clément K, Pueyo ME, Rakotoambinina B, Zouali H, Passa P, Cohen D, Robert J-J. (1992) Primary pancreatic β-cell secretory defect caused by mutations in glucokinase gene in kindreds of maturity onset diabetes of the young. *Lancet* **340**: 444–448.

Vionnet M, Stoffel M, Takeda J, Yasuda K, Bell GI, Zouali H, Lesage S, Velho G, Iris F, Passa P, Froguel P, Cohen D. (1992) Nonsense mutation in the glucokinase gene causes early-onset non-insulin-dependent diabetes mellitus. *Nature* **356**: 721–723.

Warram JH, Krolewski AS, Gottlieb MS, Kahn CR. (1984) Difference in risk of insulin-dependent diabetes in offspring of diabetic mothers and diabetic fathers. *N. Engl. J. Med.* **311**: 149–152

Wolf E, Spencer KM, Cudworth AG. (1983) The genetic susceptibility to type 1 (insulin dependent) diabetes: analysis of the HLA-DR association. *Diabetologia* **24**: 224–230.

13

Coronary artery disease and the variability gene concept:

the effect of smoking on plasma levels of high density lipoprotein and fibrinogen

Steve E. Humphries

13.1 Introduction

The critical role of genes is in coding for structural proteins and enzymes which enable the cell, organ or organism to maintain homeostasis in the face of the environmental challenges experienced. The genes fulfil this role successfully when the proteins produced enable the optimum level of chemicals and nutrients to be maintained for a long enough period to allow/permit reproduction to occur. The success of genes in maintaining homeostasis is dependent on the organism 'choosing' environments that are compatible with life and avoiding environments that are too stressful. Any organism thus has a range of environments (e.g. ambient temperature or level of particular nutrients) that can be tolerated, while experiencing more extreme environments results in accelerated cell destruction and, in the case of a complex organism/higher animal, the onset of disease and early death. Within a population, genetic variation will mean that individuals will have different abilities to maintain homeostasis when faced with a specific environmental challenge, and it is well known that the ability of cells and organisms to maintain homeostasis decreases with age, probably because of the accumulation of deleterious mutations in the nuclear and, particularly, the mitochondrial genome, which reduces the efficiency of the proteins, enzymes and energy production processes in the cell. The clinical features of any disorder with a late age of onset can therefore be thought of as being caused to some extent by the failure of the individual to maintain homeostasis, and this is particularly true for the disorder of coronary artery disease (CAD).

CAD is a multifactorial disorder, with both genetic and environmental factors being involved to varying extents in causing the disease in different individuals in the general population (Goldbourt and Neufeld, 1986). Epidemiological studies have identified a number of factors that are associated with increased risk of CAD, including high blood pressure, smoking, high dietary fat intake, obesity and as a common complication of diabetes. From these studies, a number of plasma risk factors have been identified such as elevated levels of cholesterol carried in low density lipoprotein (LDL) particles, low levels of high density lipoprotein (HDL) particles and high levels of the clotting factor fibrinogen. It is well established that genetic variation determines, in part, the levels of such lipids and proteins in the blood. Some of the genes involved have been well studied and mutations in these genes have been identified (e.g. the LDL receptor gene in patients with familial hypercholesterolaemia, see Chapter 5, and the apolipoprotein E gene, see Chapter 14). However, for any selected individual, the level of such risk factors in the blood is also due to an individual's genetically determined ability to maintain homeostasis in response to the environment being experienced. Thus, the current epidemic of CAD being seen in Western societies is not due to an increase in the frequency of mutations in important genes, but rather to an inability in some, but clearly not all, individuals to maintain optimum blood levels of these risk factor components, in the light of the environment experienced as a result of 'affluent' lifestyle changes. The most obvious environmental challenge that has changed over the last 100 years is diet, with increasing availability in Western society of all types of foods, but much interest has focused on the lipid components and particularly the amount of saturated fats. The other environmental factor which has changed in recent years is the proportion of individuals smoking cigarettes, and the subject of this chapter will be to examine the effect of genetic variation in determining the change in CAD risk factors experienced by an individual in response to smoking. The risk factors discussed will be plasma levels of HDL and its major protein constituent, apolipoprotein (apoAI), which are associated with protection from atherosclerosis, and the levels of fibrinogen, an important determinant of the thrombotic tendency of the blood.

13.2 Molecular biology of 'variability'

Failure to maintain homeostasis has been proposed previously by Berg (1987), in terms of a 'variability gene' concept, with a distinction between genes affecting the 'level' of a risk factor and other genes determining 'variability', with an individual's risk of CAD being a combination of the two. Thus, an individual with genetically determined high risk factor levels plus a 'restrictive' genotype would be at greater risk of CAD than one with a 'permissive' genotype, who could potentially lower risk by modifying lifestyle. Conversely, an individual with genetically determined low risk factor levels and a 'restrictive' genotype may not experience an elevation of risk by entering a high risk environment, whereas one who carries a 'permissive' genotype would suffer a large and clinically important increase in risk and succumb to premature CAD.

The concept of a 'variability gene' can be formulated in molecular terms in a number of different ways. One possibility is that sequences at the locus for an identified 'candidate' gene for CAD (i.e. an enzyme or protein involved in lipid metabolism or blood clotting) may themselves determine the extent of variability shown by an individual in response to changes in environment. Such variation may alter the amino acid sequence of the protein so that function, specific activity or catabolic rate of the protein is altered in the presence of a particular environmental factor (such as a drug) or dietary component (or their metabolites). Alternatively, variation in the sequences involved in control of expression of the gene, either transcription, RNA processing, translation or mRNA stability, may respond to particular environmental factors. A theoretical example of such a mechanism affecting transcription is shown in Figure 13.1. It is well established that, for many genes, the control of the rate of transcription is affected by the interaction of a large number of different *trans*-acting DNA-binding proteins, that bind close to and mainly, but not exclusively, upstream from the start of transcription of the gene. For many genes, this 'proximal promoter region' of 200–300 bp contains a number of different elements of 6–12 bp in length that are common to many genes, for example the TATA, CAAT and SP-1 boxes and the interleukin (IL)-6 element (reviewed in Jones *et al.*, 1988). Proteins that bind to these elements have been identified and, in some cases, the genes encoding them have been cloned. Experiments have shown that some of these elements have a positive effect on transcription, whilst others have a negative effect (repressor), so that the overall rate of transcription is under multifactorial control (Frankel and Kim, 1991). It is clear that the affinity of a particular *trans*-acting protein is highly dependent on the sequence of the DNA, and single base changes may reduce or abolish binding, leading to severely reduced transcription, as has been found in some examples of thalassaemia (Treisman *et al.*, 1983). Conversely, other sequence changes within the promoter may increase the strength of the promoter, for example causing hereditary persistence of fetal haemoglobin (Ronchi *et al.*, 1989). Similar conclusions have been reached on other promoters by *in vitro* expression studies using site-directed mutagenesis (e.g. reviewed in Maniatis *et al.*, 1987).

It is of relevance that many of these DNA-binding proteins function as homodimers, heterodimers or multimers, and can be visualized as such by electron microscopy (Mastrangelo *et al.*, 1991). Some of these proteins are modified post-translationally, such as by phosphorylation (Gomez and Cohen, 1991), influencing affinity for DNA or for each other and thus their activity. The model presented in Figure 13.1 shows how environmental factors may affect such modification. For example, dietary lipid composition or other components, such as drugs, may have an effect on cellular membrane fluidity or structure, which could alter the activity of membrane-associated enzymes such as protein kinases, and thus in turn alter the modification state and DNA affinity of the nuclear proteins. This modification may result in a large effect on transcription from one (polymorphic) variant promoter sequence, but not another. As well as sequence variation in the promoter region of the candidate gene itself, there is the obvious

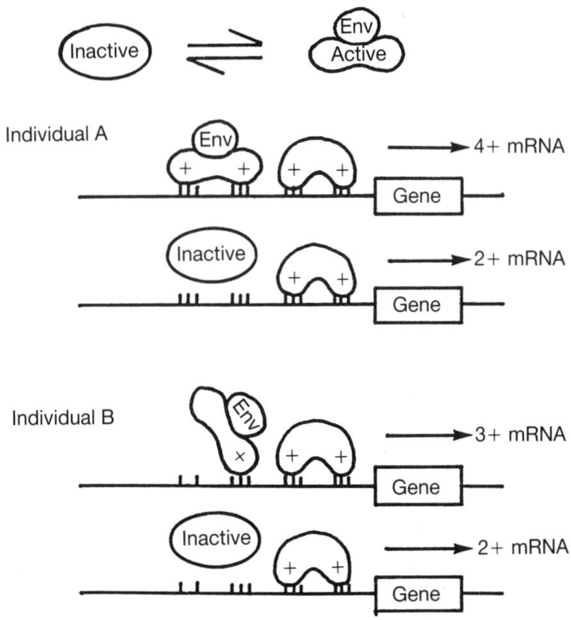

Figure 13.1. Cartoon depicting potential molecular mechanism determining greater variability in individual A compared to individual B because of sequence variation in the gene promoter. An 'inactive' (non-binding) DNA-binding protein that has a positive effect on transcription is shown modified to an active binding form by an environmental factor. Contact points between the protein and a number of specific base pairs in the DNA sequence are indicated by horizontal lines, with strong binding being denoted by +. In individual B, a sequence difference reduces the binding affinity of the DNA-binding protein and thus its positive effect on transcription. As a result, individual A changes from basal levels of 2+ mRNA to levels of 4+ mRNA in the presence of the environmental factor and thus has a doubling in the levels of the produced protein (permissive genotype). Individual B changes from the basal levels of 2+ mRNA to only 3+ mRNA, and has only a small increase in the levels of the produced protein (restrictive genotype).

possibility that changes in the levels or function of the *trans*-acting DNA-binding proteins may affect 'variability', though the effect of such changes would be likely to be wide-ranging (pleiotropy) since many of these sequence elements are common to many genes and, thus, any one of these proteins is likely to be involved in control of expression of several genes.

13.3 Smoking and coronary artery disease

There are many possible mechanisms whereby smoking may lead to CAD, with one or more of these mechanisms predominating in different individuals (for review see Oliver, 1989). Acutely, smoking is associated with a number of changes in vascular physiology, typified by an increased adrenergic state, with a rise in heart rate and blood pressure, an increase in plasma free fatty acids and myocardial oxygen consumption, and a decrease in ventricular fibrillation threshold. There is also good evidence for acute endothelial cell damage, for example as measured by an increase in circulating von Willibrand factor (Blann, 1992), which leads to a rapid increase in arterial wall permeability to fibrinogen following smoking (Allen *et al.*, 1988). Chronically, smoking increases platelet aggregation and both white blood cell count and haematocrit, and thus results in a significant increase in blood viscosity (for review see Lowe, 1993). Smoking increases fibrinogen markedly which also increases viscosity, and this viscosity effect may be particularly important where the low shear rate at bifurcations and bends in blood vessels allows the accumulation at these sites of atherogenic lipids and other constituents such as activated platelets and white cells.

Smoking is associated chronically with a number of potentially atherogenic changes in plasma lipid levels (Richmond *et al.*, 1987; Craig *et al.*, 1989), particularly with lower levels of the 'protective' HDL particles. Smokers may also be at higher risk of CAD because of associated lifestyle changes. Smokers have lower levels of linoleic acid in their adipose tissue, and low linoleic acid is strongly associated with CAD in case–control studies (Logan *et al.*, 1978). Linoleic acid is the principal polyunsaturated fatty acid consumed in the Western diet and it cannot be synthesised in the body. Smokers (and CAD-prone people) consume significantly lower amounts of linoleic acid, possibly because smoking affects taste, a suggestion which is supported by the observation that smokers also add more salt to their food. Linoleic acid has potential anti-thrombotic effects by reducing platelet aggregation, as well as by reducing plasma levels of LDL, and it may protect against ventricular fibrillation.

Many of the pathophysiological changes seen in response to smoking appear to mimic, albeit at low level, those seen in the 'acute phase' (AP) response that accompanies severe inflammation. The AP is characterized by changes in the plasma level of a number of proteins, with some positive AP reactants increasing two- to four-fold (e.g. fibrinogen and α-1-antitrypsin) and others several hundred-fold, such as C-reactive protein, and some that decrease (negative AP reactants), including albumin and apoAI (Pepys and Baltz, 1983). Monocytes play a central role in the AP, and they migrate to sites of tissue damage and respond to various external stimuli by secreting a number of cytokines and growth factors, the most important of which appear to be IL-1, IL-6 tumour necrosis factor (TNF) and transforming growth factor β (TGF-β). For smoking, it is thus believed that damage to the lung tissues (Figure 13.2a) results in the recruitment of large numbers of macrophages, and these respond by secreting IL-6, TGF-β, etc., into the blood, from where they stimulate the liver to make a 'low-grade' AP response. Hepatocytes have a specific IL-6 receptor on their surface which comprises two proteins, one of which binds IL-6 and, through interaction with the second, which is a transmembrane tyrosine kinase (Kishimoto *et al.*, 1992), stimulates the phosphorylation of specific cytoplasmic proteins. This initiates a cellular cascade of events which results among other

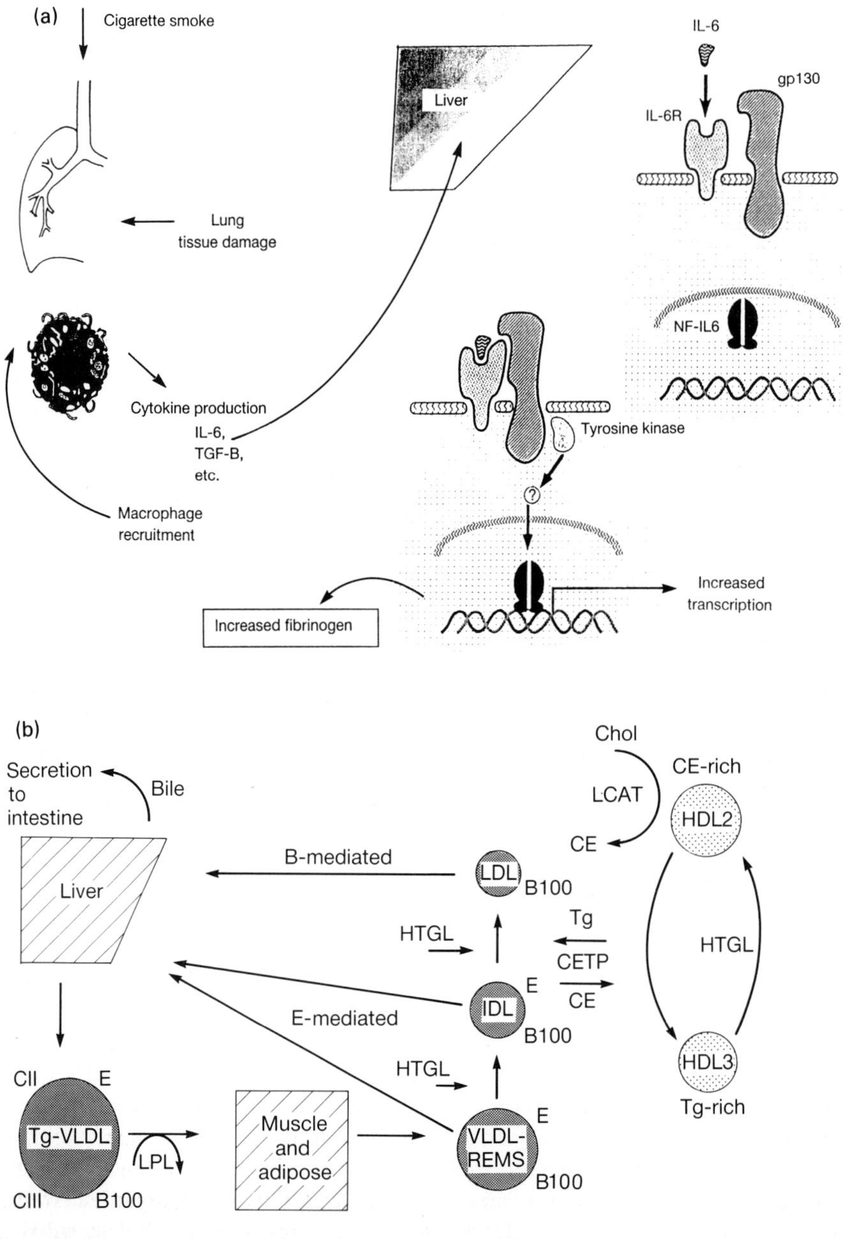

Figure 13.2. (a) The proposed relationship between smoking and elevated plasma fibrinogen levels. (b) The main features of lipid metabolism.

things in the rapid modification of a nuclear transcription factor, NF-IL-6, which significantly enhances the DNA-binding ability of the protein (Akira *et al.*, 1992). NF-IL-6 is a leucine-zipper, containing protein which has homology to the transcription factor C/EBP. The transcription of a number of liver-specific genes is controlled by C/EBP binding, due to the presence of a sequence element (consensus TGTGGAAA) in the promoter region of both positive and negative AP genes; such an element is found in both the albumin and apoAI promoter. It appears that NF-IL-6 competes for C/EBP binding in these genes and this has the effect of suppressing the transcription of negative AP proteins. By contrast, positive AP proteins have related sequence elements which are recognized only by NF-IL-6 and binding results in strong transcription; such elements have been identified in the fibrinogen gene promoter, amongst others.

Thus, smoking contributes to CAD because of direct cell-mediated acute effects on endothelial integrity and vascular physiology, because of its more chronic effects in promoting thrombosis and increasing blood viscosity, both through cellular effects and through changes in transcription of genes in the liver, and because of alterations in the plasma lipid balance to a more atherogenic profile. The molecular mechanisms by which smoking affects levels of HDL-C and fibrinogen are considered in detail below.

13.4 Lipoprotein metabolism and smoking

Cholesterol and triglycerides are transported in the blood in protein–lipid particles called lipoproteins. These particles can be distinguished by their buoyant density and by their protein constituents. Following the digestion of dietary fat, the intestine secretes large triglyceride-rich particles called chylomicrons which enter the circulation via the lymphatic system. Similar triglyceride-rich very low density lipoprotein (VLDL) particles are secreted by the liver, with both types of particles being stabilized by the large apolipoprotein apoB, and both containing other apoproteins, such as apoE and apoCI, CII and CIII. ApoCII is the essential co-factor for the enzyme lipoprotein lipase (LPL) that hydrolyses the triglyceride (TG) core of both chylomicrons and VLDL. As shown in Figure 13.2b for VLDL, the triglycerides from these particles are metabolized as an energy source in muscle tissue, or deposited in adipose tissue. Many of the triglyceride-depleted remnant particles are removed rapidly from the circulation by the liver, through interaction between apoE on the particle surface and specific receptors, but a proportion of the VLDL remnants are metabolized further, principally by the enzyme hepatic triglyceride lipase (HTGL), which hydrolyses the remaining triglyceride in the core of the particle. This is accompanied by loss of all the apoproteins except for apoB, and the final result is the cholesterol-rich LDL particle. LDL is removed from the blood by the interaction of apoB with a specific receptor, the LDL receptor, found on the surface of extra-hepatic cells and, most importantly, in the liver, where the cholesterol is metabolized to bile acids and is secreted into the gut.

The other class of circulating lipoproteins of major relevance are the HDL particles, which function in 'reverse cholesterol transport' (Reichl and Miller, 1989), accepting cholesterol from extra-hepatic cells or from other lipoproteins, and helping in the return of these lipids to the liver for further metabolism. HDL particles are of both intestinal and hepatic origin, and they contain the apoproteins apoAI and apoAII, with some particles also containing apoE and the apoC peptides (Eisenberg, 1987). ApoAI can promote cholesterol efflux from cells *in vitro* (Barbaras *et al.* 1987), and is therefore an important component of the particle. The cholesterol in an HDL particle can be esterified by the plasma enzyme lecithin–cholesterol acyl transferase (LCAT), with apoAI also acting as an obligatory co-factor for the enzyme. The HDL, now with a core of cholesterol ester, may be removed from the blood by the liver, or the core lipids may be transferred to other lipoproteins by specific transfer proteins, particularly cholesterol ester transfer protein (CETP), which mediates the exchange and transfer of cholesterol esters (CE) and triglycerides between lipoprotein particles. Therefore the LDL acquires CE by transfer from other lipoproteins including HDL, and HDL acquires TG. These triglyceride-rich particles are substrates for hepatic lipase, and small, cholesterol-poor HDL particles are regenerated which can accept more cholesterol and CE.

Several epidemiological studies have reported that levels of both HDL–cholesterol (HDL-C) and apoAI are inversely related to incidence and severity of CAD, and can independently predict the risk of CAD (reviewed in Reichl and Miller, 1989). Various environmental factors have been identified that affect plasma HDL levels, including steroid hormones, alcohol intake, stress, infection, the amount of exercise, BMI, diet, some drug therapy (such as β-blockers) and smoking. Cigarette smoking is well recognized as having a lowering effect on both HDL-C and apoAI (Wilhelmsson *et al.*, 1975; Berg *et al.*, 1979; Criqui *et al.*, 1980; Assman *et al.*, 1984; Haffner *et al.*, 1985). This effect is reversible and, within 2 weeks from cessation of smoking, the levels of HDL-C can return to normal (Stubbe *et al.*, 1982; Tuomilehto *et al.*, 1986; Feher *et al.*, 1990). The mechanism of this smoking effect is unknown, but apoAI is a negative acute phase protein and the action of growth factors, such as TGF-β, on gene expression in the liver may explain this effect, acting either through post-transcriptional mechanisms affecting mRNA stability (Morrone *et al.*, 1989) or by reducing apoAI gene transcription in the liver by competition and displacement by the induced NF-IL6 of the C/EBP positive transcription factor. Such a mechanism could be confirmed by *in vitro* or *in vivo* studies.

In spite of this large contribution of environmental factors, several studies have demonstrated that the heritability of HDL-C and apoAI is in the range of 0.43–0.66 (Hasstedt *et al.*, 1984). A strong genetic effect on levels of HDL-C and apoAI has been demonstrated by twin (Austin *et al.*, 1993) and family studies (Hamsten *et al.*, 1987a), and evidence for a major gene determining individual differences in levels of apoAI has been demonstrated by applying biometrical techniques (Moll *et al.*, 1989), although the genes involved have not been determined. In turnover studies, the major determinant of HDL-C levels appears to be its rate of removal from the blood compartment rather than the synthetic rate, suggesting that variation in receptors or enzymes may be involved. Individuals with inherited CETP deficiency have been identified who have markedly elevated levels of HDL-C (Brown *et al.*, 1989), while patients with LCAT deficiency have low levels of HDL-C and other lipoprotein abnormalities (Norum *et al.*, 1993). Because of the strong metabolic relationship between triglyceride-containing lipoproteins and HDL particles, in individuals in the general population there is an inverse relationship between plasma triglyceride levels and HDL-C levels. Because of this relationship, genetic variation that causes elevations of plasma triglycerides is associated with low levels of HDL-C, e.g. LPL deficiency, and patients with conditions which are associated with hypertriglyceridaemia, such as obesity and diabetes, usually have low HDL-C levels.

13.5 Variation at the apoAI gene locus, and levels of HDL-C

The apoAI gene itself is an obvious candidate for involvement in determining plasma levels of apoAI, and the gene for apoAI is in a cluster with those for apoCIII and apoAIV on chromosome 11. Over 10 common polymorphisms have been detected within this gene cluster, and many studies have found associations between genotypes of some of these polymorphisms and variations in levels of lipids and lipoproteins in healthy individuals and patients (reviewed in Humphries, 1988; Mehrabian and Lusis, 1992). Associations between variation in the gene cluster and levels of HDL or apoAI have been reported most consistently with the G_{-75}–A substitution in the apoAI promoter (Figure 13.3). In seven independent studies (Table 13.1) the A allele has been associated with higher levels of both HDL-C and apoAI (Jeenah *et al.*, 1990; Pagani *et al.*, 1990; Paul-Hayase *et al.*, 1992; Sigurdsson *et al.*, 1992; Xu *et al.*, 1993; Talmud *et al.*, 1994; Saha *et al.*, 1994). The size of the effect associated with the A_{-75} allele is modest, with carriers having roughly 11% higher levels of HDL-C and 9.7% higher levels of apoAI in the Icelandic study, and genotype accounting for 13.9% and 8%, respectively, of sample variance after adjustment for age and BMI (Sigurdsson *et al.*, 1992).

The A substitution at –75 bp is located between the CACAT sequence and the TAAATA box of the transcriptional start site of the apoAI gene, and creates a 6 bp perfect repeat CAGGGC-CA*GGGC. Because of its position in the promoter region of the apoAI gene, it is likely that the G–A sequence change itself may have a direct effect on the transcription of the apoAI gene, and thereby the production of apoAI either from the liver or the intestine. It is known that the promoter region between nucleotides –256 bp and –41 bp upstream from the transcription start site of the apoAI gene is necessary and sufficient for maximal levels of expression in HepG2, but that sequences between the apoAI and apoCIII gene are important for expression in the intestine (Reue *et al.*, 1988; Sastry *et al.*, 1988; Walsh *et al.*, 1993). This region contains 'enhancer-like' elements which may control transcription of the whole gene cluster as well as elements that share homology with sites for transcription factors such as C/EBP, HNF-4, NF-KB

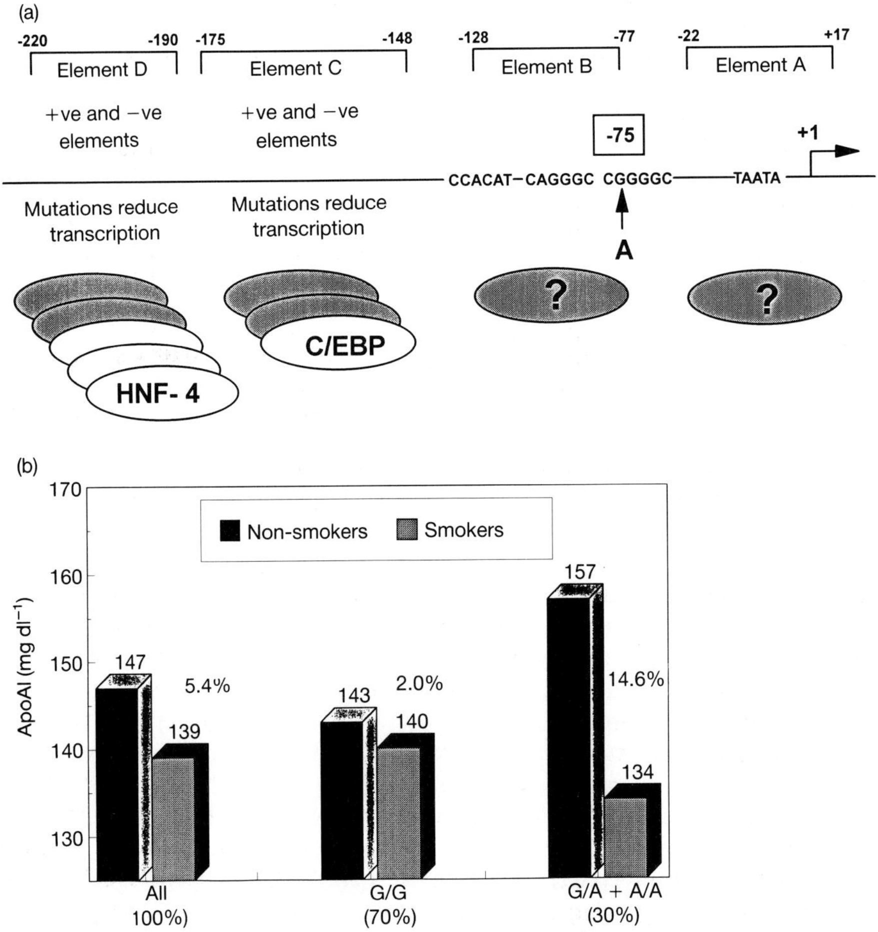

Figure 13.3. (a) The promoter region of the apoAI gene, showing the location of the G_{-75}–A substitution relative to the transcription initiation site and of the CAAT and TATA boxes that flank the substitution. Ellipses represent proteins that have been detected as binding to specific regions of the DNA, only some of which have been identified. For further details see Zannis *et al.* (1992). (b) Mean levels of HDL-C with respect to different genotypes and smoking habits in Icelandic healthy men (data from Sigurdsson *et al.*, 1992). Smoking × genotype interaction $p < 0.02$. The subjects, 149 men, 15–78 years old, were randomly selected participants in the Icelandic National Diet Survey 1990 from the south-west part of Iceland. All subjects completed a questionnaire concerning smoking habits, with smokers defined as all those who currently smoked tobacco or had ceased smoking up to 14 days before blood sampling. Those on lipid-lowering drugs, thyroxin or with diabetes were excluded. From these results we can predict that those with the A allele who smoke will have a much larger fall in apoAI levels.

Table 13.1. Mean (+SE) of apoAI and HDL-C levels in individuals with different apoAI G/A genotypes

Sample (ref.)	Genotype	No.	ApoAI (mg d l^{-1})	HDL(mmol l^{-1})
Bristol, men	GG	76	128±4	1.18±0.05
(Jeenah *et al.* 1990)	GA + AA	18	151±13†	1.43±0.13*
Brugge, boys	GG	99	120±2	1.32±0.03
(Paul-Hayase *et al.*, 1992)	GA + AA	42	125±3†	1.14±0.06
Chinese, men	GG	62	134±5	1.12±0.04
(Saha *et al.*, 1994)	GG + AA	69	152±5†	1.24±.004†
Italian, boys	GG	67	162±3	1.65±0.03
(Xu *et al.*, 1993)	GG + AA	44	173±4*	1.73±0.04
EARS, women	GG	585	159±19	1.57±0.02
(Talmud *et al.*, 1994)	GG + AA	233	165±23*	1.62±0.03
Iceland, men	GG	70	144±4	1.17±0.03
(Sigurdsson *et al.*, 1993)	GA + AA	28	157±6†	1.31±0.06‡

*$p < 0.05$, †$p < 0.025$, ‡$p < 0.005$.

and ARP-1, although the functional effect of these sequences has not been characterized in detail (reviewed in Zannis *et al.*, 1992). However, a recent study using DNase I footprint analysis on the proximal part of the apoAI promoter identified four protected areas (Papazafiri *et al.*, 1991), with one of the protein-binding regions being from –128 bp to –77 bp. This is in close proximity to the G–A sequence change at –75 bp and our hypothesis is that the G–A substitution alters the affinity for a *trans*-acting nuclear protein which affects the transcription rate of the apoAI gene and thus alters the rate of synthesis of apoAI from the liver and intestine. Recent experiments using functional tests of promoter strength (CAT assays) have observed that the A allele shows higher transcriptional activity than the G allele when using a promoter fragment of –256 to +100 bp transfected into the hepatoma HepG2 cells (Jeenah and Wells, 1992), supporting the hypothesis that the G–A substitution may be of direct functional significance. However a study using a smaller fragment of the promoter (–150 to +1) observed higher transcription from the G allele (Smith *et al.*, 1992), and the discrepancy between the two studies may be due to the differences in the size of the promoter fragment used, with small fragments missing important distal functional elements that are required for the effect of the A allele to be expressed.

13.6 ApoAI genotype and smoking

It is not feasible, for ethical reasons, to carry out longitudinal studies on HDL-C and apoAI levels where individuals of different genotype are requested to start smoking, but it is possible to make inferences about this by studying the effect associated with genotype in smokers and non-smokers drawn from the same population. In three studies (Sigurdsson *et al.*, 1992; Saha *et al.*, 1994; Talmud *et al.*, 1994) we have looked at the influence of smoking on the effect on apoAI levels associated with the A_{-75} allele in men, and have obtained very similar results to those from the Icelandic study presented in Figure 13.3. As expected, the levels of HDL-C and apoAI were lower (14% and 6%, respectively) in those who smoked compared to those who did not smoke. In men with the G allele there was essentially no apoAI-lowering effect associated with smoking, with the slightly higher levels seen in smokers presumably reflecting small differences in the prevalence of other factors affecting levels of HDL-C and apoAI, such as diet or obesity. By contrast, in those with one or more A allele, apoAI levels were 15% lower and HDL-C levels were 20% lower in smokers compared to non-smokers. Smoking thus overrides the positive effect of the A allele on levels of apoAI and HDL-C, with the result that men carrying this allele would be predicted to change from having amongst the highest levels of HDL-C and apoAI to amongst the lowest levels if they are smokers.

The mechanism of this effect is unknown, but if the A_{-75} allele is having its HDL-C-raising effect by increasing transcription of the apoAI gene, it is likely that the smoking-induced changes in expression of nuclear DNA-binding proteins such as NF-IL6 are masking or overriding this effect. Thus the uncharacterized protein identified as binding to the –128 bp to –77 bp region may be a target for smoking-induced modifications such as increased cellular levels, post-translational modification or direct competition with other factors. It cannot be ruled out that the A_{-75} change is a marker (in allelic association with) an as yet unidentified sequence change in a distant transcriptional control element, for example in the apoCIII–AIV intragenic region. These possibilities are open to experimental study at the cellular level.

The effect of this interaction on relative risk of CAD can be estimated by extrapolation from studies which have suggested that a 1% change in apoAI levels associated with a 2% change in is relative risk of CAD (Stampfer *et al.*, 1991). Based on this data, men with only the apoAI G_{-75} allele would experience no HDL-related increase in CAD risk upon smoking, while those men with one or more A_{-75} allele would have a lower risk of CAD (roughly 80%) compared to individuals with the genotype G/G but, if these same individuals smoke cigarettes, their relative risk will increase by 1.3-fold because of the large decrease (15%) in levels of apoAI. Both groups of men would be predicted to experience an increase in CAD risk of 1.8-fold because of the direct effect associated with smoking, acting through other pathways common to both groups.

13.7 Variability at the fibrinogen locus, plasma fibrinogen levels, smoking and risk of CAD

Several prospective studies have shown a direct association between plasma fibrinogen concentration and the subsequent incidence of CAD (e.g. Meade *et al.*, 1986; reviewed in Cook and Ubben, 1990). In men in the Northwick Park Heart Study (NPHS), an elevation of one standard deviation in fibrinogen (about

0.6 g l^{-1}) was associated with an 84% increase in the risk of CAD within the next 5 years. This association probably is mediated through a number of different pathophysiological processes. Individuals with elevated fibrinogen levels may have an increased propensity for coagulation, and thrombus formation in an artery that is already narrowed by atherosclerosis is a frequent cause of acute symptoms such as myocardial infarction (MI) or stroke. High levels of fibrinogen increase blood viscosity, which itself may partly be involved. Fibrinogen interacts with specific receptors on activated platelets and increases platelet aggregability *in vitro*. Elevated levels of fibrinogen may also be having a direct effect on the development of the atherosclerotic lesion, and fibrinogen and fibrinogen degradation products can be detected histochemically and immunologically in the intima of diseased artery walls, and within atherosclerotic plaques (Smith and Staples, 1981). Animal studies have also shown that intravascular fibrin deposition is related to plasma fibrinogen levels. Finally, fibrinogen has been shown to have a mitogenic effect on haemopoetic cells, through apparently specific cell surface receptors, although these studies have not been extended to endothelial cells.

Individuals who smoke have a higher risk of both CAD and stroke and many studies have observed that individuals who smoke have elevated levels of plasma fibrinogen (e.g. Wilkes *et al.*, 1988). It is very likely that a substantial proportion of the association between smoking and CAD is mediated through the plasma fibrinogen concentration (Meade *et al.*, 1987). Elevated plasma fibrinogen could be caused either by increased synthesis or by reduced removal of fibrinogen from the blood by the fibrinolytic system. Fibrinolysis occurs as a result of the action of a number of plasma proteins and *in vivo* studies have shown that the degradation products of injected radiolabelled fibrinogen accumulate at low levels in all the organs and tissues of the body (reviewed in Lane, 1986). Fibrinogen is synthesised in the liver and, since it is an acute phase protein, its plasma level is raised following infection or injury. Levels are raised by the use of oral contraceptives and following the menopause. In both men and women, the fibrinogen level increases with age and obesity and is higher in diabetics than non-diabetics. There is an inverse association between alcohol intake and the plasma fibrinogen concentration. Because of its sensitivity to these and other environmental factors, the within-individual variation of fibrinogen levels is high, accounting for up to 26% of the sample variance in standardized assays (Thompson *et al.*, 1987). The known individual and environmental factors that influence fibrinogen level account for about 20% of the population variance in fibrinogen (Meade *et al.*, 1979) although, because of the high within-individual variability in fibrinogen levels, this is probably a conservative estimate. The extent to which genetic factors may determine the plasma fibrinogen level is unclear, though path analysis has suggested an estimate for heritability of fibrinogen levels of 0.5 (Hamsten *et al.*, 1987b), twin studies a lower estimate of 0.3 (Berg and Kierulf, 1989; Reed *et al.*, 1994). To date, there have been no reports of biometrical analysis to investigate the possibility of a major gene determining fibrinogen levels.

Each plasma fibrinogen molecule is comprised of two each of the Aα-, Bβ- and γ-fibrinogen polypeptide chains, and the complex is held together by a number of inter- and intra-chain disulphide links. The cDNA sequence and gene structure for all three fibrinogen genes have now been determined. The genes are in a cluster of less than 50 kb on the long arm of chromosome 4, and each chain is synthesised as a separate mRNA, with the levels of all three mRNAs being co-ordinately controlled at least in response to acute stimulation such as de-fibrination. The rate-limiting step in the production of the mature fibrinogen molecule in the human hepatoma cell line HepG2 is the synthesis of the Bβ-polypeptide chain (Roy *et al.*, 1990), which in turn is influenced by the amount of its mRNA available. It is therefore likely that an alteration in the level of synthesis of the Bβ-chain may have an effect on the amount of fibrinogen secreted by the liver.

A cartoon of the β-gene promoter is shown in Figure 13.4a, and the region from –150 bp to the start of transcription contains a number of elements of interest (Huber *et al.*, 1987, 1990). Only in this region is there significant sequence homology between the rat and human genes, suggesting that this sequence has an important conserved function. This region also has homology with other 'acute phase' genes such as α-1-antitrypsin (Courtois *et al.*, 1987). In addition, this region has been reported to contain all the information required to act as a promoter in HepG2 cells and has been shown to bind proteins from a HepG2 cell nuclear extract. The sequence from –89 to –76 contains a conserved liver-specific transcription element which binds hepatic nuclear factor 1 (HNF1), and deletion mapping shows that just upstream lies an IL-6-responsive element, which has been identified in other genes as the motif CTGGGA (Dalmon *et al.*, 1993). It is therefore possible that sequence changes in this region of the gene may have a direct effect on the rate of transcription and thus on plasma fibrinogen levels.

In studies of the β-fibrinogen promoter, a common G/A sequence variation was detected by use of the enzyme HaeIII, with a loss of the predicted HaeIII site at position –455 in roughly 20% of alleles examined (Thomas *et al.*, 1991). In several samples of healthy individuals it has been reported that the A_{-455} allele (lack of HaeIII cutting site) is associated consistently with higher fibrinogen levels (Table 13.2), with those with one or more copies of the A_{-455} allele having higher fibrinogen levels than those with the genotype G/G (0.28 g l^{-1} higher in healthy men in the UK; Thomas *et al.*, 1991). The magnitude of this genotype effect indicates that it is likely to be of biological significance in causing an elevated risk of thrombosis and, by extrapolation from the NPHS data of the relationship between fibrinogen and CAD risk (0.6 g l^{-1} associated with 84% greater risk), men with the A allele would be at 40% higher risk of a thrombotic event. Of course, this estimate is based on healthy middle-aged men from north London, and may not be the same in other groups. However, in support of the relationship between fibrinogen genotype and risk of disease, polymorphisms at the fibrinogen locus have been reported to be associated with risk of peripheral arterial disease (Fowkes *et al.*, 1992), although not with risk of MI in a case–control study (Scarabin *et al.*, 1993).

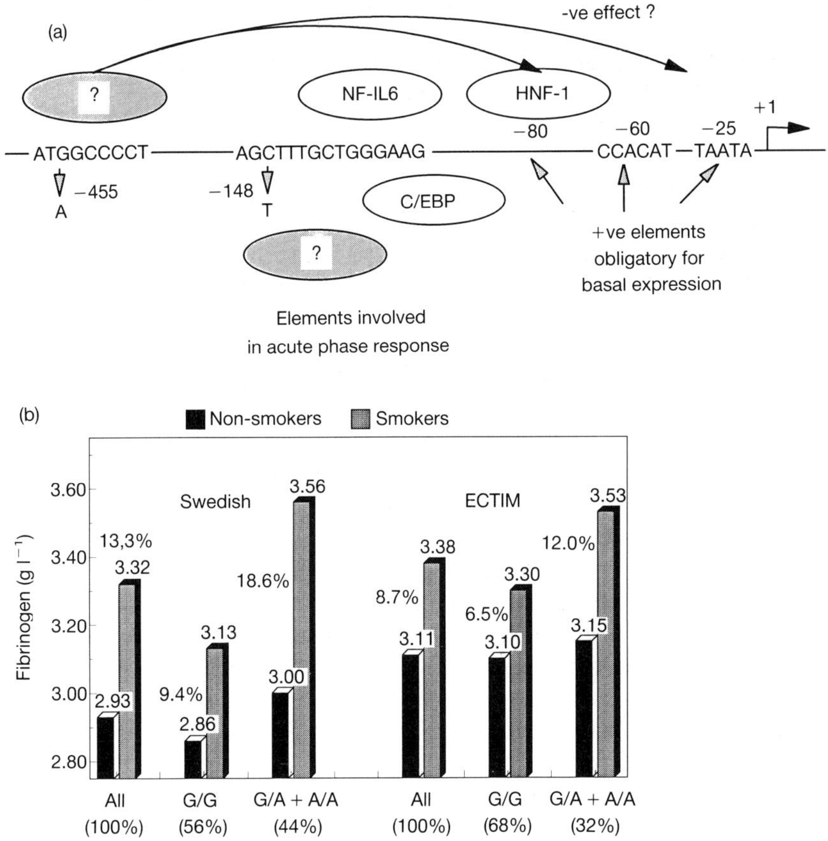

Figure 13.4. (a) Promoter region of the β-fibrinogen gene, showing the location of the G_{455}–A substitution and C_{-148}–T relative to the transcription initiation site and the CAAT and TATA boxes. (b) Mean levels of fibrinogen in men with respect to different genotypes and smoking habits. Data from Green *et al.* (1993), combining young MI patients and matched controls from Stockholm and data from the ECTIM case–control study (Scarabin *et al.*, 1993) combining MI cases and controls. Data from cases and controls were combined by calculation of the genotype-associated deviation from the sample mean for each group separately and by pooling the estimate. The results predict that those with the A allele who smoke will have twice as large a rise in fibrinogen.

Although the A_{-455} sequence is outside the region of the reported promoter sequence, it is possible that it is having a direct effect on transcription, and preliminary studies have demonstrated binding of a hepatic nuclear protein to the G but not the A sequence (Green *et al.*, unpublished observations). However, recently it has been found that the A_{-455} sequence change is in complete allelic association in all Caucasian populations studied to date with a C_{-148}–T change

Table 13.2. Mean fibrinogen levels in men with different G/A genotypes

Sample (ref.)	Genotype	No.	Fibrinogen (g l^{-1})
Healthy, UK	GG	188	2.71
(Thomas, 1991)	GA + AA	101	2.99*
PVD + healthy, UK†	GG	165	2.92
(Conner *et al.*, 1992)	GA + AA	82	2.96
MI + healthy, Sweden	GG	44	2.90
(Green *et al.*, 1993)	GG + AA	32	3.30*
Young males, EARS	GG	326	2.24
(Humphries *et al.*, 1994)	GG + AA	188	2.34*
ECTIM, control group	GG	410	2.97
(Scarabin *et al.*, 1993)	GG + AA	238	3.06*
ECTIM, MI group	GG	352	3.38
(Scarabin *et al.*, 1993)	GA + AA	181	3.54*

PVD, peripheral vascular disease; EARS, European Atherosclerosis Research Study; ECTIM, Étude Cas-Témoins sur l'Infarctes du Myocarde.
* $p < 0.025$.
† Data not adjusted for the effect of age, smoking, etc.

located close to the consensus sequence of the IL-6 element (Figure 13.4a). This raises the possibility that the G_{-455}–A change is acting as a neutral marker for the C–T change, which is the functional change working through effects on transcription of the β-fibrinogen gene that are mediated by IL-6. One possibility is that, as proposed in Figure 13.1, variation in the IL-6-responsive element in the β-promoter may increase the affinity of NF-IL-6, leading to enhanced transcription. This hypothesis is supported by recent data from our laboratory (Lane *et al.*, 1993) and from others (Baumann and Henschen, 1993) using band shift assays that the T_{-148} sequence binds a nuclear protein which the C_{-148} sequence does not. In order to test this hypothesis, experiments are in progress to insert this fragment of the gene into the appropriate vector to test promoter strength.

13.8 Interaction between smoking and genotype to determine plasma fibrinogen levels

Although the relationship between smoking and raised plasma fibrinogen is well established, it is likely that the elevation in fibrinogen experienced in response to a certain degree of smoking will vary between different individuals. In several studies, the possibility of interaction between fibrinogen genotype and smoking has been examined, and results from two of these are summarized in Figure 13.4b. In both studies, fibrinogen levels were, as expected, higher in the smokers (13% in the Swedish and 9% in the ECTIM study; (Green *et al.*, 1993, and Scarabin *et al.*, 1993, respectively) and, in both studies in the non-smokers, men with one or more A_{-455} allele had higher levels of fibrinogen than those with only the G_{-455} allele, although the raising effect associated with the A allele was small (2–5%). In those with only the G_{-455} allele, the smokers show a smaller than average elevation of fibrinogen levels (9.4% and 6.5%) while those with one or more A_{-455} allele show a roughly two-fold greater effect (elevations of 18.6% and 12.0%, respectively). Since this data is cross-sectional it must be interpreted with caution, but it suggests that individuals with the A_{-455} allele who smoke will experience a greater rise in fibrinogen-associated CAD risk than those lacking this allele. It also predicts that the A_{-455} individuals will experience a greater than average reduction in fibrinogen levels upon stopping smoking, and this prediction is testable in intervention studies. The mechanism of this genotype effect is likely to be mediated through changes in transcription of the β-fibrinogen gene. The IL-6-mediated effects on transcription of the gene act through binding of specific nuclear factors to DNA elements, and it may be that these mechanisms or interactions with adjacent elements and/or nuclear factors are disrupted by the sequence changes. If this is the case, the precise molecular effects of these polymorphisms will be amenable to study in transfection experiments or in transgenic animal model systems.

13.9 Smoking and the potential effect of combined genotypes at the apoAI and fibrinogen locus, and risk of CAD

Because the polymorphisms at the apoAI and fibrinogen loci are common, many individuals in the population will be heterozygous for more than one of the 'high risk' genotypes, which are defined here as those associated with lower apoAI or higher fibrinogen. The distribution of such multiple genotypes is easy to estimate based on the observed allele frequencies at the two loci, assuming independent segregation, which is likely since they are on different chromosomes, and no associative mating or strong selection pressures against certain genotype combinations (Hardy–Weinberg proportions). This is presented in Figure 13.5, with, for simplicity, individuals grouped for both loci into those with the G allele only or those with one or more A allele. The calculations show that 13% of the population will have the genotype fibrinogen-GG, apoAI-GA and since, when not smoking, these individuals should have the lowest fibrinogen and highest apoAI levels, their relative risk is set at 100, calculated using a simple additive model for the effect on CAD risk of the two proteins. The most frequent genotype group comprise those individuals homozygous for the common allele at both loci and, since they will have lower apoAI levels (by 9%), their predicted relative CAD risk is 18% higher. The smallest group, who have the fibrinogen-A allele as well as the apoAI-A allele, will have, compared to the lowest risk group, higher fibrinogen ($3.0\ g\ l^{-1}$ compared to $2.86\ g\ l^{-1}$, i.e. $0.14\ g\ l^{-1}$, from the Swedish study in Figure 13.4b; Green *et al.*, 1993) and thus higher CAD risk (extrapolated from NPHS data to be $0.14/0.6 \times 84\% = 20\%$). The fourth group, calculated to be 27% of the population, have the genotype predisposing to the highest fibrinogen and lowest apoAI and thus the highest CAD risk, calculated to be 138%.

However, as shown in Figure 13.5, this predicted pattern of CAD risk changes in the four genotype groups if the men are smokers. The group with the lowest non-smoking risk are predicted to experience a moderate rise in fibrinogen but a large fall in apoAI and have a calculated final CAD risk of 168%. The largest group of men experience only a moderate change in both of these factors and thus increase their predicted final risk by a similar extent, to 178%. By contrast, the other two groups are predicted to show large increases in fibrinogen levels and thus increased CAD risk, with the smallest group having the greatest change, due to an additional large fall in apoAI levels. This suggests that, overall, 66% of the population (13% + 53%; see Figure 13.5) will experience only a small smoking-associated change in CAD risk, while the rest will experience a very much larger

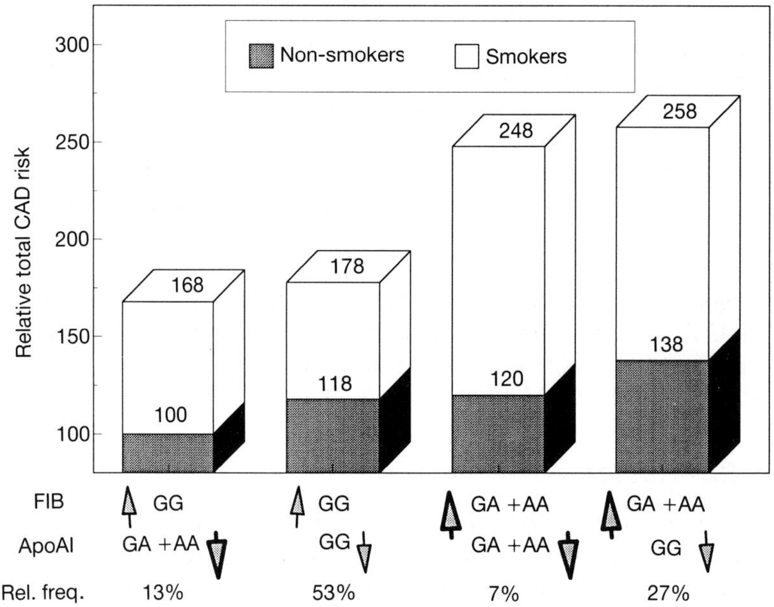

Figure 13.5. Calculated frequency of individuals in the general population having different combinations of genotypes at the apoAI and fibrinogen loci, and the predicted effect on fibrinogen levels and apoAI levels, and thus CAD risk, in men who are non-smokers (hatched bar) and smokers (non-hatched bar). For each genotype group the size of the change in apoAI or fibrinogen upon smoking is indicated by a small or a large arrow. The relative risk of the group having the lowest predicted fibrinogen and highest apoAI is taken as 100%. The change in risk associated with a 1% change in apoAI is taken to be 2% (Stampfer *et al.*, 1991) and with a 0.6 g l^{-1} change in fibrinogen to be 84% (Meade *et al.*, 1987). The predicted levels of apoAI in non-smoking and smoking men are estimated from the data on Icelandic men (Figure 13.3b; Sigurdsson *et al.*, 1993) and for fibrinogen from Swedish men (Figure 13.4b; Green *et al.*, 1993).

change, but for different reasons. From the point of view of the variability gene hypothesis, those with the most common genotype (fibrinogen-GG, apoAI-GG) will have a moderately high baseline risk of CAD because of their genotype-determined plasma level of these proteins but, because they have the 'restrictive' genotype at these loci, they will not experience a large change in these risk factors as a consequence of smoking.

These predictions are presented as an example of how variations at different loci may combine to determine an individual's risk in different ways under different environmental conditions, and the conclusions clearly must be treated with caution since they are based on several assumptions. The most important of these is that the effects on risk of the polymorphisms are additive, whereas the effects may, in fact, be synergistic or compensatory. The homeostatic abilities of the lipid metabolism and thrombotic cascade are such that large changes may induce compensatory responses, for example as are seen in the coagulation cascade in women after the menopause, where higher plasma fibrinogen levels are balanced by higher levels of anti-coagulation factors such as anti-thrombin III. Thus, those with the high risk genotypes may not experience the additive effect on CAD risk expected. However, these predictions are testable by studying a large enough sample of smokers and non-smokers to allow interaction to be detected (roughly 1000 of each). The additional important question to address is whether the smoker/non-smoker comparisons hold true in an intervention study, and of particular interest would be to investigate whether those with the fibrinogen A allele do experience, on cessation of smoking, a very large decrease in fibrinogen levels. As well as helping to understand the underlying pathophysiology of CAD in the general population, such genotype-specific information should be of use in giving people specific advice and supporting them in reducing their subsequent risk of CAD by changes in their lifestyle.

13.10 Conclusions

It is obvious that current estimation for an individual of the plasma level of CAD risk factors, such as apoAI, HDL-C or fibrinogen, is accomplished most easily and accurately by conventional direct methods rather than by genetic tests. However, it is possible that genetic tests may be much more predictive of trends over time, or changes in response to lifestyle factors, than are conventional tests, which would at least require several measures to estimate variability. Once the molecular mechanisms of the variability have been determined at the DNA level this should allow the development of tests that will have a high degree of accuracy and diagnostic potential. Individuals so identified by genetic tests could be targeted for more frequent monitoring and given appropriate advice on lifestyle changes which would reduce their possibility of entering the high risk group. Understanding the mechanism of the interaction between genetic polymorphisms and environmental factors will be important in determining the aetiology of low plasma HDL-C and apoAI levels and high plasma fibrinogen levels, and may be useful in identifying those smoking individuals who would lower their risk of

CAD substantially by cessation of smoking. Once the mechanisms controlling changes in plasma risk factors in response to personal environmental changes are better understood, it may also be possible to develop directed therapeutic strategies that will reduce risk in a genotype-specific manner, an approach which is not possible at present.

Acknowledgements

This work was supported by grants from the British Heart Foundation (RG16 and 86–77). I am particularly grateful to my colleagues, Dr Fiona Green, Dr Philippa Talmud and Rachel Peacock for helpful discussion and access to unpublished data, and to Gina Deeley for help in preparing the manuscript.

References

Akira S, Isshiki H, Nakajima T, Kinoshita S, Nishio Y, Natsuka S, Kishimoto. (1992) Regulation of expression of the interleukin 6 gene: structure and function of the transcription factor NF-IL6. In: *Polyfunctional Cytokines: IL6 and LIF. Ciba Foundation Symposium 167* (eds J March, K Widdows). John Wiley & Sons, Chichester, pp. 47–67.

Allen DR, Browse NL, Rutt DL, Butler L, Fletcher C. (1988) The effect of cigarette smoke, nicotine, and carbon monoxide on the permeability of the arterial wall. *J. Vasc. Surg.* **7:** 139–152.

Assman G, Schulte H, Schriewer H. (1984) The effects of cigarette smoking on serum levels of HDL cholesterol and HDL apolipoprotein A-I. Findings of a prospective epidemiological study on employees of several companies in Westphalia, West Germany. *J. Clin. Chem. Clin. Biochem.* **22:** 397–402.

Austin M, Newman B, Selby JV, Edwards K, Mayer EJ, Krauss RM. (1993) Genetics of LDL subclass phenotypes in women twins. Concordance, heritability, and comingling analysis. *Arterioscler. Thromb.* **13:** 687–695.

Barbaras R, Puchois P, Fruchart JC, Alihaud G. (1987) Cholesterol efflux from cultured adipose cells is mediated by LpA-I particles but not LpA-I:A-II particles. *Biochem. Biophys. Res. Commun.* **142:** 63–69.

Baumann RE, Henschen AH. (1993) Genetic variation in human Bβ fibrinogen gene promoter influences formation of a specific DNA–protein complex with the interleukin 6 response element. *Thromb. Haemost.* **69:** 1515 (abstract).

Berg K. (1987) Genetics of coronary heart disease and its risk factors. In: *Molecular Approaches to Human Polygenic Disease. Ciba Foundation Symposium 130* (eds G Bock, GM Collins). John Wiley & Sons, Chicester, pp. 14–133.

Berg K, Kierulf P. (1989) DNA polymorphisms at fibrinogen loci and plasma fibrinogen concentration. *Clin. Genet.* **36:** 229–235.

Berg K, Børresen A-L, Dahlén G. (1979) Effect of smoking on the serum levels of HDL apoproteins. *Atherosclerosis* **34:** 339–343.

Blann AD. (1992) The acute influence of smoking on endothelium. *Atherosclerosis* **96:** 249–250.

Brown ML, Inazu A, Hesler CB, Agellon LB, Mann C, Whitlock ME, Marcel YL, Milne RW, Koizumi J, Mabuchi H, Takeda R, Tall AR. (1989) Molecular basis of lipid transfer protein deficiency in a family with increased high-density lipoproteins. *Nature* **342:** 448–451.

Conner JM, Fowkes FGR, Wood J, Smith FB, Donnon PT, Lowe GDO. (1992) Genetic variation at fibrinogen loci and plasma fibrinogen levels. *J. Med. Genet.* **29:** 480–482.

Cook NS, Ubben D. (1990) Fibrinogen as a major risk factor in cardiovascular disease. *Trends Pharm. Sci.* **11:** 444–451.

Courtois G, Morgan JG, Campbell LA, Fourel G, Crabtree GR. (1987) Interaction of a liver-specific nuclear factor with the fibrinogen and α-1-antitrypsin promoters. *Science* **238:** 688–692.

Craig WY, Palomaki GE, Haddow JE. (1989) Cigarette smoking and serum lipid and lipoproteins concentrations analysis of published data. *Br. Med. J.* **298:** 284–288.

Criqui NH, Wallacee RB, Heiss G, Mishkel M, Schonfeld G, Jones GTL. (1980) Cigarette smoking and plasma high-density lipoprotein cholesterol. The Lipid Research Clinics Program Prevalence Study. *Circulation* **62** (Suppl. 4): 70–73.

Dalmon J, Laurent M, Courtois G. (1993) The human β fibrinogen promoter contains a hepatocyte nuclear factor 1-dependent interleukin-6 responsive element. *Mol. Cell Biol.* **13:** 1183–1193.

Eisenberg S. (1987) High density lipoprotein metabolism. *J. Lipid Res.* **25:** 1017–1057.

Feher MD, Rampling MW, Brown J, Robinson R, Richmond W, Cholerton S, Bain BJ, Sever PR. (1990) Acute changes in atherogenic and thrombogenic factors with cessation of smoking. *J. R. Soc. Med.* **83:** 146–148.

Fowkes FGR, Conner JM, Smith FB, Wood J, Donnan PT, Lowe GDO. (1992) Fibrinogen genotype and risk of peripheral atherosclerosis. *Lancet* **339:** 693–696.

Frankel AD, Kim PS. (1991) Modular structure of transcription factors: implications for gene regulation. *Cell* **65:** 717–719.

Goldbourt U, Neufeld HN. (1986) Genetic aspects of arteriosclerosis. *Arteriosclerosis* **6:** 357–377.

Gomez N, Cohen P. (1991) Dissection of the protein kinase cascade by which nerve growth factor activates MAP kinases. *Nature* **353:** 170–172.

Green F, Hamsten A, Blomback M, Humphries S. (1993) The role of β-fibrinogen genotype in determining plasma fibrinogen levels in young survivors of myocardial infarction and healthy controls from Sweden. *Thromb. Haemost.* **70:** 915–920.

Haffner SM, Applebaum-Bowden D, Wahl PW, Hoover JJ, Warnick GR, Albers JJ, Hazzard WR. (1985) Epidemiological correlates of high density lipoprotein subfractions, apolipoprotein AI, AII and D and lecithin cholesterol acyltransferase. Effects of smoking, alcohol and adiposity. *Arteriosclerosis* **5:** 169–177.

Hamsten A, Iselius L, Dahlen G, Faire U. (1987a) Genetic and cultural inheritance of serum lipids low and high density lipoprotein cholesterol and serum apolipoproteins AI, AII and B. *Atherosclerosis* **60:** 199–208.

Hamsten A, Iselius L, de Faire U, Blomback M. (1987b) Genetic and cultural inheritance of plasma fibrinogen concentration. *Lancet* **ii:** 998–990.

Hasstedt SJ, Albers JJ, Cheung MC, Jorde LB, Wilson DE, Edwards CQ, Cannon WN, Ash KO, Williams RR. (1984) The inheritance of high density lipoprotein cholesterol and apolipoproteins A-I and A-II. *Atherosclerosis* **51:** 21–29.

Huber P, Dalmon J, Courtois G, Laurent M, Assouline Z, Marguerie G. (1987) Characterization of the 5′-flanking region for the human fibrinogen β gene. *Nucleic Acids Res.* **15:** 1615–1625.

Huber P, Laurent M, Dalmon J. (1990) Human beta-fibrinogen gene expression: upstream sequences involved in its tissue specific expression and its dexamethasone and interleukin 6 stimulation. *J. Biol. Chem.* **265:** 5695–5701.

Humphries SE. (1988) DNA polymorphisms of the apolipoprotein genes – their use in the investigation of the genetic components of hyperlipidaemia and atherosclerosis. *Atherosclerosis* **72:** 89–108.

Humphries SE, Ye S, Talmud P, Bara L, Tiret L. (1994) European Atherosclerosis Research Study. Genotype at the fibrinogen locus (G_{-455}–A β-gene) is associated with differences in plasma fibrinogen levels in young men and women from different regions in Europe: evidence for gender–genotype–environment interaction. *Arterioscler. Thromb.*, in press.

Jeenah M, Wells C. (1992) Functional significance of a point mutation in the apoAI promoter. *Circulation* **81:** I–297.

Jeenah M, Kessling A, Miller N, Humphries SE. (1990) G to A substitution in the promoter region of the apolipoprotein AI gene is associated with elevated serum apolipoprotein AI and high density lipoprotein cholesterol concentrations. *Mol. Biol. Med.* **7:** 233–241.

Jones NC, Rigby PWJ, Ziff EB. (1988) *Trans*-acting factors and the regulation of eukaryotic trascription: lessons from studies on DNA tumor viruses. *Genes Devel.* **2:** 267–281.

Kishimoto T, Hibi M, Murakami M, Narazaki M, Saito M, Taga T. (1992) The molecular biology of interleukin 6 and its receptor. In: *Polyfunctional Cytokines: IL6 and LIF. Ciba Foundation Symposium 167* (eds GR Bock, J March, K Widdows). John Wiley & Sons, Chichester, pp. 5–23.

Lane A, Humphries SE, Green FR. (1993) Effect on transcription of two common genetic polymorphisms adjacent to the promoter region of the B-fibrinogen gene. *Thromb. Haemost.* **69:** 962.

Lane DA. (1986) Clearance and catabolism of fibrinogen and its derivatives – a review. In: *Fibrinogen and its Derivatives* (eds V Schaefers-Borchel, E Selmagi, A Henschen). Elsevier Science Publishers BV, Amsterdam, pp 181–193.

Logan RL, Thomson M, Riemersma RA, Oliver MF, Olsson AG, Rössner S, Callmer E, Walldius G, Kaijser L, Carlson LA, Lockerbie L, Lutz W. (1978) Risk factors for ischaemic heart disease in normal men aged 40. *Lancet* **i:** 949–955.

Lowe GDO. (1993) Blood viscosity and cardiovascular risk. *Curr. Opin. Lipidol.* **4:** 283–287.

Maniatis T, Goodburn S, Fischer JA. (1987) Regulation of inducible and tissue-specific gene expression. *Science* **236:** 1237–1244.

Mastrangelo IA, Courey AJ, Wall JS, Jackson SP, Hough VC. (1991) DNA looping and Sp1 multimer links: a mechanism for transcriptional synergism and enhancement. *Proc. Natl Acad. Sci. USA* **88:** 5670–5674.

Meade TW, Chakrabarti R, Haines AP, North WRS, Stirling Y. (1979) Characteristics affecting fibrinolytic activity and plasma fibrinogen concentrations. *Br. Med. J.* **i:** 153–156.

Meade TW, Mellows S, Brozovic M, Miller GJ, Chakrabarti RR, North WRS, Haines AP, Stirling Y, Imeson JD, Thompson SG. (1986) Haemostatic function and ischaemic heart disease; principal results of the Northwick Park Heart Study. *Lancet* **ii:** 533–537.

Meade TW, Imeson J, Stirling Y. (1987) Effects of changes in smoking and other characteristics on clotting factors and risk of ischaemic heart disease. *Lancet* **ii:** 986–988.

Mehrabian M, Lusis AJ. (1992) Genetic markers for studies of atherosclerosis and related risk factors. In: *Molecular Genetics of Coronary Artery Disease* (eds AJ Lusis, JI Rotter, RS Sparkes). Karger, Basel, pp. 363–418.

Moll PP, Michels VV, Weidman WH, Kottke BA. (1989) Genetic determination of plasma apolipoprotein AI in a population-based sample. *Am. J. Hum. Genet.* **44:** 124–139.

Morrone G, Cortese R, Sorrentino V. (1989) Post-transcriptional control of negative acute phase genes by transforming growth factor beta. *EMBO J.* **8:** 3767–3771.

Norum KR, Gjone E, Glomset JA. (1993) Familial lecithin : cholesterol acyltransferase deficiency, including fish eye disease. In: *The Metabolic Basis of Inherited Disease* (eds CR Scriver, AL Beaudet, WS Sly, D Valle). McGraw-Hill, New York, pp. 1181–1194.

Oliver MF. (1989) Cigarette smoking, polyunsaturated fats, linoleic acid, and coronary heart disease. *Lancet* **i:** 1241–1243.

Pagani F, Sidoli A, Giudici GA, Barenghi L, Vergani C, Baralle FE. (1990) Human apolipoprotein AI promotor polymorphism: association with hyper-alphalipoproteinemia. *J. Lipid Res.* **31:** 1371–1377.

Papazafiri P, Ogami K; Ramji DP, Nicosia A, Monaki P, Cladaras C, Zannis VI. (1991) Promoter elements and factors involved in hepatic transcription of the human apoA-I gene positive and negative regulators bind to overlapping sites. *J. Biol. Chem.* **266:** 5790–5797.

Paul-Hayase H, Rosseneu M, Robinson D, Van Bervliet JP Deslypere JP, Humphries SE. (1992) Polymorphisms in the apolipoprotein (apo) AI-CIII-AIV gene cluster: detection of genetic variation determining plasma apoAI, ApoCIII and ApoAIV concentrations. *Hum. Genet.* **88:** 439–446.

Pepys MB, Baltz ML. (1983) Acute phase proteins with special reference to C-reactive protein and related proteins (pentaxins) and serum amyloid A protein. *Adv. Immunol.* **34:** 141–212.

Reed T, Tracey RP, Fabsitz RR. (1994) Minimal genetic influences on plasma fibrinogen level in adult males in the NHLBI twin study. *Clin. Genet.*, in press.

Reichl D, Miller NE. (1989) Pathophysiology of reverse cholesterol transport: insights from inherited disorders of lipoprotein metabolism. *Arteriosclerosis* **9:** 785–797.

Reue K, Leff T, Breslow J. (1988) Human apolipoprotein CIII gene expression is regulated by positive and negative *cis*-acting elements and tissue-specific protein factors. *J. Biol. Chem.* **263:** 6857–6864.

Richmond W, Seviour PW, Teal TK, Elkeles RS. (1987) Impaired intravascular lipolysis with changes in concentration of high density lipoprotein subclasses in young smokers. *Br. Med. J.* **295:** 246–247.

Ronchi A, Nicolis S, Santoro C, Ottolenghi S. (1989) Increased SpI binding mediates erythroid-specific overexpression of a mutated (HPFH) γ-globulin promoter. *Nucleic Acids Res.* **17:** 10231–10241.

Roy SN, Mukhopadhyay G, Redman CM. (1990) Regulation of fibrinogen assembly. Transfection of HepG2 cells with Bβ cDNA specifically enhances synthesis of the three component chains of fibrinogen. *J. Biol. Chem.* **265:** 6389–6393.

Saha N, Tay JSH, Low PS, Humphries SE. (1994) G to A substitution in the promoter region of the apolipoprotein-AI gene is associated with elevated serum apolipoprotein AI levels in the Chinese. *Genet. Epidemiol.*, in press.

Sastry KN, Seedorf U, Karathanasis SK. (1988) Different *cis*-acting DNA elements control expression of the human apolipoprotein AI gene in different cell types. *Mol. Cell Biol.* **8:** 605–614.

Scarabin P-Y, Bara L, Ricard S, Poirer O, Cambou JP, Arveiler D, Luc G, Evans AE, Samama MM, Cambien F. (1993) Genetic variation at the β-fibrinogen locus in relation to plasma fibrinogen concentrations and risk of myocardial infarction. The ECTIM Study. *Arterioscler. Thromb.* **13:** 886–891.

Sigurdsson G Jr, Gudnason V, Sigurdsson G, Humphries SE. (1992) Interaction between a polymorphism of the apolipoprotein AI promoter region and smoking determines plasma levels of high density lipoprotein and apolipoprotein AI. *Arterioscler. Thromb.* **12:** 1017–1022.

Smith EB, Staples EM. (1981) Haemostatic factors in human aortic intima. *Lancet* **i:** 1171–1174.

Smith JD, Brinton EA, Breslow JL. (1992) Polymorphism in the human apolipoprotein A-I gene promoter region: association of the minor allele with decrased production rate *in vivo* and promoter activity *in vitro*. *J. Clin. Invest.* **89:** 1796–1800.

Stampfer MJ, Sacks FM, Salvini S, Willett WC, Hennekens CH. (1991) A prospective study of cholesterol, apolipoproteins, and the risk of myocardial infarction. *N. Engl. J. Med.* **325:** 373–381.

Stubbe I, Eskilsson J, Nilsson-Ehle P.(1982) High-density lipoprotein concentrations increase after stopping smoking. *Br. Med. J.* **284**: 1511–1513.

Talmud P, Ye S, Humphries S. (1994) A polymorphism in the promoter of the apolipoprotein AI gene associated with differences in apoAI levels: The European Atherosclerosis Research Study. *Genet. Epidemiol.*, in press.

Thomas A, Kelleher C, Green F, Meade TW, Humphries SE. (1991) Variation in the promoter region of the β-fibrinogen gene is associated with plasma fibrinogen levels in smokers and non-smokers. *Thromb. Haemost.* **65:** 487–490.

Thompson SG, Martin JC, Meade TW. (1987) Sources of variability in coagulation factor assays. *Thromb. Haemost.* **58:** 1073–1077.

Treisman R, Orkin SH, Maniatis T. (1983) Specific transcription and RNA splicing defects in five cloned β-thalassaemia genes. *Nature* **302:** 591–596.

Tuomilehto J, Tanskanen A, Salonen JT, Nissinen A, Koskela K. (1986) Effect of smoking and stopping smoking on serum high-density lipoprotein cholesterol levels in a representative population sample. *Preven. Med.* **15:** 35–45.

Walsh A, Azrolan N, Wang K, Marcigliano A, O'Connell A, Breslow JL. (1993) Intestinal expression of the human apoA-I gene in transgenic mice is controlled by a DNA region 3′ to the gene in the promoter of the adjacent convergently transcribed apoC-III gene. *J. Lipid Res.* **34:** 617–623.

Wilhelmsson C, Vedin JA, Elmfeldt D, Tibblin G, Wilhelmsen L. (1975) Smoking and myocardial infarction. *Lancet* **i:** 415–420.

Wilkes HC, Kelleher C, Meade TW. (1988) Smoking and plasma fibrinogen. *Lancet* **i:** 307–308.

Xu C-F, Angelico F, Del Ben M, Humphries S. (1993) Role of genetic variation at the apoAI-CIII-AIV gene cluster in determining plasma apoAI levels in boys and girls. *Genet. Epidemiol.* **10:** 113–122.

Zannis VI, Kardassis D, Cardot P, Hadzopoulou-Cladaras Zanni, EE, Cladaras C. (1992) Molecular biology of the human apolipoprotein genes: gene regulation and structure/function relationship. *Curr. Opin. Lipidol.* **3:** 96–113.

14

Genetic predisposition to dyslipidaemia and accelerated athero-sclerosis: environmental interactions and modification by gene therapy

E. Boerwinkle and L. Chan

14.1 Introduction

It is a developing paradigm in medical practice and research that both genetic and environmental factors are key contributors to disease aetiology and pathophysiology. Consideration of *both* genetic and environmental components is necessary for predicting those individuals at increased risk for disease and treating those already afflicted. However, simultaneous recognition of both is too often only rhetorical, with research efforts pursuing either genetic factors or environmental factors alone. The chronic diseases of later life such as hypertension and coronary artery disease (CAD) are common in all Westernized countries. These diseases are the result of interactions between numerous environmental and genetic factors. Disease liability is not attributable to a single factor such as a solitary foodstuff or a single genetic alteration. As a result, the contribution to disease risk of any single gene or environmental factor is likely to be small when compared to all other risk factors combined. Also, the genes affecting disease risk do not act alone; they are constantly interacting with other genes and with environmental factors. Therefore, the effect of a gene measured in one environment may be very different to its effect in another.

In the first part of this chapter, we present a simultaneous consideration of both genes and environment, and their interactions with respect to dyslipidaemia and cardiovascular disease. Then we examine alternative methods to alter the impact of risk factors for cardiovascular disease. Environmental factors can be removed or modified by altering one's lifestyle. It is only recently, however, that genetic factors in high risk individuals have been modified by direct gene transfer. We discuss the strategies involved in such an approach, and the potential problems and promise of gene therapy in treating cardiovascular disease.

14.2 The role of genes in normal lipid variation

Two basic lines of evidence document the role of genes in affecting lipid levels. First, several inborn errors of lipid metabolism have been reported and characterized. Three examples of such inborn errors are listed below. First, Brown and Goldstein (1986) have described lesions in the low density lipoprotein (LDL) receptor gene that lead to elevated LDL levels and premature atherosclerosis. Second, mutations in the lipoprotein lipase gene are the underlying cause of familial lipoprotein lipase deficiency and type I hyperlipoproteinaemia (e.g. see Langlois *et al.*, 1989; Faustinella *et al.*, 1991). Third, mutations in the gene coding for cholesterol ester transfer protein lead to an increase in plasma high density lipoprotein (HDL)–cholesterol levels (e.g. see Brown *et al.*, 1989). Even though these alterations have very large effects on an individual, their impact on the population is relatively small because of their low frequency. The second line of evidence comes from studies showing that a significant fraction of variability in lipid, lipoprotein and apolipoprotein (apo) levels is attributable to genetic variability. For example, a full one-half of the interindividual variability of plasma total and HDL–cholesterol levels is attributable to genetic factors (Hamsten *et al.*, 1986). This value is near unity for plasma lipoprotein(a) (Lp(a)) levels (Hasstedt and Williams, 1986). Of particular interest in the biometrical genetic studies of coronary heart disease risk factors has been the detection of statistical evidence for single unknown genes with large effects. Such 'major genes' with large effects have been reported for both HDL–cholesterol and apoA-I levels (Hasstedt *et al.*, 1986; Moll *et al.*, 1989). Evidence supporting the role of a single locus with a large effect on plasma Lp(a) levels is well established (e.g. see Hasstedt and Williams, 1986). Apo(a), the unique protein component of Lp(a), is closely linked and homologous to the serum protease plasminogen. Linkage analyses using a protein polymorphism in plasminogen have determined that the major gene contributing to Lp(a) levels is likely to be the apo(a) structural gene, and Boerwinkle *et al.* (1992) have estimated that 90% of the interindividual variation in plasma Lp(a) levels is attributable to length and sequence variation in the apo(a) gene.

Although traditional quantitative genetic studies indicate that genes are contributing to Lp(a) levels, they do not yield information about the identity and role of specific gene loci. Knowledge concerning individual loci is necessary for determining which individuals carry specific mutations, and for studying the interaction between genes and environments. Advances in atherosclerosis research, lipid metabolism and molecular biology have identified numerous genes and gene products that may be contributing to cardiovascular disease. Plasma lipids do not circulate free but rather are transported as lipoprotein particles stabilized by specialized proteins known as apolipoproteins. Plasma lipoproteins are produced primarily by the liver and small intestine. Once they are secreted into the circulation, various enzymes and transfer proteins alter the lipid and protein components of these particles. The lipoprotein particles may be taken up by cell surface receptors which recognize and bind to the lipoproteins which are taken up by the cells. The genes encoding these apolipoproteins, enzymes, transfer proteins and receptors are candidates for those responsible for the underlying polygenetic and major gene effects discussed above.

The gene whose effects on normal lipid variation are best understood is apoE. ApoE is a structural component of chylomicron remnants, very low density lipoproteins (VLDL), and a subset of HDL lipoproteins. ApoE plays a major role in lipid metabolism through cellular uptake of lipoprotein particles by apoE-specific and LDL (also known as apoB/E) receptors on the liver and other tissues (Mahley *et al.*, 1990). Human apoE is polymorphic, with three alleles, $\epsilon 2$, $\epsilon 3$ and $\epsilon 4$, coding for three isoforms, E2, E3 and E4, respectively.

E2-containing lipoproteins have lower affinity for hepatic lipoprotein receptors compared to lipoproteins containing E3. In addition, the conversion of VLDL to LDL in the plasma of patients with familial type III hyperlipoproteinaemia is impeded by the apoE2 isoform, but can be restored by addition of apoE3, suggesting a role for apoE in the conversion of VLDL to LDL. *In vitro*, the receptor-binding affinity of E4-containing lipoproteins may be similar to those containing the common E3 isoform but, *in vivo*, apoE4-containing lipoproteins are cleared more rapidly from the circulation than those containing apoE3.

Numerous reports have indicated that the apoE polymorphism influences plasma lipid levels. Although the frequencies of the apoE alleles are heterogeneous among populations, the effects of this gene are relatively consistent: the average effect of the $\epsilon 2$ allele is to lower plasma cholesterol levels and the average effect of the $\epsilon 4$ allele is to raise plasma cholesterol levels. Using family data, Boerwinkle and Sing (1987) directly estimated that the apoE polymorphism accounts for 12.5% of the overall polygenetic variance of total serum cholesterol levels. Boerwinkle and Utermann (1988) have investigated the effect of the apoE polymorphism on plasma apoE and total cholesterol levels in a random sample of 563 individuals. The effects of the apoE polymorphism on plasma apoE levels were opposite to those on total cholesterol levels (Table 14.1). On the basis of these and other results, they formulated a model for the effect of the apoE polymorphism on lipid metabolism and predicted that the effects of this gene will differ among populations differing in dietary fat intake. According to this model, differences among alleles in receptor binding, chylomicron clearance and subsequent regulation of LDL receptors account for the observed effects of this polymorphism on total cholesterol levels and coronary heart disease risk. Based on the results of several epidemiologic studies, an association between apoE genotypes and the prevalence of coronary heart disease has also been established (e.g. see Menzel *et al.*, 1983). The frequency of the ϵ2 allele is reduced among coronary heart disease patients, while the frequency of the ϵ4 allele generally is increased. Recently, Schächter *et al.* (1994) have implicated the apoE polymorphism in the overall longevity of humans.

Table 14.1. Average effects of the apoE polymorphism*

Phenotype	ϵ2	ϵ3	ϵ4	Variance†
Apolipoprotein E (mg dl^{-1})	0.95	−0.04	−0.19	20.2%
Total cholesterol (mg dl^{-1})	−14.2	−0.16	7.09	4.0%

* Reproduced from Boerwinkle and Utermann (1988).
† Percent of the total phenotypic variance attributable to the apoE polymorphism.

14.3 Genotype–environment interaction

The regulation of gene expression and the action of gene products are determined by the micro- and macro-environments of the organism and its cellular constituents. These environments are influenced by both the expression of other genes and by the ecology of the organism, e.g. its diet. There have been many discussions on the potential for, and effects of, genotype–environment interaction, but only a few studies have attempted to quantitate the role of genotype–environment interaction in humans.

Nestruck *et al.* (1987) reported significant interaction between the apoE polymorphism and the cholesterol-lowering drug Probucol. Probucol has both antioxidant and lipid-lowering properties, but has variable efficacy among patients. Probucol-treated type II hypercholesterolaemic individuals with an ϵ4 allele showed a larger reduction in plasma cholesterol levels than ϵ3/3 individuals. Nestruck *et al.* (1987) hypothesized that the E4 isoform and Probucol act synergistically to promote enhanced lipoprotein catabolism.

Gueguen *et al.* (1989) have investigated the interaction between the apoE polymorphism and longitudinal changes in several coronary heart disease risk factors in a sample of 158 families. The estimated frequencies of apo ϵ2, ϵ3 and ϵ4 alleles in this sample were 0.120, 0.764 and 0.116, respectively. There was no significant evidence for an effect of the apoE polymorphism on the longitudinal profile of any of the variables considered. There was a significant interaction between apoE effects and weight change on the longitudinal change of serum triglyceride levels. Accompanying weight gain, individuals with an ϵ4 allele showed a larger increase in triglyceride levels than individuals with no ϵ4 allele.

Tikkanen *et al.* (1990) and Xu *et al.* (1990) have investigated the effects of the apoE polymorphism in a sample of adults from Finland who switched from their normal diet to a diet low in total and saturated fat. They found that when the subjects were on their normal high fat diet, apoE genotype was associated with the expected and significant effects on plasma lipids and, when they switched to the low fat diet, average total serum cholesterol levels were not statistically significantly different among apoE genotypes. These results support the conjecture of Boerwinkle and Utermann (1988) that the response to changes in dietary cholesterol will differ among apoE types. A direct test of whether dietary response differs among apoE genotypes, however, did not detect significant differences (Xu *et al.*, 1990). The decrease in total serum cholesterol levels in subjects on the low total and saturated fat diet was not significantly different among apoE types (Figure 14.1).

Because of its recognized importance in remnant clearance, the apoE gene has been targeted for studying postprandial lipid response. It has been argued that prolonged postprandial lipaemia is conducive to the development of atherosclerosis (Zilversmit, 1979), and case–control studies have shown that late postprandial lipid levels are predictive of coronary heart disease, even when HDLs are considered simultaneously (Patsch *et al.*, 1992). To estimate the contribution of apoE polymorphism to distinct metabolic aspects of postprandial lipoprotein metabolism, Boerwinkle *et al.* (1994) measured multiple parameters of postpran-

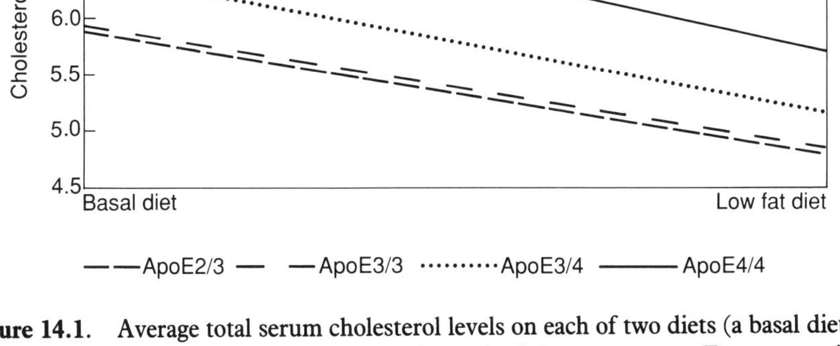

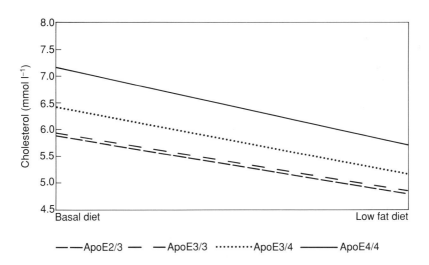

Figure 14.1. Average total serum cholesterol levels on each of two diets (a basal diet, and a diet low in total fat and saturated fat) for each of the common apoE genotypes in the Finnish dietary intervention study. Reproduced from Xu *et al.* (1990) with permission from Wiley-Liss, a division of John Wiley and Sons, Inc. Copyright © 1990 Wiley-Liss.

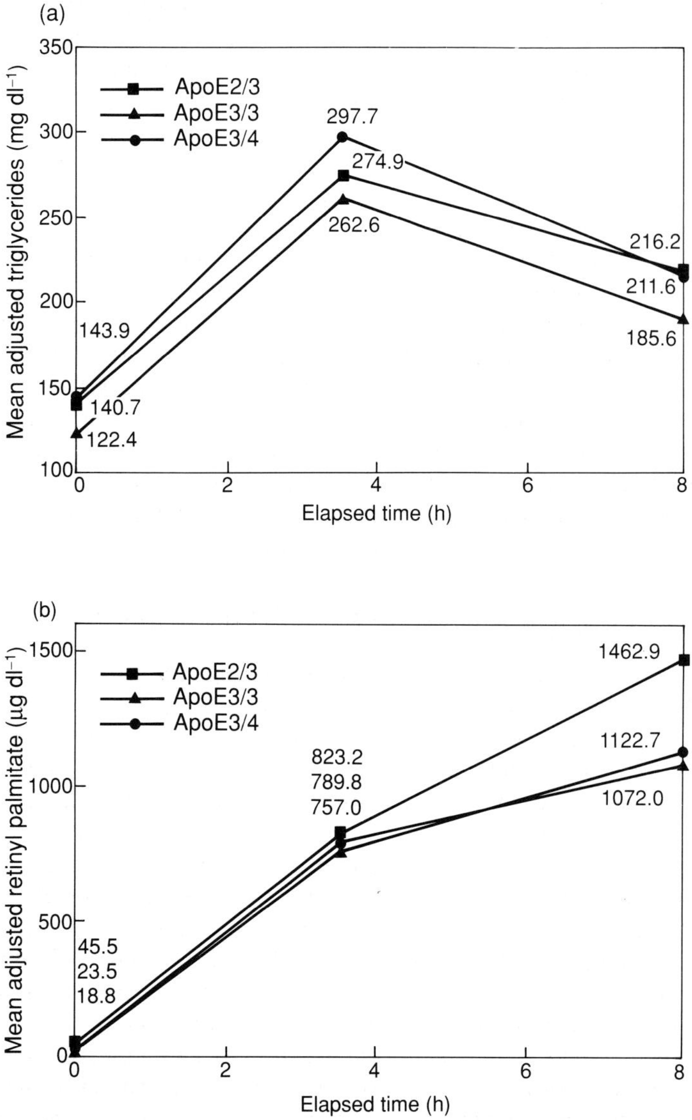

Figure 14.2. Profile of postprandial triglyceride (a) and retinyl palmitate (b) levels for each of the common apoE genotypes in 397 Caucasian subjects from the Atherosclerosis Risk in Communities Study. Reproduced from Boerwinkle *et al.* (1994) with permission from the University of Chicago Press.

dial response in a sample of 397 individuals taking part in the Atherosclerosis Risk in Communities Study. The profile of postprandial triglyceride and retinyl palmitate (which gives a measure of intestinally derived chylomicron levels) response for each of the common apoE genotypes is shown in Figure 14.2a and b, respectively. The profile of postprandial triglyceride levels was not different among the common apoE genotypes. The profile of postprandial retinyl palmitate was significantly different among apoE genotypes. In fact, the impact of the apoE polymorphism on interindividual difference in postprandial response is larger than its well-known effect on LDL–cholesterol

While the data of Boerwinkle *et al.* (1994) indicate that the final step of the remnant pathway, i.e. their removal from circulation, is delayed in subjects with the ϵ2/3 genotype, there is no evidence that the ϵ2 allele predisposes to coronary artery disease. In contrast, a moderate protective role of this allele has been suggested in some studies (Hixson *et al.*, 1991). Thus, other factors such as enhanced VLDL production by the liver or impaired lipolysis of triglyceride-rich lipoproteins, irrespective of their origin, must be implicated in the atherogenicity of remnant particles. Thus, these studies of genotype–environment interaction not only provide a reliable estimate of the magnitude of its effect on various measurements commonly used to characterize postprandial lipaemia, but also provide mechanistic insight into the effects of the apoE gene polymorphism on postprandial lipaemia and coronary heart disease.

14.4 Somatic gene therapy for hyperlipidaemia

Environmental influences can be modified as long as the individual is compliant and willing to make long-term lifestyle changes. In some cases, such changes may be difficult, requiring the assistance of psychologists and special programmes aimed at behaviour modification. From a pathophysiological standpoint, the desired changes, e.g. diet modification, cessation of smoking and exercise programmes, have been widely accepted as beneficial to one's cardiovascular health, and are readily justified. In contrast to environmental factors, which are generally viewed as modifiable, until recently we have been resigned to accepting our genetic makeup as something we cannot change. For example, patients with familial hypercholesterolaemia (FH) accept that they are born with the disease, which will stay with them for life. They hope to delay the onset of coronary artery disease by manipulating the environment, e.g. adopting a certain 'heart-healthy' lifestyle, and by potent therapeutic measures such as taking lipid-lowering drugs or some physical means of extracorporeal LDL removal. However, we now believe that one's genetic makeup can be altered and we can prevent or ameliorate many of the devastating consequences of severe hyperlipidaemia by somatic gene therapy.

14.5 Basic considerations in human gene therapy

A large number of human diseases are potential targets for somatic gene therapy (Morgan and Anderson, 1993). *Monogenic* diseases such as cystic fibrosis, adenosine deaminase deficiency, sickle cell anaemia, haemophilia A and haemophilia B are the classic targets for gene therapy because simple replacement with the normal counterpart of the diseased gene potentially should be curative. In the cardiovascular area, FH caused by LDL receptor defects is one such example. The next class of diseases potentially amenable to gene therapy are the *polygenic* disorders. Often the contribution of the individual genes involved is not clearly defined and, as discussed in the earlier part of this chapter, gene–environment interactions may play a role in these diseases which include various forms of hyperlipidaemia, diabetes, hypertension and some forms of cancer. One treatment strategy involves the over-expression of a therapeutic gene or silencing of a 'harmful' gene. Restenosis is a common complication of coronary angioplasty that can be prevented potentially by genetic manipulation of growth-related genes that modulate smooth muscle cell proliferation at the site of angioplasty. A similar strategy can also be applied systemically for the syndrome of familial combined hyperlipidaemia where, in the vast majority of cases, the underlying responsible genetic lesion has not been identified. Post-myocardial infarction patients with a serious form of this syndrome which is not responding well to conventional therapy are potential candidates for gene therapy designed to modify or lower their atherogenic lipoproteins.

14.6 Gene delivery systems

There are many ways that therapeutic genes can be delivered to target organ(s) (Mulligan, 1993). Two basic approaches can be taken:

(i) an *ex vivo* approach, whereby cells are removed from the patient, genetically modified and re-implanted in the patient, and
(ii) an *in vivo* approach, whereby the therapeutic gene is administered directly to the patient, e.g. by intravenous injection, intramuscular injection or intrabronchial installation.

The *in vivo* approach is the one preferred by the patient and clinician because of its simplicity. However, it is unclear whether gene therapy for different diseases can be optimized to be of high enough tissue selectivity and efficiency for such an approach to be applied routinely.

There are two major types of gene transfer systems for delivering therapeutic genes: (i) non-viral transfer systems, and (ii) viral transfer systems. Numerous non-viral techniques have been tried which vary markedly in efficiency. The more popular methods include direct cellular uptake of naked DNA, receptor-mediated DNA transfer (e.g. using DNA coupled to asialoorosomucoid which is taken up by the highly efficient asialoglycoprotein receptor in the liver; Wu and Wu, 1987) and liposome-mediated transfer (e.g. see Zhu *et al.*, 1993). A large number of viral vectors have been modified for gene transfer. Two systems that have received much attention recently are based on retroviruses and adenoviruses. Retroviruses can integrate into the host cell chromosome which would allow for stable expression. Special defective (i.e. replication-incompetent) retroviral vectors have been developed for this purpose. Unfortunately, retroviruses can only transduce replicating cells, and some stimulus to cell division (e.g.

partial hepatectomy for delivery to the liver) is usually required at the time the gene is introduced. Furthermore, they have limited host cell range (i.e. they infect only a few cell types) and infectious viral particles can be produced in low titres only. While research is being conducted to produce modified retroviral vectors that address these deficiencies (e.g. see Burns *et al.*, 1993), adenoviral vectors have gained considerable popularity. Their major advantage over retrovirus-based vector systems is their ability to infect non-replicating cells and the fact that they can be grown to very high titres. They have been used extensively for *in vivo* direct gene transfer. Unlike retrovirus-based systems, the adenovirus remains episomal and eventually will be lost as the host cells divide. The adenovirus-mediated expression of the LDL receptor has been accomplished in receptor-deficient mice, leading to a transient amelioration of the lipoprotein abnormality in these animals (Ishibashi *et al.*, 1993). Experience with adenovirus-based systems is limited and its safety as a gene transfer vehicle has to be tested further. The non-viral and viral gene transfer systems are not necessarily mutually exclusive and components of each can be combined for greatly enhanced efficiency.

14.7 Ethical considerations

Any new therapy must meet safety standards, and gene therapy is no exception. For therapeutic measures, which may not have been tested in large numbers of patients because of the rarity of the disease, the potential therapeutic benefits must be weighed against the potential risks of therapy. These issues have been discussed in various reviews (Walters, 1986; Morgan and Anderson, 1993) and must be re-examined periodically whenever a new therapeutic system is being developed. The concept of somatic gene therapy for serious, otherwise irreversible diseases falls within the standards of medical ethics. The whole issue of germ-line manipulation, on the other hand, is very much under debate, and will continue to be so for the forseeable future.

14.8 Current status and future of somatic gene therapy for hyperlipidaemia

FH is an autosomal dominant disorder characterized by elevated plasma and LDL–cholesterol and premature coronary artery disease (Goldstein and Brown, 1989). Heterozygotes occur in the population with a frequency of about 1 in 500. They present with plasma cholesterol and LDL–cholesterol levels of 2–3 times the normal levels and symptomatic coronary artery disease during middle age. Homozygotes occur with a frequency of about 1 in 1 000 000. Their LDL–cholesterol levels are about 4–7 times normal or higher. They usually present with angina pectoris, myocardial infarction or sudden death between the ages of 5 and 30. FH is caused by defects in the LDL receptor, and heterozygous FH generally responds to drug therapy. The response of homozygous FH patients, on the other hand, depends on whether some functional LDL receptors are still present;

receptor-negative individuals generally are quite resistant to drug or diet therapy.

The limitations of available methods of treatment for homozygous FH have made somatic gene therapy an attractive alternative. FH is a monogenic disease and its severity depends entirely on the amount of functional LDL receptor present. If the LDL receptor content of the liver can be restored toward normal, the patient should have a satisfactory clinical response. Wilson *et al.* (1988) first showed the feasibility of this approach by demonstrating that the human LDL receptor gene could be transferred to hepatocytes isolated from WHHL rabbits (an animal model of FH caused by defective LDL receptors) using retrovirus-based vectors. Following transduction *in vitro*, LDL receptor function, as measured by LDL degradation by the transduced hepatocytes, was increased compared to controls. Chowdhury *et al.* (1991) subsequently showed that autologous transplantation of WHHL hepatocytes that were transduced *ex vivo* with the retroviral vector resulted in a 30–50% decrease in total serum cholesterol that persisted for the duration of the experiment (4 months).

Based on the experience with WHHL rabbits, an experimental clinical protocol using the *ex vivo* approach for autologous transplantation of transduced hepatocytes in homozygous FH patients is currently underway (Wilson *et al.*, 1992a). Like extracorporeal LDL removal and liver transplantation, this therapeutic option is labour-intensive and expensive because it involves partial hepatectomy and the culture and transduction of large numbers of hepatocytes *in vitro* before they are transplanted back into the patient. If the LDL receptor gene can be introduced into the liver simply by an intravenous injection (i.e. using an *in vivo* approach), the method would be much more acceptable to both patients and physicians, and the much lower cost involved means that this mode of treatment would be available to many patients. Some preliminary success in the WHHL rabbit model using such a strategy has been reported. Wilson *et al.* (1992b) made use of the asialoglycoprotein receptor-mediated uptake of a retroviral LDL receptor vector. The latter was coupled to asialoorosomucoid and infusion of the DNA–protein complex into the peripheral circulation of WHHL rabbits resulted in hepatocyte-specific gene transfer and temporary amelioration of hypercholesterolaemia. Ishibashi *et al.* (1993) also accomplished transient restoration of LDL receptor function and LDL lowering in LDL receptor-deficient mice by intravenous injection of recombinant replication-defective adenovirus encoding the human LDL receptor driven by a cytomegalovirus promoter.

In addition to the treatment of the monogenic disorder FH, somatic gene therapy using the LDL receptor and other therapeutic genes (such as apoA-I, which elevates HDL, and cholesterol 7α hydroxylase, which lowers plasma cholesterol by diverting it to bile acid production) are potentially useful in the treatment of other hyperlipidaemic states such as combined familial hyperlipidaemia and polygenic hyperlipidaemia. In these cases, we take advantage of the beneficial physiological effects of enhanced expression of the therapeutic gene although it may not be itself defective in the patients being treated. Thus, somatic gene therapy has a much broader application than the gene replacement approach used for many classic monogenic diseases.

Although the *in vivo* gene therapy experiments in the FH animal models show considerable promise, the expression of the transduced gene and the cholesterol lowering effect were transient. There is active ongoing research aiming at the permanent expression of the exogenously introduced gene following *in vivo* delivery. Adenoviral vectors have problems of toxicity, and the formulation of the asialoorosomucoid–DNA complex is also subject to considerable variation such that production and standardization will be difficult. These are some of the drawbacks of currently available technology; methods to circumvent them are being investigated by various basic and clinical scientists. Alternative approaches (such as the delivery of naked DNA or of liposomes) are also being actively pursued. It is likely that, in the 21st century, somatic gene therapy will become a standard form of treatment for homozygous FH and other rare forms of dyslipidaemia that are resistant to conventional modes of treatment.

Acknowledgements

This work was supported by NIH grant HL-27341 and a grant from the March of Dimes Birth Defects Foundation (to L.C.) and NIH grant HL-40613, and contracts N01-HC55015, N01-HC55016, N01-HC55018, N01-HC55019, N01-HC55020, N01-HC55021, N01-HC55022 with the National Heart, Lung and Blood Institute (to E.B.). Eric Boerwinkle is an Established Investigator of the American Heart Association and the recipient of a Research Career Development Award from the National Institutes of Health (HL-02453).

References

Boerwinkle E, Sing CF. (1987) The use of measured genotype information in the analysis of quantitative phenotypes in man. III. Simultaneous estimation of the frequencies and effects of the apolipoprotein E polymorphism and residual polygenetic effects on cholesterol, betalipoproteins and triglyceride levels. *Ann. Hum. Genet.* **51**: 211–226.

Boerwinkle E, Utermann G. (1988) Simultaneous effects of the apolipoprotein E polymorphism on apolipoprotein E, apolipoprotein B, and cholesterol metabolism. *Am. J. Hum. Genet.* **42**: 104–112.

Boerwinkle E, Leffert CC, Lin J-P, Lackner C, Chiesa G, Hobbs HH. (1992) Apolipoprotein(a) gene accounts for greater than 90% of the variation in plasma lipoprotein(a) concentrations. *J. Clin. Invest.* **90**: 52–60.

Boerwinkle E, Brown S, Sharrett AR, Heiss G, Patsch W. (1994) Apolipoprotein E polymorphism influences postprandial retinyl palmitate but not triglyceride concentrations. *Am. J. Hum. Genet.* **54**: 341–360.

Brown MS, Goldstein JL. (1986) A receptor-mediated pathway for cholesterol homeostasis. *Science* **232**: 34–47.

Brown ML, Inazu A, Hesler CB, Agellon LB, Mann C, Whitlock ME, Marcel YL, Milne RW, Koizumi J, Mabuchi H, Takeda R, Tall AR. (1989) Molecular basis of lipid transfer protein deficiency in a family with increased high-density lipoproteins. *Nature* **342**: 448–451.

Burns JC, Friedmann T, Driever W, Burrascano M, Yee J-K. (1993) Vesicular stomatitis virus G glycoprotein pseudotyped retroviral vectors: concentration to very high titer and efficient gene transfer into mammalian and nonmammalian cells. *Proc. Natl Acad. Sci. USA* **90**: 8033–8037.

Chowdhury JR, Grossman M, Gupta S, Chowdhury NR, Baker JR, Wilson JM Jr. (1991) Long-term improvement of hypercholesterolemia after *ex vivo* gene transfer in LDLR-deficient rabbits. *Science* **254**: 1802–1805.

Faustinella F, Chang A, Van Biervliet JP, Rosseneu M, Vinaimont N, Smith LC, Chen S-H, Chan L. (1991) Catalytic triad residue mutation (Asp^{156}→Gly) causing familial lipoprotein lipase deficiency: co-inheritance with a nonsense mutation (Ser^{447}→Ter) in a Turkish family. *J. Biol. Chem.* **266**: 14418–14424.

Goldstein JL, Brown MS. (1989) Familial hypercholesterolemia. In: *The Metabolic Basis of Inherited Disease* (eds CR Scriver, AL Beaudet, WS Sly, D Valle). McGraw-Hill, New York, pp. 1215–1250.

Gueguen R, Visvikis S, Steinmetz J, Siest G, Boerwinkle E. (1989) An analysis of genotype effects and their interactions by using the apolipoprotein E polymorphism and longitudinal data. *Am. J. Hum. Genet.* **45**: 793–802.

Hamsten A, Iselius L, Dahlen G, de Faire U. (1986) Genetic and cultural inheritance of serum lipids, low and high density lipoprotein cholesterol and serum apolipoproteins AI, AII, and B. *Atherosclerosis* **60**: 199–208.

Hasstedt SJ, Williams RR. (1986) Three alleles for quantitative Lp(a). *Genet. Epidemiol.* **3**: 3–55.

Hasstedt SJ, Ash KO, Williams RR. (1986) A reexamination of major locus hypotheses for high density lipoprotein cholesterol levels using 2170 persons screened in 55 Utah pedigrees. *Am. J. Med. Genet.* **24**: 57–66.

Hixson JE, PDAY Research Group. (1991) Apolipoprotein E polymorphisms affect atherosclerosis in young males. *Arterioscler. Thromb.* **11**: 1237–1244.

Ishibashi S, Brown MS, Goldstein JL, Gerard RD, Hammer RE, Herz J. (1993) Hypercholesterolemia in low density lipoprotein receptor knockout mice and its reversal by adenovirus-mediated gene delivery. *J. Clin. Invest.* **92**: 883–893.

Langlois S, Deeb S, Brunzell JD, Kastelein JJ, Hayden MR. (1989) A major insertion accounts for a significant proportion of mutations underlying human lipoprotein lipase deficiency. *Proc. Natl Acad. Sci. USA* **86**: 948–952.

Mahley RW, Innerarity TL, Rall SC Jr, Weisgraber KH, Taylor JM. (1990) Apolipoprotein E: genetic variants provide insights into its structure and function. *Curr. Opin. Lipidol.* **1**: 87–95.

Menzel HJ, Kladetzky RG, Assmann G. (1983) Apolipoprotein E polymorphism and coronary disease. *Arteriosclerosis* **3**: 310–315.

Moll PP, Michels VV, Weidman WH, Kottke BA. (1989) Genetic determination of plasma apolipoprotein AI in a population-based sample. *Am. J. Hum. Genet.* **44**: 124–139.

Morgan RA, Anderson WF. (1993) Human gene therapy. *Annu. Rev. Biochem.* **62**: 191–217.

Mulligan RC. (1993) The basic science of gene therapy. *Science* **260**: 926–932.

Nestruck AC, Bouthillier D, Sing CF, Davignon J. (1987) Apolipoprotein E polymorphism and plasma cholesterol response to probucol. *Metabolism* **36**: 743–747.

Patsch JR, Miesenbock G, Hopferwieser T, Muhlberger, Knapp E, Dunn JK, Gotto AM Jr, Patsch W. (1992) Relation of triglyceride metabolism and coronary artery disease. Studies in the postprandial state. *Arterioscler. Thromb.* **12**: 1336–1345.

Schächter F, Favre-Delanef L, Guenet F, Roujer H, Frojuel P, Lesseven-Ginet L, Cohen D. (1994) Genetic associations with human longevity at the apoE and ACE loci. *Nature Genetics* **6**: 29–32.

Tikkanen MJ, Huttunen JK, Ehnholm C, Pietinen P. (1990) Apolipoprotein E4 homozygosity predisposes to serum cholesterol elevation during a high-fat diet. *Arteriosclerosis* **10**: 285–288.

Walters L. (1986) The ethics of human gene therapy. *Nature* **320**: 225–227.

Wilson JM, Johnston DE, Jefferson DM, Mulligan RC. (1988) Correction of the genetic defect in hepatocytes from the Watanabe heritable hyperlipidemic rabbit. *Proc. Natl Acad. Sci. USA* **85**: 4421–4425.

Wilson JM, Grossman M, Raper SE, Baker JR, Newton RS, Thoene JG. (1992a) *Ex vivo* gene therapy of familial hypercholesterolemia. *Hum. Gene Ther.* **3**: 179–222.

Wilson JM, Grossman M, Wu CH, Chowdhury NR, Wu GY, Chowdhury JR. (1992b)

Hepatocyte-derived gene transfer *in vivo* leads to transient improvement of hypercholesterolemia in low density lipoprotein receptor-deficient rabbits. *J. Biol. Chem.* **267**: 963–967.

Wu GY, Wu CH. (1987) Receptor-mediated *in vitro* gene transformation by a soluble DNA carrier system. *J. Biol. Chem.* **262**: 4429–4432.

Xu CF, Boerwinkle E, Tikkanen MJ, Huttunen JK, Humphries S, Talmud P. (1990) Genetic variation at the apolipoprotein gene loci contribute to response of plasma lipids to dietary change. *Genet. Epidemiol.* **7**: 261–275.

Zhu N, Liggitt D, Liu Y, Debs R. (1993) Systemic gene expression after intravenous DNA delivery into adult mice. *Science.* **261**: 209–211.

Zilversmit DB. (1979) Atherogenesis: a postprandial phenomenon. *Circulation* **60**: 473–485.

Index